PATHWAYS TO EXPLORATION
RATIONALES AND APPROACHES FOR A U.S. PROGRAM OF HUMAN SPACE EXPLORATION

Committee on Human Spaceflight

Aeronautics and Space Engineering Board
Space Studies Board
Division on Engineering and Physical Sciences

Committee on National Statistics
Division of Behavioral and Social Sciences and Education

NATIONAL RESEARCH COUNCIL
OF THE NATIONAL ACADEMIES

THE NATIONAL ACADEMIES PRESS
Washington, D.C.
www.nap.edu

THE NATIONAL ACADEMIES PRESS   **500 Fifth Street, NW**   **Washington, DC 20001**

NOTICE: The project that is the subject of this report was approved by the Governing Board of the National Research Council, whose members are drawn from the councils of the National Academy of Sciences, the National Academy of Engineering, and the Institute of Medicine. The members of the committee responsible for the report were chosen for their special competences and with regard for appropriate balance.

This report is based on work supported by Contract NNH10CC48B between the National Academy of Sciences and the National Aeronautics and Space Administration. Any opinions, findings, conclusions, or recommendations expressed in this publication are those of the authors and do not necessarily reflect the views of the agency that provided support for the project.

International Standard Book Number-13: 978-0-309-30507-5
International Standard Book Number-10: 0-309-30507-1
Library of Congress Control Number: 2014950546

Cover: Design by Tim Warchocki.

Copies of this report are available free of charge from

Aeronautics and Space Engineering Board
National Research Council
The Keck Center of the National Academies
500 Fifth Street, NW
Washington, DC 20001

Additional copies of this report are available from the National Academies Press, 500 Fifth Street, NW, Keck 360, Washington, DC 20001; (800) 624-6242 or (202) 334-3313; http://www.nap.edu.

THE NATIONAL ACADEMIES
Advisers to the Nation on Science, Engineering, and Medicine

The **National Academy of Sciences** is a private, nonprofit, self-perpetuating society of distinguished scholars engaged in scientific and engineering research, dedicated to the furtherance of science and technology and to their use for the general welfare. Upon the authority of the charter granted to it by the Congress in 1863, the Academy has a mandate that requires it to advise the federal government on scientific and technical matters. Dr. Ralph J. Cicerone is president of the National Academy of Sciences.

The **National Academy of Engineering** was established in 1964, under the charter of the National Academy of Sciences, as a parallel organization of outstanding engineers. It is autonomous in its administration and in the selection of its members, sharing with the National Academy of Sciences the responsibility for advising the federal government. The National Academy of Engineering also sponsors engineering programs aimed at meeting national needs, encourages education and research, and recognizes the superior achievements of engineers. Dr. C. D. Mote, Jr., is president of the National Academy of Engineering.

The **Institute of Medicine** was established in 1970 by the National Academy of Sciences to secure the services of eminent members of appropriate professions in the examination of policy matters pertaining to the health of the public. The Institute acts under the responsibility given to the National Academy of Sciences by its congressional charter to be an adviser to the federal government and, upon its own initiative, to identify issues of medical care, research, and education. Dr. Victor J. Dzau is president of the Institute of Medicine.

The **National Research Council** was organized by the National Academy of Sciences in 1916 to associate the broad community of science and technology with the Academy's purposes of furthering knowledge and advising the federal government. Functioning in accordance with general policies determined by the Academy, the Council has become the principal operating agency of both the National Academy of Sciences and the National Academy of Engineering in providing services to the government, the public, and the scientific and engineering communities. The Council is administered jointly by both Academies and the Institute of Medicine. Dr. Ralph J. Cicerone and Dr. C. D. Mote, Jr., are chair and vice chair, respectively, of the National Research Council.

www.nationalacademies.org

COMMITTEE ON HUMAN SPACEFLIGHT

MITCHELL E. DANIELS, JR., Purdue University, *Co-Chair*
JONATHAN LUNINE, Cornell University, *Co-Chair*
BERNARD F. BURKE, Massachusetts Institute of Technology (emeritus professor)
MARY LYNNE DITTMAR, Dittmar Associates Inc.
PASCALE EHRENFREUND, George Washington University
JAMES S. JACKSON, University of Michigan
FRANK G. KLOTZ,[1] Council on Foreign Relations
FRANKLIN D. MARTIN, Martin Consulting, Inc.
DAVID C. MOWERY, University of California, Berkeley (emeritus professor)
BRYAN D. O'CONNOR, Independent Aerospace Consultant
STANLEY PRESSER, University of Maryland
HELEN R. QUINN, SLAC National Accelerator Laboratory (emeritus professor)
ASIF A. SIDDIQI, Fordham University
JOHN C. SOMMERER, Johns Hopkins University (retired)
ROGER TOURANGEAU, Westat, Inc.
ARIEL WALDMAN, Spacehack.org
CLIFF ZUKIN, Rutgers University, The State University of New Jersey

Staff

SANDRA GRAHAM, Senior Program Officer, *Study Director*
MICHAEL H. MOLONEY, Director, Aeronautics and Space Engineering Board and Space Studies Board
ABIGAIL SHEFFER, Associate Program Officer
AMANDA R. THIBAULT, Research Associate
DIONNA J. WILLIAMS, Program Coordinator
F. HARRISON DREVES, Lloyd V. Berkner Space Policy Intern, Summer 2013
JINNI MEEHAN, Lloyd V. Berkner Space Policy Intern, Fall 2013
CHERYL MOY, Christine Mirzayan Science and Technology Policy Graduate Fellow, Fall 2012
SIERRA SMITH, Lloyd V. Berkner Space Policy Intern, Fall 2013
PADAMASHRI SURESH, Christine Mirzayan Science and Technology Policy Graduate Fellow, Winter 2014

PUBLIC AND STAKEHOLDER OPINIONS PANEL

ROGER TOURANGEAU, Westat, Inc., *Chair*
MOLLY ANDOLINA, DePaul University
JENNIFER L. HOCHSCHILD, Harvard University
JAMES S. JACKSON, University of Michigan
ROGER D. LAUNIUS, Smithsonian Institution
JON D. MILLER, University of Michigan
STANLEY PRESSER, University of Maryland
CLIFF ZUKIN, Rutgers, the State University of New Jersey

Staff

KRISZTINA MARTON, Senior Program Officer, Committee on National Statistics
CONSTANCE CITRO, Director, Committee on National Statistics
JACQUELINE R. SOVDE, Program Associate, Committee on National Statistics

---

[1] General Klotz resigned from the committee on April 10, 2014, to take up an appointment as under secretary of energy for nuclear security and administrator of the National Nuclear Security Administration.

TECHNICAL PANEL

JOHN C. SOMMERER, Johns Hopkins University (retired), *Chair*
DOUGLAS S. STETSON, Space Science and Exploration Consulting Group, *Vice Chair*
ARNOLD D. ALDRICH, Aerospace Consultant
DOUGLAS M. ALLEN, Independent Consultant
RAYMOND E. ARVIDSON, Washington University in St. Louis
RICHARD C. ATKINSON, University of California, San Diego (professor emeritus)
ROBERT D. BRAUN, Georgia Institute of Technology
ELIZABETH R. CANTWELL, Lawrence Livermore National Laboratory
DAVID E. CROW, University of Connecticut (professor emeritus)
RAVI B. DEO, EMBR
ROBERT S. DICKMAN, RD Space, LLC
DAVA J. NEWMAN, Massachusetts Institute of Technology
JOHN ROGACKI, Florida Institute for Human and Machine Cognition (Ocala)
GUILLERMO TROTTI, Trotti and Associates, Inc.
LINDA A. WILLIAMS, Wyle Aerospace Group

Staff

ALAN C. ANGLEMAN, Senior Program Officer, Aeronautics and Space Engineering Board
DIONNA J. WILLIAMS, Program Coordinator, Space Studies Board

Preface

The mandate to carry out this study originated in the National Aeronautics and Space Administration (NASA) Authorization Act of 2010, which, as shown below, required that NASA ask the National Academies to perform a human spaceflight study that would review "the goals, core capabilities, and direction of human space flight." The language of the act reflected concerns that—in the absence of an accepted and independent basis for the establishment of long-term goals—political cycles and other factors would continue to drive instability in the human spaceflight program.

National Aeronautics and Space Administration Authorization Act of 2010 (P.L. 111-267), Section 204

SEC. 204. INDEPENDENT STUDY ON HUMAN EXPLORATION OF SPACE.

(a) IN GENERAL.—In fiscal year 2012 the Administrator shall contract with the National Academies for a review of the goals, core capabilities, and direction of human space flight, using the goals set forth in the National Aeronautics and Space Act of 1958, the National Aeronautics and Space Administration Authorization Act of 2005, and the National Aeronautics and Space Administration Authorization Act of 2008, the goals set forth in this Act, and goals set forth in any existing statement of space policy issued by the President.

(b) ELEMENTS.—The review shall include—
  (1) a broad spectrum of participation with representatives of a range of disciplines, backgrounds, and generations, including civil, commercial, international, scientific, and national security interests;
  (2) input from NASA's international partner discussions and NASA's Human Exploration Framework Team;
  (3) an examination of the relationship of national goals to foundational capabilities, robotic activities, technologies, and missions authorized by this Act;
  (4) a review and prioritization of scientific, engineering, economic, and social science questions to be addressed by human space exploration to improve the overall human condition; and
  (5) findings and recommendations for fiscal years 2014 through 2023.

In the decade or so leading up to the request, the human spaceflight program in the United States had experienced considerable programmatic turbulence, with frequent and dramatic changes in program goals and mission plans in response to changes in national policies. The changes had a high cost in program resources and opportunities and imposed what many feared was an intolerable burden on already constrained human exploration budgets. Because of the effects of continuing volatility in the exploration program, stakeholders in human spaceflight—including those in the government—had been seeking a means of stabilizing the program for some time. Many studies have been conducted by respected members of the research, policy, and commercial communities, but the resulting changes in human exploration policy have often been limited. In particular, uncertainty among policy planners about the fundamental rationale for and future of the U.S. human spaceflight program remained.

Since the Apollo era, the space science community has had considerable success in selecting and setting priorities among suites of missions by using decadal surveys prepared by the National Research Council (NRC). In large part because of that process, NASA's space science programs have achieved remarkable stability in the long term. The pursuit of the goal of a similar level of long-term stability in human exploration led to the mandate for the present study in the NASA Authorization Act of 2010. The language of the act made it clear that a broad array of perspectives and expertise must be represented in the study and that a wide array of benefits, including societal benefits, must be examined.

After the law's enactment, the NRC's Aeronautics and Space Engineering Board (ASEB) and Space Studies Board (SSB) discussed the requested study and possible approaches to its execution at their fall and spring meetings in 2010 and 2011. As a result of those discussions, a small working group made up of members of the two boards and the Division on Engineering and Physical Sciences (DEPS) chair was assembled to formulate a statement of task, which became the basis of discussion with NASA that went on from November 2010 through February 2012. The extended discussion included the NASA administrator, deputy administrator, associate administrator for human exploration and operations, and other key NASA staff. Consultations were also conducted with key House of Representatives and Senate staff. Consultations were also held with the Office of Management and Budget on the wording of the statement of task. It became clear during those discussions that the study committee would need to look well beyond scientific and technical issues and extend its inquiries into such fields as sociology, economics, and political science. A collaboration was therefore formed between the NRC's DEPS and the Division of Behavioral and Social Sciences and Education (DBASSE). DBASSE selected its Committee on National Statistics to serve as a partner with the NRC's leading space boards—SSB and ASEB—in DEPS to carry out the study. ASEB was to have the lead role in the project. In addition, the DBASSE Board on Behavioral, Cognitive, and Sensory Sciences (BBCSS) played an extensive consulting role during study development.

A statement of task describing the study was agreed on in early 2012, and funding became available for the activity in the second half of 2012. The final version of the statement of task assigned to the committee reads as follows:

In accordance with Section 204 of the NASA Authorization Act 2010, the National Research Council (NRC) will appoint an ad hoc committee to undertake a study to review the long-term goals, core capabilities, and direction of the U.S. human spaceflight program and make recommendations to enable a sustainable U.S. human spaceflight program.

The committee will:

1. Consider the goals for the human spaceflight program as set forth in (a) the National Aeronautics and Space Act of 1958, (b) the National Aeronautics and Space Administration Authorization Acts of 2005, 2008, and 2010, and (c) the National Space Policy of the United States (2010), and any existing statement of space policy issued by the president of the United States.

2. Solicit broadly-based, but directed, public and stakeholder input to understand better the motivations, goals, and possible evolution of human spaceflight—that is, the foundations of a rationale for a compelling and sustainable U.S. human spaceflight program—and to characterize its value to the public and other stakeholders.

3. Describe the expected value and value proposition of NASA's human spaceflight activities in the context of national goals—including the needs of government, industry, the economy, and the public good—and in the context of the priorities and programs of current and potential international partners in the spaceflight program.

4. Identify a set of high-priority enduring questions that describe the rationale for and value of human exploration in a national and international context. The questions should motivate a sustainable direction for the long-term exploration of space by humans. The enduring questions may include scientific, engineering, economic, cultural, and social science questions to be addressed by human space exploration and questions on improving the overall human condition.

5. Consider prior studies examining human space exploration, and NASA's work with international partners, to understand possible exploration pathways (including key technical pursuits and destinations) and the appropriate balance between the "technology push" and "requirements pull". Consideration should include the analysis completed

by NASA's Human Exploration Framework Team, NASA's Human Spaceflight Architecture Team, the Review of U.S. Human Spaceflight Plans (Augustine Commission), previous NRC reports, and relevant reports identified by the committee.

6. Examine the relationship of national goals to foundational capabilities, robotic activities, technologies, and missions authorized by the NASA Authorization Act of 2010 by assessing them with respect to the set of enduring questions.

7. Provide findings, rationale, prioritized recommendations, and decision rules that could enable and guide future planning for U.S. human space exploration. The recommendations will describe a high-level strategic approach to ensuring the sustainable pursuit of national goals enabled by human space exploration, answering enduring questions, and delivering value to the nation over the fiscal year (FY) period of FY2014 through FY2023, while considering the program's likely evolution in 2015-2030.

A clear outcome of the wide-ranging consultations carried out by ASEB and SSB in advance of this study's formal beginning—and arguably a requirement of the language in the congressional mandate—was that the committee appointed to carry out the statement of task should contain a breadth of backgrounds spanning expertise in not only human exploration but also such fields as space science, science more broadly, sociology, the science of public polling, political science and history, and economics. In that regard, the Committee on Human Spaceflight looks different from committees that have carried out many previous studies related to human spaceflight by the NRC or other organizations, and it provides a fresh independent perspective on the issues involved in this much-studied endeavor. Because the committee's membership was so broadly based, the NRC decided to appoint two panels of subject-matter experts to assist in providing an independent assessment of the technical challenges of human spaceflight and to provide expert analysis of decades of public polling and of the stakeholder opinions solicited for this study. The Technical Panel facilitated a robust and independent understanding of the technical and engineering aspects of the study, and the Public and Stakeholder Opinions Panel obtained and examined public and stakeholder data and analyses to help the committee to understand the motivations, goals, and possible evolution of human spaceflight. (The panels were responsible for the development of Chapters 4 and 3, respectively, of this report, but it should be noted that the whole report has been adopted by the committee on a consensus basis.) In addition, the committee and its panels were assisted by contractors who had extensive experience in mission technical and cost assessments and in the development and conduct of surveys.

The committee and its panels engaged in extensive data-gathering activities throughout the course of the study. They included review of an extensive database of literature on human spaceflight that included several decades of blue-ribbon studies and the writings of diverse stakeholders. The committee and panel deliberations were informed by invited speakers who had a variety of backgrounds and included representatives of NASA, international space agencies, the aerospace industry, congressional staff, and academe. During the course of the study, the committee disseminated widely a call for interested parties to submit papers that described their own ideas on the role of human spaceflight and their vision of a suggested future, and about 200 responses were received and reviewed by the committee (a list is provided in Appendix H). To broaden the scope of the study's outreach further, the committee turned to social media and held a 1-day Twitter event that allowed any interested parties an opportunity to provide less formal input. Those activities were separate from the formal stakeholder survey and public poll analysis conducted by the Public and Stakeholder Opinions Panel during the course of the study. Various members of the committee gathered information and input from relevant U.S. and international conferences during the study and conducted information-gathering visits to NASA's Johnson Space Center, Kennedy Space Center, and Marshall Space Flight Center.

The committee is grateful to the many people who participated and provided input into this study through all those activities. The Public and Stakeholder Opinions Panel would like to thank the National Opinion Research Center at the University of Chicago and Westat interns Reanne Townsend and Kay Ricci for their help in the stakeholder survey and the graduate survey research class of the Bloustein School of Planning and Public Policy of Rutgers, The State University of New Jersey for assistance in compiling the public opinion data. The committee also acknowledges the vital analytic support that Randy Persinger and Torrey Radcliffe, of the Aerospace Corporation, provided to the committee and the Technical Panel and the considerable leadership shown by NRC staff in assisting the committee and panels in their work.

Acknowledgment of Reviewers

This report has been reviewed in draft form by individuals chosen for their diverse perspectives and technical expertise, in accordance with procedures approved by the National Research Council's (NRC's) Report Review Committee. The purpose of this independent review is to provide candid and critical comments that will assist the institution in making its published report as sound as possible and to ensure that the report meets institutional standards for objectivity, evidence, and responsiveness to the study charge. The review comments and draft manuscript remain confidential to protect the integrity of the deliberative process. We wish to thank the following individuals for their review of this report:

Norman M. Bradburn, University of Chicago,
Erik L. Burgess, Burgess Consulting, Inc.,
David C. Byers, Independent Consultant, Las Vegas, Nevada,
Eileen M. Collins, Space Presentations, LLC,
Ian A. Crawford, Birkbeck College, University of London,
Edward F. Crawley, Massachusetts Institute of Technology,
Donald A. Dillman, Washington State University,
Irwin Feller, Pennsylvania State University,
James W. Head III, Brown University,
Gerda Horneck, Institute of Aerospace Medicine, German Aerospace Center, DLR,
Kathleen Hall Jamieson, University of Pennsylvania,
John M. Logsdon, George Washington University,
James Clay Moltz, Naval Postgraduate School,
Simon Ostrach, Case Western Reserve University,
Andy Peytchev, Research Triangle Institute,
Joseph H. Rothenberg, Swedish Space Corporation,
Carol E. Scott-Conner, University of Iowa Hospitals and Clinics,
Marcia S. Smith, Space and Technology Policy Group, LLC, and
Patricia G. Smith, Patti Grace Smith Consulting, LLC.

Although the reviewers listed above have provided many constructive comments and suggestions, they were not asked to endorse the conclusions or recommendations, nor did they see the final draft of the report before its release. The review of this report was overseen by Louis J. Lanzerotti, New Jersey Institute of Technology. Appointed by the NRC, he was responsible for making certain that an independent examination of this report was carried out in accordance with institutional procedures and that all review comments were carefully considered. Responsibility for the final content of this report rests entirely with the authoring committee and the institution.

Contents

SUMMARY   1

1 OVERVIEW OF ANALYSIS AND FINDINGS   8
   1.1 Introduction, 8
   1.2 U.S. Space Policy Past and Present, 11
   1.3 International Context, 20
   1.4 Enduring Questions and Rationales, 26
        1.4.1 Enduring Questions, 26
        1.4.2 Rationales, 27
        1.4.3 Value and Value Propositions, 28
   1.5 Public and Stakeholder Opinion, 29
        1.5.1 Analysis of Public Opinion Polls, 29
        1.5.2 Stakeholder Views, 32
   1.6 A Strategic Approach to a Sustainable Program of Human Spaceflight, 34
        1.6.1 Horizon Goal: Mars, 34
        1.6.2 Stepping Stones, 35
        1.6.3 Pathway Principles and Decision Rules, 38
        1.6.4 Two Examples of Futures for Human Spaceflight: The Fiscal Challenge Ahead, 39
        1.6.5 Risk Tolerance in a Sustained Program of Human Space Exploration, 41
   1.7 Summary: A Sustainable U.S. Human Space Exploration Program, 41

2 WHY DO WE GO THERE?   44
   2.1 Introduction, 44
   2.2 Outreach Efforts, 44
   2.3 Enduring Questions, 46
   2.4 Rationales for Human Spaceflight, 48
        2.4.1 Economic and Technology Impacts, 48
        2.4.2 National Security and Defense, 53
        2.4.3 National Stature and International Relations, 57

2.4.4 Education and Inspiration, 58
2.4.5 Scientific Exploration and Observation, 61
2.4.6 Survival, 66
2.4.7 Shared Human Destiny and Aspiration, 67
2.5 Assessment of Rationales, 70
2.6 Value Propositions, 70
2.6.1 The Problem with Value Propositions, 70
2.6.2 Stakeholder Value and the Impacts of Ending Human Spaceflight, 76
2.7 Conclusions on the Benefits of Human Spaceflight, 81

3 PUBLIC AND STAKEHOLDER ATTITUDES   83
3.1 Public Opinion, 83
3.1.1 Interest in Space Exploration and the Attentive Public, 84
3.1.2 Support for Spending on Space Exploration, 86
3.1.3 Trends in Support for Specific Human Spaceflight Missions, 88
3.1.4 Human Versus Robotic Missions, 90
3.1.5 NASA's Role, International Collaboration, and Commercial Firms, 90
3.1.6 Rationales for Support of Space Exploration, 91
3.1.7 Correlates of Support for Space Exploration, 92
3.1.8 Summary of Findings on Public Opinion, 93
3.2 Stakeholder Survey, 95
3.2.1 Characteristics of the Respondents, 96
3.2.2 Rationales for Space Exploration and Human Space Exploration, 98
3.2.3 Views on a Course for the Future, 101
3.2.4 Other Findings, 105
3.2.5 Correlates of Support for Human Spaceflight, 106
3.2.6 Summary of Findings of the Stakeholder Survey, 106

4 TECHNICAL ANALYSIS AND AFFORDABILITY ASSESSMENT OF HUMAN   109
   EXPLORATION PATHWAYS
4.1 Introduction and Overview, 109
4.2 Technical Requirements, 112
4.2.1 Possible Destinations in the Context of Foreseeable Technology, 112
4.2.2 Design Reference Missions, 114
4.2.3 Potential Pathways, 119
4.2.4 Drivers and Requirements of Key Mission Element Groups, 120
4.2.5 Contribution of Key Mission Elements to the Pathways, 125
4.2.6 Challenges in Developing Key Capabilities, 130
4.2.7 Affordability, 151
4.2.8 Assessment of Pathways Against Desirable Pathway Properties, 163
4.3 Technology Programs, 168
4.3.1 NASA Technology Programs, 169
4.3.2 Human Exploration and Operations Mission Directorate, 169
4.3.3 Commercial Programs, 171
4.3.4 Department of Defense, 172
4.3.5 International Activities, 172
4.3.6 Robotic Systems, 173
4.4 Key Results from the Panel's Technical Analysis and Affordability Assessment, 175

APPENDIXES

A   Statement of Task   179
B   Methodological Notes About the Public Opinion Data   181
C   Stakeholder Survey Methods   188
D   Stakeholder Survey Mail Questionnaire (Version A)   193
E   Frequency Distributions of Responses to the Stakeholder Survey by Respondent Group   201
F   Acronyms and Abbreviations   231
G   List of Briefings to the Committee and Panels   233
H   List of Input Papers   238
I   Committee, Panel, and Staff Biographies   243

Summary

The United States has publicly funded its human spaceflight program continuously for more than a half-century. Today, the United States is the major partner in a massive orbital facility, the International Space Station (ISS), that is a model for how U.S. leadership can engage nations through soft power and that is becoming the focal point for the first tentative steps in commercial cargo and crewed orbital spaceflights. Yet, a national consensus on the long-term future of human spaceflight beyond our commitment to the ISS remains elusive.

The task for the Committee on Human Spaceflight originated in the National Aeronautics and Space Administration [NASA] Authorization Act of 2010, which required that the National Academies perform a human-spaceflight study that would review "the goals, core capabilities, and direction of human space flight." The explicit examination of rationales, along with the identification of enduring questions set the task apart from numerous similar studies performed over the preceding several decades, as did the requirement that the committee bring broad public and stakeholder input into its considerations. The complex mix of historical achievement and uncertain future made the task faced by the committee extraordinarily challenging and multidimensional. Nevertheless, the committee has come to agree on a set of major conclusions and recommendations, which are summarized here.

ENDURING QUESTIONS

Enduring questions are questions that serve as motivators of aspiration, scientific endeavors, debate, and critical thinking in the realm of human spaceflight. The questions endure in that any answers available today are at best provisional and will change as more exploration is done. Enduring questions provide motivations that are immune to external forces and policy shifts. They are intended not only to stand the test of time but also to continue to drive work forward in the face of technological, societal, and economic constraints. Enduring questions are clear and intrinsically connect to broadly shared human experience. On the basis of the analysis reported in Chapter 2, the committee asserts that the enduring questions motivating human spaceflight are these:

- **How far from Earth can humans go?**
- **What can humans discover and achieve when we get there?**

RATIONALES FOR HUMAN SPACEFLIGHT AND THE PUBLIC INTEREST

All the arguments that the committee heard in support of human spaceflight have been used in various forms and combinations to justify the program for many years. In the committee's view, these rationales can be divided into two sets. Pragmatic rationales involve economic benefits, contributions to national security, contributions to national stature and international relations, inspiration for students and citizens to further their science and engineering education, and contributions to science. Aspirational rationales involve the eventual survival of the human species (through off-Earth settlement) and shared human destiny and the aspiration to explore. In reviewing the rationales, the committee concluded as follows:

- *Economic matters.* There is no widely accepted, robust quantitative methodology to support comparative assessments of the returns on investment in federal R&D programs in different economic sectors and fields of research. Nevertheless, it is clear that the NASA human spaceflight program, like other government R&D programs, has stimulated economic activity and has advanced development of new products and technologies that have had or may in the future generate significant economic impacts. It is impossible, however, to develop a reliable comparison of the returns on spaceflight versus other government R&D investment.
- *Security.* Space-based assets and programs are an important element of national security, but the direct contribution of human spaceflight in this realm has been and is likely to remain limited. An active U.S. human spaceflight program gives the United States a stronger voice in an international code of conduct for space, enhances U.S. soft power, and supports collaborations with other nations; thus, it contributes to our national interests, including security.
- *National stature and international relations.* Being a leader in human space exploration enhances international stature and national pride. Because the work is complex and expensive, it can benefit from international cooperative efforts. Such cooperation has important geopolitical benefits.
- *Education and inspiration.* The United States needs scientists and engineers and a public that has a strong understanding of science. The challenge and excitement of space missions can serve as an inspiration for students and citizens to engage with science and engineering although it is difficult to measure this. The path to becoming a scientist or engineer requires much more than the initial inspiration. Many who work in space fields, however, report the importance of such inspiration, although it is difficult to separate the contributions of human and robotic spaceflight.
- *Scientific discovery.* The relative benefits of robotic versus human efforts in space science are constantly shifting as a result of changes in technology, cost, and risk. The current capabilities of robotic planetary explorers, such as Curiosity and Cassini, are such that although they can go farther, go sooner, and be much less expensive than human missions to the same locations, they cannot match the flexibility of humans to function in complex environments, to improvise, and to respond quickly to new discoveries. Such constraints may change some day.
- *Human survival.* It is not possible to say whether human off-Earth settlements could eventually be developed that would outlive human presence on Earth and lengthen the survival of our species. That question can be answered only by pushing the human frontier in space.
- *Shared destiny and aspiration to explore.* The urge to explore and to reach challenging goals is a common human characteristic. Space is today a major physical frontier for such exploration and aspiration. Some say that it is human destiny to continue to explore space. While not all share this view, for those who do it is an important reason to engage in human spaceflight.

As discussed in Chapter 2, the pragmatic rationales have never seemed adequate by themselves, perhaps because the benefits that they argue for are not unique to human spaceflight. Those that are—the aspirational rationales related to the human destiny to explore and the survival of the human species—are also the rationales most tied to the enduring questions. Whereas the committee concluded from its review and assessment that no single rationale alone seems to justify the costs and risks of pursuing human spaceflight, the aspirational rationales, when supplemented by the practical benefits associated with the pragmatic rationales, do, in the committee's judgment, argue for a continuation of our nation's human spaceflight program, provided that the pathways and decision rules recommended by the committee are adopted (see below).

The level of public interest in space exploration is modest relative to interest in other public-policy issues such as economic issues, education, and medical or scientific discoveries. Public opinion about space has been generally favorable over the past 50 years, but much of the public is inattentive to space exploration, and spending on space exploration does not have high priority for most of the public.

HORIZON GOAL

The technical analysis completed for this study shows clearly that for the foreseeable future the only feasible destinations for human exploration are the Moon, asteroids, Mars, and the moons of Mars. Among that small set of plausible goals for human space exploration,[1] the most distant and difficult is a landing by human beings on the surface of Mars; it would require overcoming unprecedented technical risk, fiscal risk, and programmatic challenges. Thus, the "horizon goal" for human space exploration is Mars. All long-range space programs, by all potential partners, for human space exploration converge on that goal.

POLICY CHALLENGES

A program of human space exploration beyond low Earth orbit (LEO) that satisfies the pathway principles defined below is not sustainable with a budget that increases only enough to keep pace with inflation. As shown in Chapter 4, the current program to develop launch vehicles and spacecraft for flight beyond LEO cannot provide the flight frequency required to maintain competence and safety, does not possess the "stepping-stone" architecture that allows the public to see the connection between the horizon goal and near-term accomplishments, and may discourage potential international partners.

Because policy goals do not lead to sustainable programs unless they also reflect or change programmatic, technical, and budgetary realities, the committee notes that those who are formulating policy goals will need to keep the following factors in mind:

- Any defensible calculation of tangible, quantifiable benefits—spinoff technologies, attraction of talent to scientific careers, scientific knowledge, and so on—is unlikely ever to demonstrate a positive economic return on the massive investments required by human spaceflight.
- The arguments that triggered the Apollo investments—national defense and prestige—seem to have especially limited public salience in today's post-Cold War America.
- Although the public is mostly positive about NASA and its spaceflight programs, increased spending on spaceflight has low priority for most Americans. However, although most Americans do not follow the issue closely, those who pay more attention are more supportive of space exploration.

INTERNATIONAL COLLABORATION

International collaboration has become an integral part of the space policy of essentially all nations that participate in space activities around the world. Most countries now rarely initiate and carry out substantial space projects without some foreign participation. The reasons for collaboration are multiple, but countries, including the United States, cooperate principally when they benefit from it.

It is evident that near-term U.S. goals for human exploration are not aligned with those of our traditional international partners. Although most major spacefaring nations and agencies are looking toward the Moon, specifically the lunar surface, U.S. plans are focused on redirection of an asteroid into a retrograde lunar orbit where astronauts

---

[1] Although there is no strictly defined distinction between human spaceflight and human space exploration, the committee takes the latter to mean spaceflight beyond low Earth orbit, in which the goal is to have humans venture into the cosmos to discover new things.

would conduct operations with it. It is also evident that given the rapid development of China's capabilities in space, it is in the best interests of the United States to be open to its inclusion in future international partnerships. In particular, current federal law that prevents NASA from participating in bilateral activities with the Chinese serves only to hinder U.S. ability to bring China into its sphere of international partnerships and substantially reduces the potential international capability that might be pooled to reach Mars. Also, given the scale of the endeavor of a mission to Mars, contributions by international partners would have to be of unprecedented magnitude to defray a significant portion of the cost. This assessment follows from the detailed discussion in Chapter 4 of what is required for human missions to Mars.

RECOMMENDATIONS FOR A PATHWAYS APPROACH

NASA and its international and commercial partners have developed an infrastructure in LEO that is approaching maturity—that is, assembly of the ISS is essentially complete. The nation must now decide whether to embark on human space exploration beyond LEO in a sustained and sustainable fashion. Having considered past and current space policy, explored the international setting, articulated the enduring questions and rationales, and identified public and stakeholder opinions, the committee drew on all this information to ask a fundamental question: What type of human spaceflight program would be responsive to these factors? The committee argues that it is a program in which humans operate beyond LEO on a regular basis—a sustainable human exploration program beyond LEO.

A sustainable program of human deep-space exploration requires an ultimate horizon goal that provides a long-term focus that is less likely to be disrupted by major technological failures and accidents along the way or by the vagaries of the political process and the economic scene. There is a consensus in national space policy, international coordination groups, and the public imagination for Mars as a major goal for human space exploration. NASA can sustain a human space-exploration program that pursues the horizon goal of a surface landing on Mars with meaningful milestones and simultaneously reasserts U.S. leadership in space while allowing ample opportunity for substantial international collaboration—but only if the program has elements that are built in a logical sequence and if it can fund a frequency of flights sufficiently high to ensure the maintenance of proficiency among ground personnel, mission controllers, and flight crews. In the pursuit of that goal, NASA needs to engage in the type of mission planning and related technology development that address mission requirements and integration and develop high-priority capabilities, such as entry, descent, and landing for Mars; radiation safety; and advanced in-space propulsion and power. Progress in human exploration beyond LEO will be measured in decades with costs measured in hundreds of billions of dollars and significant risk to human life.

In addition, the committee has concluded that the best way to ensure a stable, sustainable human-spaceflight program that pursues the rationales and enduring questions that the committee has identified is to develop a program through the rigorous application of a set of pathway principles. The committee's highest-priority recommendation is as follows:

NASA should adopt the following pathway principles:
I.   **Commit to designing, maintaining, and pursuing the execution of an exploration pathway beyond low Earth orbit toward a clear horizon goal that addresses the "enduring questions" for human spaceflight.**
II. **Engage international space agencies early in the design and development of the pathway on the basis of their ability and willingness to contribute.**
III. **Define steps on the pathway that foster sustainability and maintain progress on achieving the pathway's long-term goal of reaching the horizon destination.**
IV. **Seek continuously to engage new partners that can solve technical or programmatic impediments to progress.**
V.   **Create a risk-mitigation plan to sustain the selected pathway when unforeseen technical or budgetary problems arise. Such a plan should include points at which decisions are made to move to a less ambitious pathway (referred to as an "off-ramp") or to stand down the program.**

VI. **Establish exploration pathway characteristics that maximize the overall scientific, cultural, economic, political, and inspirational benefits without sacrificing progress toward the long-term goal, namely,**
  a. **The horizon and intermediate destinations have profound scientific, cultural, economic, inspirational, or geopolitical benefits that justify public investment.**
  b. **The sequence of missions and destinations permits stakeholders, including taxpayers, to see progress and to develop confidence in NASA's ability to execute the pathway.**
  c. **The pathway is characterized by logical feed-forward of technical capabilities.**
  d. **The pathway minimizes the use of dead-end mission elements that do not contribute to later destinations on the pathway.**
  e. **The pathway is affordable without incurring unacceptable development risk.**
  f. **The pathway supports, in the context of available budget, an operational tempo that ensures retention of critical technical capability, proficiency of operators, and effective use of infrastructure.**

The pathway principles will need to be supported by a set of operational decision rules as NASA, the administration, and Congress face inevitable programmatic challenges along a selected pathway. The decision rules that the committee has developed provide operational guidance that can be applied when major technical, cost, and schedule issues arise as NASA progresses along a pathway. Because many decisions will have to be made before any program of record is approved and initiated, the decision rules have been designed to provide the framework for a sustainable program through the lifetime of the selected pathway. They are designed to allow a program to stay within the constraints that are accepted and developed when the pathway principles are applied. The committee recommends that,

Whereas the overall pathway scope and cost are defined by application of the pathway principles, once a program is on a pathway, technical, cost, or schedule problems that arise should be addressed by the administration, NASA, and Congress by applying the following decision rules:
  A. **If the appropriated funding level and 5-year budget projection do not permit execution of a pathway within the established schedule, do not start down that pathway.[2]**
  B. **If a budget profile does not permit the chosen pathway, even if NASA is well along on it, take an "off-ramp."**
  C. **If the U.S. human spaceflight program receives an unexpected increase in budget for human spaceflight, NASA, the administration, and Congress should not redefine the pathway in such a way that continued budget increases are required for the pathway's sustainable execution; rather, the increase in funds should be applied to rapid retirement of important technology risks or to an increase in operational tempo in pursuit of the pathway's previously defined technical and exploration goals.**
  D. **Given that limitations on funding will require difficult choices in the development of major new technologies and capabilities, give high priority to choices that solve important technological shortcomings, that reduce overall program cost, that allow an acceleration of the schedule, or that reduce developmental or operational risk.**
  E. **If there are human spaceflight program elements, infrastructure, or organizations that are no longer contributing to progress along the pathway, the human spaceflight program should divest itself of them as soon as possible.**

---

[2] The committee recognizes that budget projections are unreliable, but they are also indispensable. One way to make such projections more robust would be for NASA to conduct sensitivity analysis and evaluate plans against a range of possible 5-year budget projections that may vary by 10 percent or more. The analysis and evaluation might be undertaken as part of the risk-mitigation plan.

RECOMMENDATIONS FOR IMPLEMENTING A SUSTAINABLE PROGRAM

The committee was not charged to recommend and has not recommended any particular pathway or set of destination targets. The recommended pathways approach combines a strategic framework with practical guidance that is designed to stabilize human space exploration and to encourage political and programmatic coherence over time.

If the United States is to have a human space-exploration program, it must be worthy of the considerable cost to the nation and great risk of life. The committee has found no single practical rationale that is uniquely compelling to justify such investment and risk. Rather, human space exploration must be undertaken for inspirational and aspirational reasons that appeal to a broad array of U.S. citizens and policy-makers and that identify and align the United States with technical achievement and sophistication while demonstrating its capability to lead or work within an international coalition for peaceful purposes. Given the expense of any human spaceflight program and the substantial risk to the crews involved, it is the committee's view that the only pathways that fit those criteria are ones that ultimately place humans on other worlds.

Although the committee's recommendation to adopt a pathways approach is made without prejudice as to which particular pathway might be followed, it was clear to the committee from its independent analysis of several pathways that a return to extended surface operations on the Moon would make substantial contributions to a strategy ultimately aimed at landing people on Mars and would probably provide a broad array of opportunities for international and commercial cooperation. No matter which pathway is selected, the successful implementation of any plan developed in concert with a pathways approach and decision rules will rest on several other conditions. In addition to its highest-priority recommendation of the pathways approach and decision rules, the committee offers the following priority-ordered recommendations as being the ones that are most critical to the development and implementation of a sustainable human space-exploration program.

NASA should
1. **Commit to design, maintain, and pursue the extension of human presence beyond low Earth orbit (LEO). This step should include**
   a. **Committing NASA's human spaceflight asset base, both physical and human, to this effort.**
   b. **Redirecting human spaceflight resources as needed to include improving program-management efficiency (including establishing and managing to appropriate levels of risk), eliminating obsolete facilities, and consolidating remaining infrastructure where possible.**
2. **Maintain long-term focus on Mars as the horizon goal for human space exploration, addressing the enduring questions for human spaceflight: How far from Earth can humans go? What can humans do and achieve when we get there?**
3. **Establish and implement the pathways approach so as to maximize the overall scientific, cultural, economic, political, and inspirational benefits of individual milestones and to conduct meaningful work at each step along the pathway without sacrificing progress toward long-term goals.**
4. **Vigorously pursue opportunities for international and commercial collaboration in order to leverage financial resources and capabilities of other nations and commercial entities. International collaboration would be open to the inclusion of China and potentially other emerging space powers in addition to traditional international partners. Specifically, future collaborations in major new endeavors should seek to incorporate**
   a. **A level of overall cost-sharing that is appropriate to the true partnerships that will be necessary to pursue pathways beyond LEO.**
   b. **Shared decision-making with partners, including a detailed analysis, in concert with international partners, of the implications for human exploration of continuing the International Space Station beyond 2024.**
5. **Engage in planning that includes mission requirements and a systems architecture that target funded high-priority technology development, most critically**
   a. **Entry, descent, and landing for Mars.**
   b. **Advanced in-space propulsion and power.**
   c. **Radiation safety.**

In this report the committee has provided guidance on how a pathways approach might be successfully pursued and the likely costs of the pathways if things go well. However, the committee also concludes that if the resulting plan is not appropriately financed, it will not succeed. Nor can it succeed without a sustained commitment on the part of those who govern the nation—a commitment that does not change direction with succeeding electoral cycles. Those branches of government—executive and legislative—responsible for NASA's funding and guidance are therefore critical enablers of the nation's investment and achievements in human spaceflight, commissioning and financing plans and then ensuring that the leadership, personnel, governance, and resources are in place at NASA and in other federally funded laboratories and facilities to advance it.

1

Overview of Analysis and Findings

1.1 INTRODUCTION

The United States has publicly funded its human spaceflight program continuously for more than a half-century—through three wars and a half-dozen recessions—from the early Mercury and Gemini suborbital and Earth orbital missions, to the Apollo lunar landings, and then to the first reusable winged, crewed spaceplane, which the United States operated for 3 decades. Today, the United States is the major partner in a massive orbital facility, the International Space Station (ISS), which is becoming the focal point for the first tentative steps in commercial cargo and crewed orbital spaceflights. Yet, the long-term future of human spaceflight beyond the ISS is unclear.

Pronouncements by multiple presidents of bold new U.S. ventures to the Moon, to Mars, and to an asteroid in its native orbit (summarized in Section 1.2 of this chapter) have not been matched by the same commitment that accompanied President Kennedy's now fabled 1961 speech, namely, the substantial increase in NASA funding needed to make them happen. In the view of many observers, the human spaceflight program conducted by the U.S. government today has no strong direction and no firm timetable for accomplishments.

The complex mix of historic achievement and uncertain future made the task[1] faced by this committee extraordinarily difficult. In responding to the task, the committee assessed the historically stated rationales for human spaceflight as rigorously as possible given the available knowledge base with the intent of identifying a set of "enduring questions" akin to the ones that motivate strategic plans for scientific disciplines. The committee also sought to describe the value and "value proposition" of the program, to solicit and interpret stakeholder and public opinion, and to provide conclusions, recommendations, and decision rules that can guide future human spaceflight programs pursued or led by this country. The fruits of the committee's labors[2] as presented here provide a map leading to a human spaceflight program that can avoid some of the ills and false starts of the past. However, to set course on such an endeavor, the nation will need its investment in the human spaceflight program to grow each year in the coming decades. To continue on the present course—pursuit of an exploration system to go beyond low Earth orbit (LEO) while simultaneously operating the ISS through the middle of the next decade as the major partner, all

---

[1] The committee's statement of task is given in Appendix A.

[2] Including six committee meetings, four Technical Panel meetings, three Public and Stakeholder Opinion Panel meetings, visits to Johnson Space Center, Kennedy Space Center, and Marshall Space Flight Center (the key NASA human spaceflight centers), a call to the public for white papers, stakeholder interviews, a public discussion on Twitter, and participation in international conferences.

under a budget profile that fails even to keep pace with inflation[3]—is to invite failure, disillusionment, and the loss of the longstanding international perception that human spaceflight is something that the United States does best.

Throughout this chapter, which gives an overview of the committee's most important conclusions and historical background related to the relevant issues, the recommendations and conclusions are in boldface type. The committee justifies and quantifies them in the later chapters. The essential cornerstones of the committee's findings can be summarized as follows:

- *The rationales for human spaceflight are a mix of the aspirational and the pragmatic.* The primary rationale for the Apollo program was to demonstrate in an unambiguous but peaceful way the technological supremacy of the United States over the Soviet Union by beating it to the Moon.[4] The rationale for Apollo took place not only against a backdrop of Cold War potential for nuclear war but also in the midst of an existential conflict between two fundamentally different economic systems—a conflict that is now over. Quantification of the value of human spaceflight to the nation today, in terms of economic return or increased quality of life, is difficult. That does not mean that there are no benefits: W.B. Cameron wrote that "not everything that can be counted counts, and not everything that counts can be counted."[5]
- *The level of public interest in space exploration is modest relative to interest in other public-policy issues.* As Chapter 3 documents, public opinion about space has been generally favorable over the past 50 years, but much of the public is inattentive to space exploration, and spending on space exploration does not have high priority for most of the public.
- *The horizon goal for human space exploration is Mars.* In Chapter 4, the committee shows that there is a small set of plausible goals for human space exploration in the foreseeable future, of which the most distant and difficult is a landing by human beings on the surface of Mars. All long-range space programs, by all potential partners, for human space exploration converge on this goal.
- *A program of human space exploration beyond LEO that satisfies the pathway principles defined below is not sustainable with a human spaceflight budget that increases only enough to keep pace with inflation.* As shown in Chapter 4, the current program to develop launch vehicles and spacecraft for flight beyond LEO cannot be sustained with constant buying power over time, in that it cannot provide the flight frequency required to maintain competence and safety, does not possess the "stepping-stone" architecture that allows the public to see the connection between the horizon goal and near-term accomplishments, and may discourage potential international partners. In the section "Pathway Principles and Decision Rules" below, the committee outlines a "pathways" approach, which requires the United States to settle on a definite pathway to the horizon goal and adhere to principles and decision rules to get there.

In the course of developing its findings, the committee identified some important issues that the nation will need to grapple with if it chooses to embark on a renewed effort in deep-space exploration involving humans:

- *The nation's near-term goals for human exploration beyond LEO are not aligned with those of our traditional international partners.* The committee heard from representatives of international partners that their near-term goal for human spaceflight was lunar-surface operations. They also made it clear that they could not undertake such a program on their own and were relying on the United States to play a leadership role in human exploration of the Moon. Although those partners expressed interest in aspects of the asteroid-redirect mission (ARM), the committee detected a concern that ARM would divert U.S. resources and attention from an eventual return to the Moon. Of the several pathways examined, the one that

---

[3] In this report, future inflation is projected to be 2.5 percent per year, which is consistent with "2013 NASA New Start Inflation Index for FY14," http://www.nasa.gov/sites/default/files/files/2013_NNSI_FY14(1).xlsx.

[4] President Kennedy, 2 days after Yuri Gagarin's historic flight into space on April 12, 1961, said in a meeting to NASA and cabinet officers, "If somebody can just tell me how to catch up . . . there's nothing more important." H. Sidey, *John F. Kennedy, President,* Atheneum Press, New York, 1963, pp. 122-123.

[5] W.B. Cameron, *Informal Sociology: A Casual Introduction to Sociological Thinking,* Random House, New York, 1963.

does not include a meaningful return to the Moon—that is, extended operations on the lunar surface—has higher development risk than the others.

- *Continued operation of the ISS beyond 2020 will have a near-term effect on the pace that NASA can sustain in exploration programs beyond LEO, but it also affords an opportunity for extended studies related to long-term exposure to microgravity.* The United States and its international partners are committed to operating the ISS jointly through 2020, and the United States recently proposed extension to 2024. In its presentations to the committee, NASA made clear its desire to operate the complex facility through 2028. In addition, the United States has designated half its ISS space and resources a national laboratory, which proponents say will require extension of the ISS beyond 2020 or even 2024 to increase the probability that the research and development (R&D) conducted there will provide substantial returns, including promised commercial benefits. Continued operation will compete with new programs beyond LEO, and this will aggravate the problems of funding. At the same time, the committee recognizes that much work remains to determine physiological tolerance to and countermeasures for the microgravity environments that will be experienced for long periods on flights to and from Mars, and the ISS is the platform on which to do this work. There is thus a tension between moving beyond the ISS to the exploration of deep space at a safe and sustainable pace and conducting the medical studies in LEO needed to execute human exploration of Mars 2 or 3 decades from now.

- *The prohibition of NASA's speaking to Chinese space authorities has left open opportunities for collaboration that are being filled by other spacefaring nations.* The recent docking of a piloted Chinese vessel to a new orbital module and the first robotic rover operations by China on the Moon are the latest developments in a program that marches steadily and strategically toward what might eventually become a lead role among the nations in spaceflight. In contrast with the failure-prone early histories of the U.S. and USSR human spaceflight programs, China has proceeded methodically, deliberately, and with little in the way of visible failure. The U.S. government's response to that has been inconsistent: regarding China as a potential partner in some activities and as a threat in others. The committee is concerned that current U.S. law is impeding the nation's ability to collaborate with China when appropriate whereas traditional U.S. international partners have not imposed such restrictions on themselves.

The remainder of this chapter looks at past, present, and prospective human spaceflight from a number of perspectives, all of which inform the committee's final considerations of a sustainable program. The next section begins with a historical trajectory of space policy with regard to human spaceflight and ends with the current situation. Section 1.3 then assesses the international context in terms of current partnerships and the capabilities of nonpartner nations, such as China. Section 1.4 summarizes the enduring questions and rationales for human spaceflight offered over time and is followed by a summation of the opinions of the U.S. public and stakeholders in human spaceflight.[6] Section 1.6 discusses strategic approaches to a sustainable human spaceflight program beyond LEO that is based on what the committee calls a pathways approach. The chapter concludes by summarizing the requirements for undertaking such an effort and the consequences of embarking on a new program of deep-space exploration without adequate funding. The committee formulates a set of pathways principles and decision rules to guide the program and offers two examples of how the pathways approach might be used to design a human spaceflight program. The report reviews and rearticulates why the nation might choose to move forward and lays out an approach that is responsive to the enduring questions and rationales that are developed and analyzed here.

Although the statement of task mentions two time horizons—one extending to 2023 and the other to 2030—the committee has not attempted to separate recommendations for the two horizons. The pathways approach described here requires integrated programs that will span the entire period up to 2030, so any attempt to divide the time window into a "before" and "after" 2023 is artificial. In fact, in developing and exercising the pathways approach, the committee of necessity considered a time horizon extending into the middle of this century, well beyond the year 2030 specified in the task statement. The committee therefore acknowledges the possibility that

---

[6] Sections 1.2–1.5 summarize the extensive and detailed analyses of enduring questions, value propositions, and stakeholder and public opinions that are offered in Chapters 2 and 3.

over the half-century considered, advances in science and technology in bioengineering, artificial intelligence, and other fields may come far more quickly and unpredictably than the advances contemplated for the human spaceflight pathways proposed in this report. Breakthroughs in these other realms could serve to solve many of the large obstacles to exploration beyond LEO. In particular, the line between the human and the robotic may be blurred more profoundly than simple linear extrapolations predict. In such an eventuality, exploration of the "last frontier" of space might well occur in a more rapid and far-reaching way than is envisioned in this report; indeed, whether it would still be accurately described as human exploration[7] is unknowable.

1.2 U.S. SPACE POLICY PAST AND PRESENT

The U.S. Space and Rocket Center near NASA's Marshall Space Flight Center includes an exhibit that is titled "Great Nations Dare" and is an immersion into the history of exploration.[8] It is a fitting reminder, at the place where the massive Saturn V moon rocket was developed and built, that history is replete with examples of nations and societies that were at the forefront of exploration for brief periods and sooner or later lost their momentum. Although the specific reasons for exploration have varied (expansion of power, trade routes, precious metals, spreading religion, and so on), there has almost always been a nationalistic competitive element that helped in obtaining resources for these expensive adventures.

The committee was charged to consider the goals of NASA as set forth in its founding legislation and the legislative acts and policy directives that followed. The committee provides below a brief history of U.S. human spaceflight efforts and brings the story up to the present day. The goal is not to chronicle every major accomplishment but rather to highlight the principal changes and shifts in civil space policy—especially as related to human spaceflight—that drove the program at the highest level.

Early space exploration was driven largely by competition between nations. The program's effective birth can be traced back to the National Aeronautics and Space Act, which was signed on July 29, 1958, and led to the formation of NASA.[9] In response to the shock of the launch of the Soviet *Sputnik* satellite on October 4, 1957, U.S. policy-makers were mobilized into creating and consolidating a federal infrastructure in support of space activities. Once NASA officially opened its doors on October 1, 1958, its initial activities were guided by the original act and by the Eisenhower administration's preliminary U.S. policy for outer space.[10] The Space Act laid out eight objectives of the U.S. civilian space program:

- The "expansion of human knowledge of phenomena in the atmosphere and space."
- Improving the "usefulness, performance, speed, safety, and efficiency" of rockets and spacecraft.
- The "development and operation of [robotic and crewed] vehicles."
- The "establishment of long-range studies of" (a) the benefits to be gained (b) opportunities for, and (c) problems involved, in the use aerospace activities "for peaceful and scientific purposes."
- "The preservation of the role of the United States as a leader in aeronautical and space science and technology."
- Cooperation with the Department of Defense as required.
- "Cooperation by the United States with other nations and groups of nations in work done pursuant to this Act and in peaceful application of the results, thereof."

---

[7] For the purposes of this report, human exploration of space is defined as flight into regions of space beyond LEO in which humans are in the vehicles. Here, *regions* can refer to position, to orbital energy (velocity), or both. The committee defines as human exploration ambiguous cases in which humans are in martian or lunar orbit telerobotically conducting surface operations because of the astronauts' proximity to the target and their remoteness from Earth.

[8] "'Great Nations Dare' Exploration Technology Exhibit," http://www.nasa.gov/centers/marshall/news/exhibits/great_nations.html, accessed January 19, 2014.

[9] National Aeronautics and Space Act of 1958, Public Law 85-568, 72 Stat., 426. Signed by the president on July 29, 1958, reproduced in full in John M. Logsdon, editor, with L.J. Lear, J. Warren-Findley, R.A. Williamson, and D.A. Day, *Exploring the Unknown: Selected Documents in the History of the U.S. Civil Space Program, Volume I: Organizing for Exploration*, NASA, Washington, D.C., 1995, pp. 334-345.

[10] "National Security Council, NSC 5814, 'U.S. Policy on Outer Space,' June 20, 1958" in *Exploring the Unknown, Volume I*, 1995, pp. 345-359.

- To make effective use of the scientific and engineering resources in the country in cooperation with interested agencies.

On the basis of those objectives, NASA prepared a formal long-range plan in December 1959 ("The Long Range Plan of the National Aeronautics and Space Administration"). The original plan featured a balanced program of science, applications, and human space exploration with the possibility of human flight to the Moon "beyond 1970."[11] Hopes for a long and stable space policy were, however, thrown into doubt with the continuing successes of the Soviet space program, in particular, the launch of the first human being into space in 1961. Cosmonaut Yuri Gagarin's historic flight on April 12, 1961, set into motion a series of events that culminated in a major policy speech on May 25, 1961, in which President Kennedy called for landing an American on the Moon before the end of the decade and returning him safely to Earth. Driven by the need to reassert U.S. confidence in the arena of space, President Kennedy's decision (and considerable congressional support for it) set NASA on a crash program to achieve the Moon-landing goal.[12] In the following 2 years, NASA's budget increased by 89 percent and 101 percent.[13] Over the course of a decade, NASA became a large federal bureaucracy with an associated contractor workforce whose primary (although not sole) goal was to develop the capabilities for human spaceflight with the proximate and primary objective of landing humans on the Moon—only to see its share of the federal budget and horizons shrink with the end of Apollo.

Through the 1960s, NASA implemented the highly successful Mercury and Gemini programs as a lead-up to the Apollo project. The first American astronaut to enter orbit, John Glenn, was launched aboard the Mercury *Friendship 7* vehicle (Figure 1.1). With Mercury and Gemini, NASA centers gained critical experience in performing increasingly complex human space missions that involved extravehicular activity, rendezvous and docking, and long-duration flight. Despite the setback of the tragic *Apollo 1* fire in early 1967, when three astronauts were killed during a ground test at Cape Kennedy, the program progressed by leaps and bounds. It culminated in the landing of *Apollo 11* astronauts Neil Armstrong and Edwin "Buzz" Aldrin on the Moon on July 20, 1969, and thus fulfilled President Kennedy's mandate. After five further landings (and *Apollo 13*, which failed to land), the Apollo program ended in 1972 (Figure 1.2). Equipment left over from Apollo was used for the long-duration Skylab project in 1973–1974 and the Apollo-Soyuz Test Project (ASTP) in 1975.

By the time ASTP was implemented, NASA was heavily invested in a new program whose origins date back to the report of the Space Task Group (STG) headed by Vice President Agnew. In its report, *The Post-Apollo Space Program: Directions for the Future*, which was delivered to President Nixon in September 1969, the STG endorsed a NASA proposal for "a balanced manned and unmanned space program conducted for the benefit of all mankind" that could include missions to Mars, a space station, and the construction of a space shuttle for routine access to Earth orbit (Figure 1.3).[14] In response, however, President Nixon issued a major statement on the U.S. space program in March 1970 that downgraded the STG's objectives. Arguing against the very high priority afforded Apollo, President Nixon contended that "space expenditures must take their place within a rigorous system of national priorities." In effect, such an approach has guided U.S. civilian space policy for the past four decades. By the time President Nixon left the White House, the NASA budget had fallen from its peak of nearly 4 percent of the total federal budget to less than 1 percent, which is essentially where it has remained.

In January 1972, President Nixon announced the development of a partially reusable crewed vehicle, the space shuttle. That decision was the outcome of a complicated set of negotiations over post-Apollo goals in human spaceflight that included a possible Earth-orbiting space station, a Mars mission, and a space shuttle.[15] Because the

---

[11] "NASA Long Range Plan, 1959," http://www.senate.gov/artandhistory/history/resources/pdf/ NASALongRange1959.pdf.

[12] The classic work on the Kennedy decision is John M. Logsdon's *The Decision to Go to the Moon: Project Apollo and the National Interest* (MIT Press, Cambridge, Mass., 1970). See also the more recent *John F. Kennedy and the Race to the Moon* (Palgrave Macmillan, New York, 2010).

[13] Figures taken from Logsdon, *John F. Kennedy and the Race to the Moon,* 2010.

[14] Space Task Group, *The Post-Apollo Space Program: Directions for the Future,* September 1969, available in NASA Historical Reference Collection, History Office, NASA Headquarters, Washington, D.C., http://www.hq.nasa.gov/office/pao/History/taskgrp.html.

[15] Roger D. Launius, "NASA and the Decision to Build the Space Shuttle, 1969-72," *The Historian* 57.1 (September 1994): 17-34.

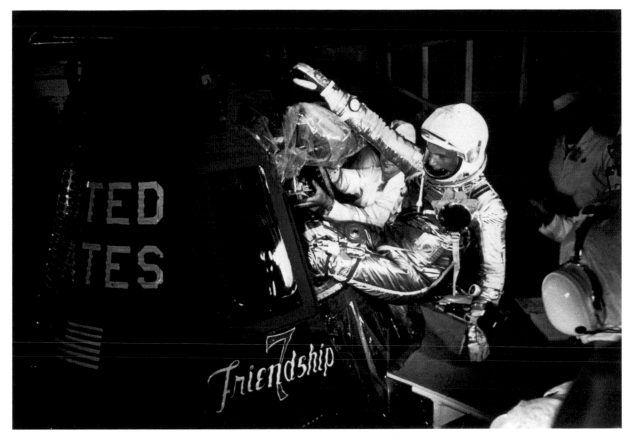

FIGURE 1.1 John Glenn climbs into the Mercury capsule, which he dubbed Friendship 7, on February 20, 1962, before launching into space. SOURCE: Courtesy of NASA, "Glenn Launch Highlighted Changing World," February 17, 2012, http://www.nasa.gov/topics/history/features/Glenn-50thKSC.html#.U34ERyhhu6I.

administration lacked enthusiasm for the station and the Mars option proved too ambitious and expensive, leading NASA officials believed, in the words of space-policy scholar John Logsdon, as follows:

> NASA had to get a go-ahead for the shuttle in 1971 if NASA were to maintain its identity as a large development organization with human spaceflight as its central activity. The choice of whether or not to approve the space shuttle thus became the *de facto* policy decision on the kind of civilian space policy and program the United States would pursue during the 1970s and beyond.[16]

(Two decades later, the Columbia Accident Investigation Board [CAIB] would call that approach "straining to do too much with too little."[17]) With further budget cuts and compromises, the original fully reusable space shuttle concept was downgraded, by the time of President Nixon's announcement, to a partially reusable, more expensive, and, as would become evident later, less safe system.

After some delays, the space shuttle began flying on April 12, 1981, when the first orbiter, *Columbia*, lifted off on a successful 2-day mission with astronauts John Young and Robert Crippen (Figure 1.4). With the launch of

---

[16] John M. Logsdon, "The Evolution of U.S. Space Policy and Plans," in *Exploring the Unknown,* 1995, p. 384.

[17] "NASA—Report of Columbia Accident Investigation Board," http://www.nasa.gov/columbia/home/CAIB_Vol1.html. (See especially, pp. 102-105, and 209). CAIB's comment echoed a similar comment made by the Augustine commission in 1990 that NASA "is trying to do too much and allowing too little margin for the unexpected."

FIGURE 1.2 Eugene Cernan, Apollo 17 commander, was the last human to walk on the Moon, finishing up the third of three moonwalks on December 13, 1972. SOURCE: Courtesy of NASA.

135 missions from 1981 to 2011, the Space Shuttle Program saw the use of five orbiters: *Columbia*, *Challenger*, *Discovery*, *Atlantis*, and *Endeavour*. Key achievements of the program included the launch of numerous satellites and interplanetary probes, deployment of the Hubble Space Telescope, a vast array of scientific experiments in Earth orbit, and several crucial servicing missions to Hubble. In its later years, the space shuttle served as a ferry vehicle (both up and down) for crews and supplies for the Russian space station *Mir* and later the ISS. More than 350 astronauts—most from NASA but also from other nations, agencies, and corporations—flew on the space shuttle. Despite those successes, the Space Shuttle Program was plagued by two fatal disasters, involving STS-51L

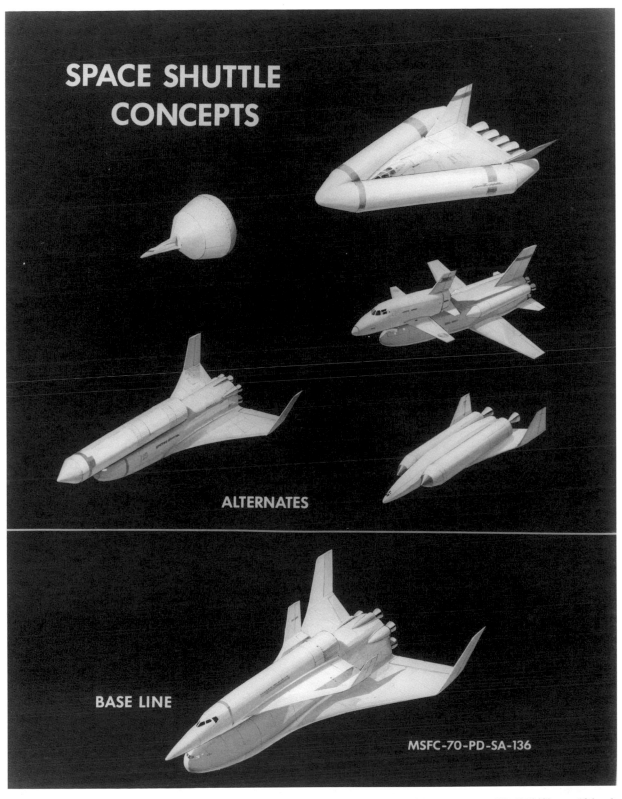

FIGURE 1.3 Space shuttle concepts. SOURCE: Courtesy of NASA; available at http://history.nasa.gov/SP-4219/Chapter12.html.

FIGURE 1.4 Space shuttle *Columbia* launch. SOURCE: Courtesy of NASA. STS-1 Shuttle Mission Imagery, S81-30462 (April 12, 1981), http://spaceflight.nasa.gov/gallery/images/shuttle/sts-1/html/s81-30462.html.

Challenger in 1986 and STS-107 *Columbia* in 2003, that killed all seven crew members on each mission. After *Challenger*, President Reagan announced that all commercial and Department of Defense payloads would be shifted off the space shuttle; this relieved the program of one of its original rationales as an all-purpose launch system for satellites.[18] Safety concerns eventually came to dominate discussions of the potential continuation of the Space Shuttle Program. In January 2004, President George W. Bush effectively set into motion the process by which the program came to a definitive end in 2011.

Although the Space Shuttle Program was finally concluded, the United States continued to have a permanent presence in space through its participation in the ISS. The roots of the station program date back to the administration of President Reagan, when NASA leadership advanced the idea of a large space station in Earth orbit as a way to underscore U.S. leadership in space activity and to exploit the commercial potential of space. In January 1984, in his State of the Union speech, President Reagan directed "NASA to develop a permanently manned space station and to do it within a decade." The design of the new station evolved through a number of iterations in the 1980s, driven largely by substantial cost overruns, changing requirements, and the repercussions of the *Challenger* accident in 1986.[19]

In May 1986, the National Commission on Space, chartered by Congress, issued a major report, *Pioneering the Space Frontier,* that recommended "a pioneering mission for 21st century America" and emphasized U.S. leadership in space activities, including missions to the Moon (by about 2005) and Mars (by about 2015). The report for the first time acknowledged the importance of maintaining U.S. leadership in a global economy with rising economic powers in Asia.[20] The *Challenger* disaster, however, interrupted that expectation, and yet another task force followed, this one commissioned by NASA and chaired by astronaut Sally K. Ride. In its 1987 report titled *Leadership and America's Future in Space,* the commission recommended a "strategy of evolution and natural progression . . . [that] would begin by increasing our capabilities in transportation and technology—not goals in themselves, but as the necessary means to achieve our goals in science and exploration." With a focus on capabilities for the first time, the goal for the United States would once again be human missions to the Moon and Mars to be carried out individually or in collaboration with other nations. The objective was unequivocally stated: "There is no doubt that exploring, prospecting, and settling Mars should be the ultimate objective of human space exploration. But America should not rush headlong toward Mars; we should adopt a strategy to continue an orderly expansion outward from Earth."[21] However, lukewarm support from Congress and events outside the United States changed the landscape of U.S. space policy before those recommendations could even be fully considered.

As the Cold War came to a close, President George H. W. Bush announced the Space Exploration Initiative (SEI) in July 1989, on the 20th anniversary of the first Moon landing. The SEI called for continuing investment in what, after *Challenger*, was named Space Station Freedom, which would be followed by human missions to the Moon but, in the words of President Bush, "this time, back to stay" and later "a manned mission to Mars." The president suggested that such missions had historical precedent, such as the voyages of Columbus and the Oregon Trail.[22] Lack of congressional support, however, left the SEI dead by the time President George H. W. Bush left the White House. A new advisory report in December 1990, the so-called Augustine Report, *Report of the Advisory Committee on the Future of the U.S. Space Program,* said that NASA was currently "overcommitted in terms of program obligations relative to resources available—in short, it [was] trying to do too much."[23]

The end of the Cold War provided a new set of opportunities. With the Freedom space station program over budget and on the verge of cancellation, NASA proposed combining its elements with elements of the post-Soviet (Russian) space station program, known as *Mir*, to create an international space station. A December 1992 study, *A*

[18] "Statement on the Building of a Fourth Shuttle Orbiter and the Future of the Space Program, August 15, 1986," from Public Papers of Ronald Reagan, 1986, http://www.reagan.utexas.edu/archives/speeches/1986/081586f.htm.

[19] W.D. Kay, Democracy and super technologies: The politics of the space shuttle and Space Station Freedom, *Science, Technology and Human Values* 19.2(April):131-151, 1994.

[20] National Commission on Space, *Pioneering the Space Frontier: The Report of the National Commission on Space: An Exciting Vision of Our Next Fifty Years in Space,* 1986, http://history.nasa.gov/painerep/begin.html.

[21] *Leadership and America's Future in Space: A Report to the Administrator, August 1987,* http://history.nasa.gov/riderep/cover.htm.

[22] NASA, "Space Exploration Initiative," http://history.nasa.gov/sei.htm, accessed January 19, 2014.

[23] *Report of the Advisory Committee on the Future of the U.S. Space Program,* December 1990, http://history.nasa.gov/augustine/racfup1.htm.

Post Cold War Assessment of U.S. Space Policy,[24] called for the United States to "develop a 'cooperative strategy' as a central element of its future approach to overall space policy." International collaboration with the Russians had already moved to the fore by this time, driven largely by geopolitical issues, especially the need to prevent Russian engineers from working for hostile nations and having Russia join the Missile Technology Control Regime. A joint station was perceived as an ideal vehicle to achieve those aims.[25] In his State of the Union address in January 1994, President Clinton announced the plan to build such a station—which became the ISS—with Russian partners.[26] As a result, NASA implemented the shuttle–*Mir* program, in which the space shuttle carried Russian cosmonauts into orbit. These missions culminated in docking and visiting missions to *Mir* starting with STS-71 in 1995. Soon, U.S. astronauts, beginning with Norman E. Thagard, spent long tours aboard *Mir*. Although marred by a number of accidents (including a fire and a collision, both unrelated to the space shuttle), the experience proved critical to the beginning of joint operations with Russia on the ISS.

The ISS, whose first element was launched into orbit in 1998, is a joint project among the space agencies of the United States, Russia, Europe (collectively and through individual nations), Japan, and Canada. Since the arrival of Expedition 1 at the ISS on November 2, 2000, it has been continuously occupied, constituting the longest continuous human presence in space; astronauts from at least 15 different nations have visited the station since then and completed more than 35 expeditions in orbit. For most of the first decade or so, the bulk of station servicing was carried out by the space shuttle, but since the end of the Space Shuttle Program cargo and crew deliveries have been taken over by vehicles from Russia (Soyuz TMA and Progress M), Europe (the Automated Transfer Vehicle, ATV), Japan (the H-II Transfer Vehicle, HTV), and commercial contractors (Dragon and Cygnus). Since the summer of 2011, only Soyuz can carry crews to the ISS, and in fact the United States has no independent capability to launch crews into orbit—an outcome set into motion by the space shuttle *Columbia* accident in 2003.

In investigating the larger structural, institutional, and cultural causes of the accident, the CAIB noted that there had been "a failure of national leadership" in not replacing the aging space shuttles and lamented a lack of "strategic vision" in civilian space activities. As a result, in January 2004, President George W. Bush, in a major speech, outlined a plan to extend the "human presence across the solar system, starting with a human return to the Moon by the year 2020, in preparation for human exploration of Mars and other destinations."[27] Despite considerable investments in the new initiative, called Project Constellation, it did not enjoy across-the-board support. A new Augustine Commission conducted a major review of human spaceflight and issued a report, *Seeking a Human Spaceflight Program Worthy of a Great Nation,* in October 2009 that noted that the Constellation program, as defined, could not be executed without substantial increases in funding. Eventually, in February 2010, President Obama announced that the program would be canceled. The NASA Authorization Act of 2010, signed on October 11, 2010, effectively terminated the program.

It is worth noting that all the blue ribbon and advisory panels formed to recommend a course of action for human spaceflight (or, more broadly, U.S. space policy) have focused on a set of key goals that are surprisingly uniform over the decades, especially after 1969. They all include a space program that would advance long-range technologies (for daily life and ultimately for human missions to Mars); develop sensible mission architectures (with a proper balance of human and robotic systems); promote science, technology, engineering, and mathematics education; maintain U.S. leadership in a competitive global environment; open commercial investment opportunities; improve affordability; support national security; and expand international collaboration. All the panels have suggested that Mars be the ultimate goal of human spaceflight with return to the lunar surface as an intermediate step.

---

[24] Vice President's Space Policy Advisory Board, *A Post Cold War Assessment of U.S. Space Policy: A Task Group Report,* December 1992, available at http://history.nasa.gov/33080.pt1.pdf and http://history.nasa.gov/33080.pt2.pdf.

[25] In June 1992, President George H.W. Bush and Boris Yeltsin signed an agreement calling for, among other things, increased collaboration between the United States and Russia. NASA and the Russian Space Agency soon ratified a 1-year contract that included consideration of increased Russian participation in a U.S. space station program.

[26] The United States and Russia agreed to cooperate in human spaceflight on September 2, 1993. A formal agreement signed on November 1, 1993, brought the Russian Space Agency as a partner with NASA on the new station. See *Space Station: Impact of the Expanded Russian Role on Funding and Research,* GAO Report to the Ranking Minority Member, Subcommittee on Oversight of Government Management, Committee on Governmental Affairs, U.S. Senate, June 1994, http://archive.gao.gov/t2pbat3/151975.pdf.

[27] NASA, "President Bush Offers New Vision for NASA," http://www.nasa.gov/missions/solarsystem/bush_vision.html, January 14, 2004.

The Obama administration issued a new national space policy in June 2010.[28] The document adheres to six broad goals: energize domestic industry, expand international collaboration, strengthen stability in space, increase resilience of mission-essential functions, perform human and robotic missions, and improve capabilities to conduct science and study Earth's resources. With respect to "space science, exploration, and discovery" and *human spaceflight* in particular, the document notes that the administrator of NASA shall

- Set far-reaching exploration milestones. By 2025, begin crewed missions beyond the moon, including sending humans to an asteroid. By mid-2030s, send humans to orbit Mars and return them safely to Earth.
- Continue the operation of the [ISS] ... likely to 2020.
- Seek partnerships with the private sector to enable safe, reliable, and cost-effective commercial spaceflight capabilities and services for the transport of crew and cargo to and from the ISS.

The National Space Policy offers general guidelines, but de facto work on human spaceflight relies on the considerations laid out in three consecutive NASA Authorization Acts issued in 2005, 2008, and 2010, each of which added to, clarified, and updated many of NASA's immediate goals in light of the winding down of the Space Shuttle Program, the end of construction of the ISS, and plans for human exploration beyond LEO. The committee notes below only the provisions of the acts that are related to human spaceflight.

The 2005 Act (signed into law on December 30, 2005) codified President George W. Bush's Vision for Space Exploration, specifically, its call for a sustained human presence on the Moon that would begin with precursor robotic missions. It also authorized collaboration with international partners. At the time, the goal was to launch a new crewed spacecraft, the Crew Exploration Vehicle, in 2014, continue work on the ISS during the ensuing decade, and return U.S. astronauts to the Moon by 2020. The act formally stipulated that the "United States segment of the ISS [would be] designated a national laboratory."[29]

In the 2008 act (signed into law on October 15, 2008), Congress emphasized that "developing United States human space flight capabilities to allow independent American access to the International Space Station, and to explore beyond low Earth orbit, is a strategically important national imperative." A second clause, which noted that "all prudent steps should . . . be taken to bring the Orion Crew Exploration Vehicle and Ares I Crew Launch Vehicle to full operational capability as soon as possible and to ensure the effective development of a United States heavy lift launch capability for missions beyond low Earth orbit" has been partly invalidated given the cancellation of Project Constellation, of which Ares I was a fundamental part. As for long-term exploration goals, the NASA Authorization Act reiterated two intertwined goals: that the United States should participate in concert with other nations and that it should explore beyond Earth orbit with a view to establishing a "human-tended" lunar outpost designated the Neil A. Armstrong Lunar Outpost. Congress affirmed its support for "the broad goals of the space exploration policy of the United States, including the eventual return to and exploration of the Moon and other destinations in the solar system and the important national imperative of independent access to space."[30]

The 2010 act (signed into law on October 11, 2010) reiterates provisions dating back to the 1958 act and updating the provisions of the 2005 and 2008 acts.[31] At the time of this writing, it is the most recent authorization act approved by Congress that addresses U.S. national space policy, and it is essentially a statement of purpose that takes into account the cancellation of Constellation and a reorientation from work toward a lunar infrastructure. The act notes that the "commitment to human exploration goals is essential for providing the necessary long-term focus and programmatic consistency and robustness of the United States civilian space program" and emphasizes the U.S. commitment to a full spectrum of activities on board the ISS, the promotion of commercial partnerships to sustain activities on board the ISS, and the maintenance of agreed-on international partnerships. Noting the impending end of the Space Shuttle Program and the lack of independent access for humans to LEO, the act reiterates that "it is . . . essential that a United States capability be developed as soon as possible." Although conceding that such

---

[28] *National Space Policy of the United States of America,* June 28, 2010, http://www.whitehouse.gov/sites/default/files/national_space_policy_6-28-10.pdf.

[29] Public Law 109-155, December 20, 2005.

[30] Public Law 110-422, October 15, 2008.

[31] Public Law 111-267, October 11, 2010.

capabilities could be provided by commercial entities, the act emphasizes that "it is in the United States national interest to maintain a government operated space transportation system for crew and cargo delivery to space."

With respect to activities *beyond* LEO, the 2010 act notes that "the United States must develop as rapidly as possible replacement vehicles capable of providing both human and cargo launch capability to low-Earth orbit and to destinations beyond low-Earth orbit." More specifically, it acknowledges that human space missions beyond LEO will "drive developments in emerging areas of space infrastructure and technology" and that "a long term objective for human exploration of space should be the eventual international exploration of Mars." The strategy to achieve that ultimate goal would incorporate international partners but also adopt a "pay-as-you-go approach." Furthermore, "requirements in new launch and crew systems . . . should be scaled to the minimum necessary to meet the core national mission capability needed to conduct cislunar missions. These initial missions, along with the development of new technologies and in-space capabilities can form the foundation for missions to other destinations." As for the specific architecture of space vehicles, the act committed NASA to develop the Space Launch System (SLS). It would be able to deliver 70–100 metric tons (MT) to LEO in its initial version and 100 to 130 MT later. NASA was also to continue development of "a multi-purpose crew vehicle to be available as practicable, and no later than for use with the Space Launch System," which evolved into an updated Orion.

Those stipulations and the more general guidelines of the original National Aeronautics and Space Act of 1958 continue to guide NASA activities to this day, at a time when its human spaceflight program is at a critical crossroads, especially with regard to plans to extend beyond LEO.

1.3 INTERNATIONAL CONTEXT

Collaboration has a long-established tradition in space activities, and the committee was explicitly charged to include consideration of the programs and priorities of current and potential international partners when examining the value of human spaceflight. Collaboration has become an integral part of the space policy of essentially all nations that are participating in space; most nations now do not execute substantial space projects without some foreign participation. The reasons for collaboration are multiple, but in a survey of cooperative agreements between international partners,[32] it was found that nations cooperate principally when it benefits themselves. It is generally understood that international collaboration taps into the resources of multiple countries to increase the scope of programs beyond the capabilities of individual participants.

The benefits of collaboration are numerous. Collaboration gives nations the opportunity to enlarge their spectrum of mission possibilities by coordinating development and optimizing planning and resources. There is an assumption that if the partners contribute capability, the whole can be greater than the sum of the parts, and the cost can be shared among the partners. A space project is potentially more affordable for each individual partner, and the pool of scientific and technological expertise brought to bear on the project is enriched. However, the evidence that international collaboration reduces costs for the lead partner, such as the United States, remains inconclusive.[33] Indeed, senior NASA officials reported to the present committee that international collaboration does not reduce costs. The scale and scope of a human mission to Mars would have to include major systems and subsystems developed by multiple mission partners to affect costs for the major contributor. Such collaborations, especially if partners are brought in during early stages and with firm cost-sharing arrangements in place, might make an effort of this scale less vulnerable to disruption and sudden changes.

---

[32] N. Peter, The changing geopolitics of space activities, *Space Policy* 22:100-109, 2006.

[33] There have been some studies on the cost effects of collaboration in large-scale scientific and technologic projects, but their conclusions are largely ambivalent. See National Research Council, *Assessment of Impediments to Interagency Collaboration on Space and Earth Science Missions* (The National Academies Press, Washington, D.C., 2011); Office of Technology Assessment, U.S. Congress, *International Partnerships in Large Science Projects,* 1998. It is worth noting that all the attendees at a roundtable for the leading space agencies—among them NASA, the European Space Agency, the Japan Aerospace Exploration Agency, and the Canadian Space Agency—held in late 2013 noted the benefits of international collaboration but cautioned that cost was *not* one of them. See "Many Benefits from International Cooperation, But Not Cost Savings, Says Panel," November 20, 2013, SpacePolicyOnline.com, http://www.spacepolicyonline.com/news/many-benefits-from-international-cooperation-but-not-cost-savings-says-panel.

The advantages of collaboration are usually more marked under particular conditions, such as when the division of labor among participants reduces the inefficiency inherent in a management structure that involves multiple nations. But it is essential to underscore that successful collaboration requires satisfaction of the core interests and needs of *all* partners.

As discussed by Nicolas Peter,

> Not all countries regard international cooperation equally; several counties actively solicit, establish, and work to maintain partnerships with spacefaring countries, while others have a more nationalistic and individual approach. Moreover, cooperation is not static, but highly dynamic, and has an intrinsic reverberating process in which partners adapt to the other, which implies in turn an adaptation to the new situation by other stakeholders.[34]

Thus, collaboration is intrinsically more problematic in programs that have decadal timescales than in programs that can be accomplished within a few years, because longer timescales greatly increase the probability of diverging political realities. Reciprocity is therefore an important source of stability. The feasibility of cooperative projects will, as a matter of practice, be based on the prior behavior of the potential partners.

The noticeable proliferation in the number of spacefaring countries has increased possibilities for collaboration. Space agencies are now looking to a variety of partners as they plan their future programs. Moreover, as the number of successful examples of working together grows, nations that are new to space activities will find collaboration helpful in mitigating the inevitable risks and softening the barriers to entry associated with space initiatives. The record shows that all recent newcomers to space—including China—have sought to establish cooperative agreements with more experienced spacefaring countries to benefit from potential transfers of technology and experience. At the same time, experienced spacefaring countries, in particular the United States, have on occasion viewed the potential transfers of technology and knowledge as threatening. Nonetheless, international collaboration is still thought to provide resilience to long-term, large-scale programs and space missions, such as the ISS.

In January 2014, for the first time the U.S. Department of State hosted an International Space Exploration Forum (ISEF), which was attended by ministers and government representatives from more than 30 nations. Although the focus of the meeting included both robotic and human space exploration activities, sponsorship of the meeting and the level of involvement of the U.S. government—Deputy Secretary of State William Burns attended—reflected an awareness of space exploration and human spaceflight in particular in developing and strengthening international relationships. The official Department of State announcement on the ISEF noted that "many of the spaceflight achievements of the past half-century would not have been possible without international collaboration" but conceded that "competition-driven innovation at the industrial and scientific levels is also an important element for the evolution of space exploration."[35]

Discussions about future cooperative missions, including missions beyond Earth orbit, are ongoing. One of the major forums for these discussions is the International Space Exploration Coordination Group, the work of representatives of 14 space agencies, which in August 2013 published an updated version of *The Global Exploration Roadmap* (GER).[36] The cooperative work that was required to develop the plans has built a network among the space agencies of the nations involved that contributes, with many other such networks, to peaceful relationships between the nations. The GER serves as a framework to coordinate planning for future international missions with the ultimate goal of sending humans to Mars. Each agency envisions contributions ranging from large-scale systems, such as Orion and SLS, to robotic missions, mission planning, landers, and cargo vehicles. Not all agencies will participate in every element of the roadmap, but competences will be built on and leveraged to produce a more effective program of work.

For now, the focus of international partnerships and collaboration in human spaceflight is the ISS. All the major partners are committed to using and operating the ISS to 2020. After that time, differing national objectives,

---

[34] N. Peter, The changing geopolitics of space activities, 2006, p. 101.

[35] U.S. Department of State, "International Space Exploration Forum Summary," January 10, 2014, http://www.state.gov/r/pa/prs/ps/2014/01/219550.htm.

[36] Lunar and Planetary Institute, "ISECG Posts Update to the Global Exploration Roadmap," August 20, 2013, https://www.globalspaceexploration.org/news/2013-08-20.

funding profiles, and evolving space-policy postures make continuation of the ISS with all the original partners less certain. Both with regard to the ISS and with regard to exploration beyond Earth orbit, NASA continues to play the leadership role (as demonstrated by its proposal to extend ISS operations to 2024), in part because of its larger expenditures on these programs relative to those of other nations. However, the nature of U.S. leadership varies on the basis of the nature of U.S. relationships with different agencies. Many look to the United States— that is, they depend on the continuation of NASA's human spaceflight programs to ensure collaboration with the United States, which underpins funding for their own programs. Others are more internally focused in setting their funding, planning, and objectives.

A growing number of nations have substantial human spaceflight programs, including continuing system or technology development efforts that might contribute directly to future NASA human space activities. Programs of greatest note are those of Russia, the European Space Agency (ESA), Japan, and China. In addition, Canada is an active partner in the ISS program, and India has announced a long-term plan for human spaceflight.

Public perceptions of spaceflight vary among nations. For emerging space nations, such as China and India, "space exploration represents one of a constellation of important ways with which to announce their 'arrival' as global powers," and it announces their emergence into an elite club of spacefaring powers.[37] The Chinese acclaim their astronauts, or yǔhángyuán (also referred to as taikonauts in the Western press), as embodiments of Chinese history, culture, and technological prowess.[38] In India, accomplishments in space represent national aspirations to become a global power.[39]

The Russian human spaceflight program is centered entirely around the ISS, with its several critical modules and its capacity to deliver crews and cargo to the facility. However, building on decades of space station activity, Russia continues to advance conceptual studies for follow-on programs, including proposals for human activity beyond LEO.[40] In a space-policy statement issued by the Russian Federal Space Agency (Roskosmos) in April 2013, the Russian government noted that "space activities are one of the primary factors determining the level of development and influence of Russia in the modern world."[41] Despite that acknowledgement, Russia has struggled to expand its human spaceflight program beyond the ISS, and robotic missions to the Moon and planets have been almost nonexistent in the last 20 years. The Russian government has continued to press forward with the development of a follow-on spacecraft to the Soyuz, identified with the generic designation PTK NP, that is comparable in size and mission with NASA's Orion.[42] The vehicle is expected to carry four cosmonauts on missions beyond LEO— principally to lunar orbit—lasting about a month.[43] The Russians expect to begin flight testing of the spacecraft with crews in 2018, although this will be a challenging deadline to meet. There are long-term plans for the exploration of both the Moon and Mars, but, given fiscal realities, these would be folded into any global initiative involving at the very least the Europeans—a point underscored in the 2013 Russian space-policy statement that noted that the Russian state interests "support the possibility of full-scale participation in projects of the international community in the research, mastery and use of space, including that of the Moon, Mars and other bodies in the solar system."[44]

China's first launch of a human into space in 2003 resulted in undeniably enhanced regional prestige, both in the eyes of a domestic audience and internationally. Since then, China has conducted four human space missions

---

[37] A.A. Siddiqi, "Spaceflight in the National Imagination," in *Remembering the Space Age* (S.J. Dick, ed.), NASA History Division, Washington, D.C., 2008, p. 27.

[38] James R. Hansen, "The Great Leap Upward: China's Human Spaceflight Program and Chinese National Identity" in *Remembering the Space Age*, 2008, p. 109-120.

[39] See, for example, the recent book by a former head of the Indian Space Research Organization: U.R. Rao, *India's Rise as a Space Power*, Cambridge University Press India, New Delhi, 2014.

[40] The current Russian orbital segment consists of five modules (*Zarya*, *Zvezda*, *Pirs*, *Poisk*, and *Rassvet*). In addition, Russians provide the Soyuz TMA and Progress-M supply vehicles.

[41] "The Principles of Space Policy of the Russian Federation in the Area of Space Activities in the Period up to 2030 and Future Perspectives Approved by the President of the Russian Federation on 19 April 2013 No. Pr-906" (in Russian), http://www.federalspace.ru/media/files/docs/3/osnovi_do_2030.doc.

[42] PTK NP stands for New-Generation Piloted Transport Ship.

[43] "New Spacecraft Will Cost $10 Billion," October 23, 2013, Interfax.ru, http://www.interfax.ru/russia/txt/336877.

[44] "The Principles of State Policy of the Russian Federation."

of increasing technical complexity.[45] Those missions have involved a Soyuz-like transport vehicle named *Shenzhou* and a small space station known as *Tiangong*. More ambitious missions in Earth orbit are impending. A larger modular space-station program is in development, and the first "experimental" core component of the station is to be launched in 2018. Two more modules are to be orbited and attached to the core by 2020. The station is to be completed by 2022 with the attachment of three more modules, giving it a size and scope similar to the deorbited Russian *Mir* space station. Some of the modules are meant specifically for international visitors or cooperative ventures with non-Chinese partners.[46]

At the end of 2011, China released a white paper detailing its philosophy and programs, updating progress since 2006, and laying out goals for the next 5 years. With regard to human spaceflight, China's goals through 2016 include plans to "launch space laboratories, manned spaceship and space freighters; make breakthroughs in and master space station key technologies, including astronauts' medium-term stay, regenerative life support and propellant fueling; conduct space applications to a certain extent and make technological preparations for the construction of space stations."[47] As for a human lunar landing, there appears to be no formal government approval of a dedicated project, but the Chinese are exploring various options for such a mission.[48] Despite claims by some in the West, there is still no indication of when such an event might be realistically implemented by the Chinese.

China also has a growing robotic program. It landed a soft-lander, *Chang'e-3*, on the Moon in December 2013 and deployed a rover named *Yùtù* that was a partial success. The Chinese are planning a lunar-sample return mission for the 2017 timeframe with the *Cháng'é-5* lunar probe.

Although China wants to engage cooperatively with the United States in human spaceflight, participation in a joint program appears to be unlikely in the near future because of security concerns, particularly in the U.S. Congress. In the meantime, China's plans do not depend on what the United States does with regard to human spaceflight. Furthermore, China has stated clearly that its space program is open to cooperative engagement with other nations, including Russia and the ESA—with or without the United States (Figure 1.5).

As discussed by Peter,

> European openness towards international cooperation with the new space actors, through the establishment of strategic partnerships by ESA, the various national space agencies, or more recently by the European Union (EU), provide a variety of possibilities for Europe to cooperate with the rest of the world, illustrating that Europe's diversity is definitively a "multiplier factor" for international cooperation. At different European space levels numerous cooperative ventures have been started with partners that, in some cases, have never been traditional partners of the West—e.g., China.[49]

That partly explains Europe's important role in international space collaboration. Peter goes on to observe that "the center of gravity of space collaboration may be drifting from the United States to Europe in certain areas; and new centers in Asia (China, Japan, India) or in Latin America (Brazil, Argentina) are growing in importance."[50]

The human spaceflight program of ESA, which accounts for about 10 percent of the agency's total budget, is focused almost entirely on supporting the ISS.[51] ESA's principal contribution to the ISS is the large, pressurized *Columbus* module, delivered to the station in 2008. ESA also provides the ATV (Figure 1.6), launched by the

---

[45] *Shenzhou-6* in 2005 (a 5-day mission with two astronauts), *Shenzhou-7* in 2008 (a mission that included extravehicular activity), *Shenzhou-9* in 2012 (which with a female astronaut docked with the *Tiangong-1* small space station), and *Shenzhou-10* in 2013 (a 2-week mission to *Tiangong-1*).

[46] Information based on papers presented at the 64th International Astronautical Congress held in Beijing, September 23-27, 2013.

[47] "Full Text: China's Space Activities in 2011," December 2011, http://english.sina.com/technology/2011/1228/427254.html; "Space plan from China broadens challenge to U.S.," *New York Times,* December 29, 2011.

[48] This judgment is based on several papers at the 64th International Astronautical Congress (IAC), Beijing, September 2013. See, for example, Lin-li Guo (China Academy of Space Technology, CAST), "Key Technology of Manned Lunar Surface Landing, Liftoff and Operating," IAC-13.A5.1.3; Yang Liu (Beijing Special Engineering Design and Research Institute), "Study on Technical Approach for Manned Deep-Space Exploration," IAC-13.A5.4-D2.8.6; Li Guoai (China Academy of Launch Vehicle Technology), "Long March Family Launch Vehicles for Deep Space Exploration," IAC-13.A3.1.11.

[49] N. Peter, The changing geopolitics of space activities, 2006, p. 103.

[50] Ibid.

[51] European Space Agency (ESA), "ESA budget by domain for 2013" [image], released January 24, 2013, http://www.esa.int/spaceinimages/Images/2013/01/ESA_budget_by_domain_for_2013_M_Million_Euro.

FIGURE 1.5 Shenzou 10 crew after return to Earth from China's fifth human spaceflight mission. June 26, 2013. From left to right, Zhang Xiaoguang, mission commander Nie Haisheng, and Wang Yaping. At present, only two nations—Russia and China—have vehicles capable of sending humans into space and returning them to Earth. SOURCE: Reuters.

Ariane 5 rocket that delivers cargo to and takes trash away from the station. The first was launched in 2008, and the plan is for a total of five to be launched over 6 years. ESA maintains its own independent astronaut corps, ground infrastructure, and science experiments program centered on the ISS.

ESA has not developed an independent capability to deliver humans into orbit, but its capabilities and accomplishments ensure that the agency—led by the leading partners, France, Germany, and Italy—will have a place in any large international endeavor in human spaceflight beyond the ISS, although its contribution will rely on partners that have an independent capability, such as the United States and Russia, and possibly in the future on China.[52] ESA has recently committed to developing a service module based on the ATV for NASA's Orion Multi-Purpose Crew Vehicle for at least one mission of the Orion, set for 2017, with a possibility of a second one.[53] The agreement puts European hardware in the critical path of development of U.S. hardware designed for operation beyond LEO, much as Russian hardware did for the ISS. ESA's work on Orion is closely linked to ESA's barter payment for the common systems operations costs on the ISS that is covered by the United States. At the same time, the Europeans appear to be open in the longer term to broadened potential partnerships with China—including partnerships in human spaceflight—although this will face some potential bureaucratic and political obstacles.[54]

---

[52] Thomas Reiter, "ESA Views on Human Spaceflight and Exploration," presentation to the committee, April 23, 2013.

[53] ESA, "European Ministers Decide to Invest in Space to Boost Europe's Competitiveness and Growth," Press Release PR 37 2012, November 21, 2012, http://www.esa.int/About_Us/Ministerial_Council_2012/European_ Ministers_decide_to_invest_in_space_to_boost_ Europe_s_competitiveness_and_growth; ESA, "ESA Workhorse to Power NASA's Orion Spacecraft," January 16, 2013, http://www.esa.int/ Our_Activities/Human_Spaceflight/ Research/ESA_workhorse_to_power_NASA_s_Orion_spacecraft.

[54] "Shifting Constellations: Europe Eyes China in Space Race," February 8, 2013, *Spiegel Online,* http://www.spiegel.de/international/europe/ esa-mulls-new-alliance-as-china-becomes-space-leader-a-882212.html.

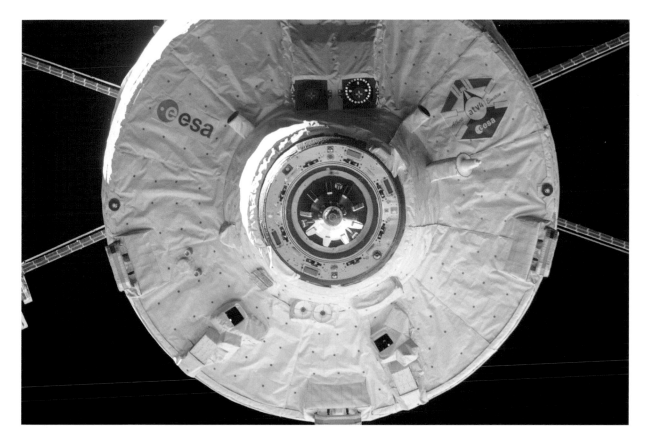

FIGURE 1.6 Automated Transfer Vehicle 4 (ATV-4) undocking. SOURCE: Courtesy of ESA and NASA, http://www.esa.int/spaceinimages/Images/2013/11/ATV-4_undocking.

Like ESA, Japan is a major partner in ISS operations and also does not have an independent capacity to deliver and recover humans from space. Japan's material contribution to the ISS is represented by the Japanese Experimental Module, *Kibo*, which is the largest pressurized module attached to the ISS. Four crewmembers can work simultaneously in the assembled module. Japan also provides the HTV, a robotic spacecraft to deliver supplies to and remove waste from the ISS. Japan maintains a small astronaut corps and, like ESA, is unlikely to engage in any human spaceflight program that is independent of an international framework or agreements at the corporate level, such as the one that provides the proximity operations system for Orbital Sciences Corporation's Cygnus cargo capsule.[55]

India has collaborated with the United States for many years in space, as well as with other international space agencies. The Indian rocket and space programs that began 50 years ago, have grown substantially since inception, and launched the country's first lunar probe in 2008 after decades of successful launch-vehicle and satellite development and deployment. *Chandrayaan-1* successfully reached lunar orbit in November 2008 and included multiple payloads that were produced jointly or independently by NASA, ESA, the United Kingdom, Germany, Poland, and Bulgaria. Among other achievements, the mission is notable for demonstrating the presence of water molecules near the Moon's south pole—a finding that has implications for in situ utilization of lunar resources.[56]

---

[55] Misuzu Onuki, "Profile: Naoki Okumura, President, Japan Aerospace Exploration Agency," *Space News*, December 9, 2013, http://www.spacenews.com/article/features/38565profile-naoki-okumura-president-japan-aerospace-exploration-agency.

[56] Water was identified by the Moon Mineralogy Mapper, one of two NASA-sponsored instruments on board *Chandrayaan-1*. See C.M. Pieters et al., Character and spatial distribution of OH/H_2O) on the surface of the Moon seen by M^3 on Chandrayaan-1, *Science* 326(5952):568-572, 2009.

More recently, India launched the Mars Orbiter Mission in November 2013 with the main objective of developing the technologies required for design, planning, management, and operation of an interplanetary mission. If successful, it will make India the fourth nation or international consortium to reach Mars.[57] India has been considering a human spaceflight program for some time and has initiated basic research studies although as of early 2014 it has yet to commit to the program officially.[58]

The committee has several findings related to the future of international collaborations beyond LEO:

- *It is evident that near-term U.S. goals for human exploration are not aligned with those of our traditional international partners.* Although most major spacefaring nations and agencies are looking toward the Moon, specifically the lunar surface, U.S. plans are focused on redirection of an asteroid into a retrograde lunar orbit where astronauts would conduct operations with it. NASA officials have asserted to the committee that the United States can support landed operations by other nations from such an orbit, but it is not clear whether this approach has been vetted with potential international partners. Although the United States is not expected to follow the desires of other nations blindly in shaping its own exploration program, there are a number of advantages for the United States in being a more active player in lunar surface operations, as discussed in Chapter 4.
- *Given the rapid development of China's capabilities in space, it is in the best interests of the United States to be open to future international partnerships.* In particular, current federal law that prevents NASA from participating in bilateral activities with the Chinese serves only to hinder U.S. ability to bring China into its sphere of international partnerships and substantially reduces the potential international capability that might be pooled to reach Mars.
- *Given the scale of the endeavor of a mission to Mars, contributions by international partners would have to be of unprecedented magnitude to defray a significant portion of the cost.* This finding follows from the detailed discussion later in the report of what is required for human missions to Mars. The United States will need to increase the budget for human spaceflight by substantially more than the rate of inflation over the coming decades, international contributions would have to be unprecedented in scale, or both.

1.4 ENDURING QUESTIONS AND RATIONALES

1.4.1 Enduring Questions

The committee was tasked with identifying enduring questions that can motivate a sustainable direction for human spaceflight. Enduring questions are those that serve as motivators of aspiration, scientific endeavors, debate, and critical thinking in the realm of human spaceflight. The questions endure in that any answers available today are at best provisional and will change as more exploration is done. Enduring questions should provide motivations that are immune to external forces and policy shifts. They are intended not only to stand the test of time but also to continue to drive work forward in the face of technologic, societal, and economic constraints. Enduring questions should be clear and should intrinsically connect to broadly shared human experience.

On the basis of the analysis reported in Chapter 2, the committee asserts that the enduring questions motivating human spaceflight are these:

- **How far from Earth can humans go?**
- **What can humans discover and achieve when we get there?**

---

[57] Prior first-time Mars encounters were by the U.S. *Mariner-4* (1965), the Soviet *Mars-2* (1971), and the European *Mars Express* (2003). A Japanese probe, *Nozomi*, flew by Mars in 2003 in a failed bid to enter orbit.

[58] Some Iranian sources have claimed that Iran is also planning a human spaceflight program but firm timetables for such a project are unknown.

1.4.2 Rationales

All the arguments that the committee heard for supporting human spaceflight have been used in various forms and combinations to justify the program for many years. The committee identified the following general set of rationales for human spaceflight: economic benefits, contributions to national security, contributions to national stature and international relations, inspiration of students and citizens to further their science and engineering education, contributions to science, the eventual survival of the human species (through off-Earth settlement), and shared human destiny and the aspiration to explore. The first five of these rationales can be considered pragmatic in that human space exploration is seen as benefiting a goal outside its mission of exploration. The committee classifies the last two as aspirational. (Human survival is of course a pragmatic human goal, but the timeline and uncertainties make the possibility of off-Earth human settlement an aspiration, one that is deeply linked to answers to the enduring questions.) Each of the rationales is discussed in more detail in Chapter 2. All the various arguments for human spaceflight—which the committee has found in prior documents, in presentations to the committee, and in the committee's various outreach and survey efforts—come down to some combination of these elements. The committee concluded from its review and assessment that

No single rationale seems to justify the value of pursuing human spaceflight.

The aspirational rationales, human destiny and human survival, are typically invoked in arguing for the unique value of the human spaceflight program and then supported by reference to one or more of the more pragmatic rationales, particularly when the question of the spending of public funds comes up. As discussed in Chapter 2, the pragmatic rationales have never seemed adequate by themselves, perhaps because the benefits that they argue for are not unique to human spaceflight. The ones that are unique to human spaceflight—the aspirational rationales related to the human destiny to explore and the survival of the human species—are also the ones most closely tied to the enduring questions.

The committee's conclusions related to each of the rationales are offered below. For the pragmatic rationales, the conclusions begin from the position that the goals implicit in these rationales—be they economic growth, contributions to space-based science, or student interest in science study—are, in and of themselves, important national goals. But the committee's conclusions on the pragmatic rationales are therefore about the impact of human spaceflight on those important national goals, as discussed in some detail in Chapter 2. However for the two aspirational rationales, the committee's conclusions are more directly connected to the associated goals. The committee's conclusions on the rationales are as follows:

Economic matters. **There is no widely accepted, robust quantitative methodology to support comparative assessments of the returns on investment in federal R&D programs in different economic sectors and fields of research. Nevertheless, it is clear that the NASA human spaceflight program, like other government R&D programs, has stimulated economic activity and has advanced development of new products and technologies that have had or may in the future generate significant economic impacts. It is impossible, however, to develop a reliable comparison of the returns on spaceflight versus other government R&D investment.**

Security. **Space-based assets and programs are an important element of national security, but the direct contribution of human spaceflight in this realm has been and is likely to remain limited. An active U.S. human spaceflight program gives the United States a stronger voice in an international code of conduct for space, enhances U.S. soft power, and supports collaborations with other nations; thus, it contributes to our national interests, including security.**

National stature and international relations. **Being a leader in human space exploration enhances international stature and national pride. Because the work is complex and expensive, it can benefit from international cooperative efforts. Such cooperation has important geopolitical benefits.**

Education and inspiration. **The United States needs scientists and engineers and a public that has a strong understanding of science. The challenge and excitement of space missions can serve as an inspiration for students and citizens to engage with science and engineering although it is difficult to measure this. The path to becoming a scientist or engineer requires much more than the initial inspiration. Many who work in space fields, however, report the importance of such inspiration, although it is difficult to separate the contributions of human and robotic spaceflight.**

Scientific discovery. **The relative benefits of robotic versus human efforts in space science are constantly shifting as a result of changes in technology, cost, and risk. The current capabilities of robotic planetary explorers, such as Curiosity and Cassini, are such that although they can go farther, go sooner, and be much less expensive than human missions to the same locations, they cannot match the flexibility of humans to function in complex environments, to improvise, and to respond quickly to new discoveries. Such constraints may change some day.**

Human survival. **It is not possible to say whether human off-Earth settlements could eventually be developed that would outlive human presence on Earth and lengthen the survival of our species. That question can be answered only by pushing the human frontier in space.**

Shared destiny and aspiration to explore. **The urge to explore and to reach challenging goals is a common human characteristic. Space is today a major physical frontier for such exploration and aspiration. Some say that it is human destiny to continue to explore space. While not all share this view, for those who do it is an important reason to engage in human spaceflight.**

1.4.3 Value and Value Propositions

The committee was tasked with describing "the expected value and value proposition of NASA's human spaceflight activities in the context of national goals." While most people can straightforwardly define what is meant by the "value" of an object or endeavor, the term *value proposition* is much less familiar. In business, a value proposition is a statement of the benefits or experiences being delivered by an organization to recipients and of the price or description of the resources expended for them. It is a concept whose applicability to a government program like NASA's space exploration program might reasonably be questioned.[59] However, the value-proposition approach to the assessment of public programs is rooted in a widespread observation that there are large differences between private- and public-sector organizations in defining objectives and in the feasibility of measuring outcomes. Put simply, there is no obvious "bottom line" for most public programs, which by definition are conducted as not-for-profit activities.

It has been argued that public programs should be evaluated in terms of their ability to achieve a broad set of objectives (or "values") and of the efficiency with which the objectives are accomplished. The effectiveness of public programs in achieving a broad set of objectives forms the core of value-proposition analysis as applied to public-sector activities. The committee's review of the value-proposition analyses of public agencies in general—and of NASA's human space exploration efforts in particular—reveals that such a value approach lacks clear definition of objectives and lacks the formulation and tracking of appropriate metrics to measure the performance of any public agency along the path to meeting these objectives. Such analyses remain largely theoretical, and the notion that one can aggregate a variety of measures of outcomes, efficiency, and progress into any single equivalent of a business value proposition remains very difficult to realize.

Chapter 2 presents a novel and detailed analysis of how value propositions might be developed for the publicly funded U.S. space program by looking at how stakeholders (both narrowly and broadly defined) derive benefits from the program and specifically at what opportunities would no longer be available if human spaceflight were

---

[59] The committee is not questioning the applicability of the value-proposition approach to private space ventures, or "new space," in which companies have customers, business models, and bottom lines as in any other commercial endeavor.

discontinued. In the end, a rigorous analysis of the value propositions for NASA human spaceflight at the national level, akin to what might be done for a large business venture, is beyond the scope of this report. It may well be beyond the scope of *any* report in that such an approach may not provide sufficient insight into whether and how future human exploration programs should be conducted to be truly useful in developing strategies for moving beyond LEO. However, as an alternative perspective to the standard listing of rationales for conducting human spaceflight, the committee provides a detailed value-proposition analysis in Chapter 2.

1.5 PUBLIC AND STAKEHOLDER OPINION[60]

1.5.1 Analysis of Public Opinion Polls

1.5.1.1 Interest in Space Exploration as Revealed Through Public Opinion

At any given time, a relatively small proportion of the U.S. public pays close attention to space exploration. Survey data collected over the years indicate that an average of about one-fourth of U.S. adults tend to say that they have a high level of interest in space exploration (Figure 1.7). Interest in space exploration is lower than that in other policy issues. For example, the 2010 General Social Survey (GSS),[61] which placed the estimate of those who are "very interested" in space exploration at 21 percent, found that this issue was at the bottom of a list of 10 issues that it asked about, trailing such related topics as new technologies and inventions (38 percent) and new scientific discoveries (39 percent).

As in the case of most other policy issues, far fewer people feel well informed about space exploration than hold a high level of interest in it, and this is important because citizen engagement depends on a combination of interest in the topic and a sense of being adequately informed about it. Estimates of the "attentive public," those who are both very interested in and well informed about space exploration, during the past few decades have been under 10 percent.

1.5.1.2 Support for Spending on Space Exploration

Although the public has a favorable view of NASA and most of the spaceflight missions that it has sponsored (Figure 1.8), most Americans do not favor increased spending on space exploration. According to data from the GSS over the past 40 years, higher percentages say that we are spending too much on space exploration than say that we are spending too little (see Figure 3.3 in Chapter 3), but the percentages have gotten closer in recent years. Questions about spending on federal programs are often quite sensitive to whether the question mentions the cost of the program, and questions that note that NASA's budget is a very small part of the federal budget tend to find more support for increased funding. The GSS questions do not mention cost, but the resulting data show that in comparison to other spending priorities, space exploration ranks near the bottom.

1.5.1.3 Trends in Support for Human Spaceflight Missions

Despite its reservations about increased funding for space exploration, the public has consistently reported positive views about specific human spaceflight missions, including the Space Shuttle Program, the Moon landing, and sending astronauts to Mars (see Figure 1.8). A majority of the public seems to have positive views about those missions, and views about the Apollo program seem to have grown more positive over time.

---

[60] Through its Public and Stakeholder Opinions Panel, the committee was able to obtain a detailed analysis of attitudes toward the space program among the general public and a carefully selected set of stakeholders. The complete analysis is given in Chapter 3. As described in further detail in Chapter 3, findings about public opinion are based on a review of data collected over the years by the nation's major polling organizations, whereas the stakeholder views also summarized here are based on a survey conducted as part of the present study.

[61] See the General Social Survey website of the National Opinion Research Center at the University of Chicago at http://www3.norc.org/GSS+Website/.

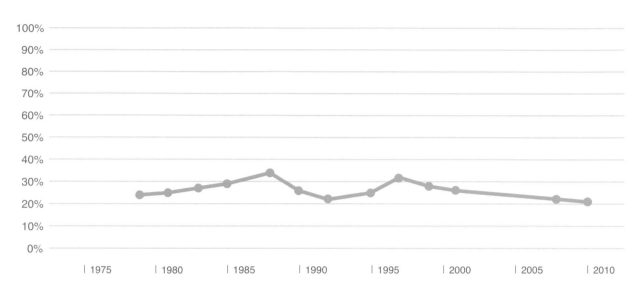

FIGURE 1.7 Percentage "very interested" in space exploration, 1979–2010. SOURCE: National Science Foundation Survey of Public Attitudes Toward and Understanding of Science and Technology; 2008, 2010: General Social Survey.

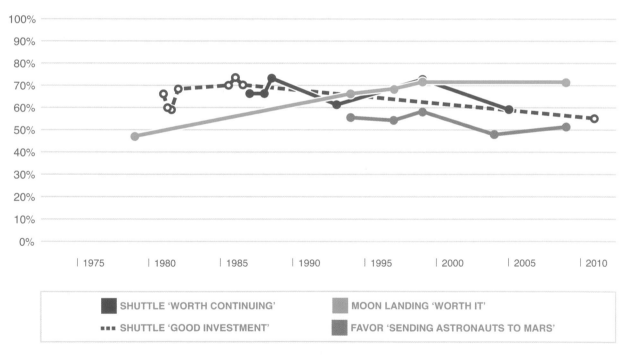

FIGURE 1.8 Public support for the space shuttle, Moon landing, and Mars mission, 1979–2011. SOURCE: Shuttle continuation: CBS/NYT (1987, 1988), CBS (1993, 1999, 2005); Shuttle investment: NBC/AP (1981,1982), NBC/WSJ (1985,1986), Pew (2011); Moon landing: CBS/NYT (1979, 1994), CBS (1997, 1999, 2009); Mars: CBS/NYT (1994), CBS (1997, 1999, 2004, 2009).

1.5.1.4 Human versus Robotic Missions

Apparent levels of support for space exploration programs depend in part on whether and how the question refers to the cost of the programs. In the same vein, preferences for human versus robotic space exploration tend to be influenced by whether cost is mentioned. For example, the Gallup Organization in 2003 asked this: "Some people feel the U.S. space program should concentrate on unmanned missions like Voyager 2, which will send back information from space. Others say the U.S. should concentrate on maintaining a human spaceflight program like the space shuttle. Which comes closer to your view?" Human space exploration was preferred over robotic missions by a margin of 52 to 37 percent. But in an AP/IPSOS poll in the next year that prefaced the question with "Some have suggested that space exploration on the Moon and Mars would be more affordable using robots than sending humans . . .," answers tilted heavily in the other direction—a preference for robots by a margin of 57 to 38 percent.

1.5.1.5 NASA's Role, International Collaboration, and Commercial Firms

When asked whether they thought it was essential for the United States to continue to be a world leader in space exploration, slightly over half (58 percent) of respondents said it was essential in a 2011 Pew Research Center survey. The percentage seems to have fluctuated. When the question asks about concern about other countries pulling ahead of the United States, fewer people seem to say that they are very concerned about this than when the question is presented in the context of maintaining leadership. Another survey conducted by CNN/ORC at almost exactly the same time, July 2011, found just 38 percent saying that it was very important for the United States "to be ahead of the Russians and other countries in space exploration."[62] Few recent surveys have explored international collaboration in depth, but the available data suggest that the public is generally positive about international collaboration.

There is little in the survey literature about the public's views on the roles of government and the private sector in the exploration of space or human spaceflight, and this reflects both the low salience of space exploration and the relatively recent emergence of private space activities. The available data[63] suggest that the public may be becoming more receptive to private commercial activity in space.

1.5.1.6 Rationales for Space Exploration

A relatively small number of surveys have probed rationales for public support of space exploration. Most of them offered a list of rationales to respondents in closed-ended questions. Not surprisingly, the apparent level of support for many rationales is higher when they are explicitly mentioned in the question than when the question is open ended. However, the summary finding from the data seems to be that no particular rationale for space exploration consistently garners agreement from a clear majority of the U.S. public.

1.5.1.7 Correlates of Support for Space Exploration

A closer examination of the survey results indicates that support for space exploration is higher in some segments of the public than in others. To explore these patterns, the committee's Public and Stakeholder Opinion Panel examined answers to a number of questions by age, sex, race, education, and partisanship from a recent study conducted by the Pew Research Center in 2011. The questions included whether the space shuttle was a good investment, whether it is essential for the United States to play a leadership role in space exploration, and whether the space program "contributes a lot" to national pride and patriotism, to scientific advances that all Americans use, and to encouraging interest in science and technology. In this survey, sex and race were the strongest predictors of support for the space program: men tended to be more positive about the space program than women, and whites more than blacks.

The committee carried out a similar analysis to explore differences among groups in support of space exploration on the basis of the 2010 GSS question on spending priorities: "We are faced with many problems in this

---

[62] Poll conducted by CNN/Opinion Research International, July 19-20, 2011.
[63] See "The Role of the Private Sector" in Chapter 3.

country, none of which can be solved easily or inexpensively. I'm going to name some of these problems, and for each one I'd like you to tell me whether you think we're spending too much money on it, too little money, or about the right amount . . . the space exploration program?" Men were more likely to say we are spending too little on space exploration than women, whites more likely than members of other races, and college graduates more likely than respondents with less education.

1.5.1.8 Summary of Public Opinion Findings

The level of public interest in space exploration is modest relative to that in other public policy issues. At any given time, a relatively small proportion of the U.S. public pays close attention to this issue, and an even smaller proportion feels well informed about it. Space exploration fares relatively poorly among the public compared to other spending priorities. No particular rationale for space exploration appears to attract support consistently from a clear majority of the public. Those trends—generally positive views of space exploration and human spaceflight but low support in terms of funding and low levels of public engagement—have held true over the past few decades, during a time when the nation developed, flew, and retired a winged, reusable space vehicle and led a consortium of nations in building a large, orbiting research facility.

1.5.2 Stakeholder Views

The Public and Stakeholder Opinions Panel conducted a survey of key stakeholder groups. For the purposes of the study, stakeholders were defined as those who may reasonably be expected to have an interest in NASA's programs and to be able to exert some influence over its direction. The stakeholder groups included in the survey were in industry, the space science and engineering community, higher education, security, defense, foreign policy, writing and popularization, and space advocacy. Further detail about the stakeholders and the data-collection process and tables of raw and analyzed results are included in Chapter 3.

1.5.2.1 Stakeholder Views on Rationales for Space Exploration and Human Spaceflight

One of the primary goals of the survey was to understand stakeholder views about rationales for space exploration and human spaceflight. There was a clear consensus on the rationale for space exploration in general but less agreement about the rationale for human spaceflight.

For space exploration, whether the question asked how important each reason was, which reason was the most important, or simply asked respondents to list the reasons they found important, "expanding knowledge and scientific understanding" was the top choice of a majority of the respondents (see Tables 3.8–3.10 in Chapter 3). It was picked by 58 percent of respondents as the most important reason for space exploration. This rationale was equally dominant among those who were involved in space-related work and those not so involved.

None of the rationales traditionally given for *human* spaceflight received support from a majority of the respondents. When asked which of the rationales traditionally given for human spaceflight was the most important, "satisfying a basic human drive to explore new frontiers" was selected by the highest number of respondents, but fewer than one-fourth agreed that it was the most important reason, and the rest of the responses were split among a number of other rationales.

When asked to describe the reasons for human spaceflight in the form of an open-ended question, about one-third of the respondents provided reasons that can be summarized as "humans can accomplish more than robots in space." Somewhat fewer than one-third argued that the reason was to satisfy a basic human drive to explore new frontiers. "Expanding knowledge and scientific understanding" was mentioned by about one-fourth as a reason for human spaceflight as well. Those who were involved in space-related work were more likely to provide additional rationales for human spaceflight as a response to the open-ended question, but these rationales were endorsed by small percentages of respondents overall (even among those who were involved in space-related work).

To provide an additional perspective on the reasons for human spaceflight, respondents were asked in an open-ended format what they thought would be lost if NASA's human spaceflight program were terminated. There was no majority agreement when the question was asked that way either. The most commonly mentioned loss was national prestige, which was mentioned by only one-fourth of respondents. And 15 percent said that nothing would be lost.

1.5.2.2 Stakeholder Views on a Course for the Future

Respondents were asked to consider what goals a worthwhile and feasible U.S. human space exploration program might work toward over the next 20 years. They were presented with a list of possible projects that NASA could pursue and asked to indicate how strongly they favored or opposed each of them. The options were presented with approximate overall costs to provide some context for the scale of the projects.

Overall, the option that received the most "strongly favor" responses was continuing with LEO flights until 2020 (the survey was done before the U.S. administration's proposal to its partners to continue ISS operations through 2024). A majority of respondents favored four programs (either "strongly" or "somewhat"): LEO flights to the ISS until 2020, extending the ISS to 2028, conducting orbital missions to Mars to teleoperate robots on the surface, and returning to the Moon to explore more of it with short visits. Those who said that they were involved in space-related work, and in particular those who were involved in human spaceflight-related work, were generally more likely to favor most programs strongly than those who were not so involved, but continuing with LEO flights to the ISS until 2020 was the option that received the most "strongly favor" responses in all three groups.

Those who were under 40 years old were generally more likely to favor most projects strongly than those who were 40 years old and over. Continuing with LEO flights to the ISS until 2020 and extending the ISS to 2028 were the two options with the most "strongly favor" responses among respondents who were under 40 years old, and they were followed by establishing outposts on the Moon and landing humans on Mars.

Respondents were asked to rate the importance of several possible projects or activities for NASA over the next 20 years. The stakeholders were grouped into three categories—those who were not involved in space-related work, those who were involved in nonhuman space-related work, and those who were involved in human space-related work. The three groups differed in their views about what NASA should be doing over the next 20 years. For example, the stakeholders who were involved in human space-related work were more likely than stakeholders in the other two categories to say that it was very important to make the investment needed to sustain a vigorous program of human spaceflight (see Table 3.18). Still, majorities in all three groups thought it very important to make the investment needed to sustain a vigorous program of robotic space exploration. The two groups of stakeholders who were not involved in human spaceflight were more likely than those who were involved in human spaceflight to support improving orbital technologies, such as weather and communication satellites. Finally, substantial minorities of all three stakeholder groups (at least 40 percent) rated expanding space collaboration with other countries as very important.

The stakeholders had clear views about the role of NASA versus the private sector. More than 90 percent thought that NASA should take the lead in space exploration for scientific research, but nearly 85 percent said that the private sector should take the lead in space travel by private citizens, and more than two-thirds said that the private sector should take the lead in economic activities in space. Nearly half thought that NASA should take the lead in establishing an off-planet human presence, but almost one-third said that neither NASA nor the private sector should do this.

1.5.2.3 Summary of Stakeholder Survey Results

For space exploration in general, "expanding knowledge and scientific understanding" emerged as the rationale that was shared by the overwhelming majority of the respondents. However, when restricted to human spaceflight, no single rationale garnered agreement from a majority of the respondents. Support for human spaceflight appears to decline steadily with age, although it is important to note that due to the panel's intentional focus on policy leaders, the respondents to the survey tended to be older overall than the general population. Support for human spaceflight goes up with involvement in work related to human space exploration. There are also differences among the stakeholder groups in their support for human spaceflight: support was lowest among the nonspace scientists and highest among the advocates and popularizers.

1.6 A STRATEGIC APPROACH TO A SUSTAINABLE PROGRAM OF HUMAN SPACEFLIGHT[64]

Having laid out past and current space policy, explored the international setting, articulated the enduring questions and rationales, and examined public and stakeholder opinions, the committee draws a fundamental question: What type of human spaceflight program would be responsive to those factors? The committee argues that it is a program in which humans operate beyond LEO on a routine basis, in other words, a sustainable human exploration program beyond LEO. It is not the role of the committee to recommend whether the nation should move forward with such a program at this time. However, it is important to recognize that the assembly of the ISS is now essentially complete and that it has a finite lifetime. If the nation does not decide soon whether to embark on human space exploration beyond LEO, it will de facto begin ramping down its human spaceflight activities in the early 2020s as preparations for the closeout of ISS begin. More important, because major new spaceflight programs have lead times of years (sometimes a decade) between a decision to pursue a program and first flight, delaying a decision until near the end of the ISS's lifetime will guarantee a long gap in any human spaceflight activity—just as the termination of the Space Shuttle has led to a hiatus in U.S. capability to take astronauts up and bring them back to Earth. Chapter 4 argues that SLS and Orion flight rates that are too far below historical norms will not be sustainable over the course of an exploration pathway that spans decades. That will be the case for the first two launches of SLS, which are the only ones scheduled.[65] Hence, the committee has concluded the following:

If the nation deems continuity in human spaceflight to be a desirable national objective, it must decide now on whether to pursue a sustainable program of human space exploration and on the nature of such a program.

In what follows, the committee outlines the essential nature of a sustainable program.

1.6.1 Horizon Goal: Mars

Within the limits of foreseeable technologies, there are a small number of places that humans can go beyond LEO, and only two that have significant gravitational wells:[66] the Moon and Mars. Mars is the farthest practical exploration "horizon" for the foreseeable future—the most distant goal that is consistent with human physiological limits with likely future technologies (Chapter 4). Mars is a goal most compatible with the committee's enduring questions, and the intrinsic fascination that Mars has held in the popular imagination for well over a century makes it an attractive target. Such a program in any realistic funding scenario requires a sustained effort for several decades, and it would address the enduring questions. There is a consensus in national space policy, international coordination groups, and the public imagination that Mars is the horizon goal of human space exploration. The committee has concluded the following:

A sustainable program of human deep-space exploration must have an ultimate, "horizon" goal that provides a long-term focus that is less likely to be disrupted by major technological failures and accidents or by the vagaries of the political process and economic scene.

---

[64] Much of the technical discussion in this section is derived from the deliberations of the Technical Panel, which are described in detail in Chapter 4.

[65] The first two launches of SLS are planned to occur 4 years apart, in 2017 and 2021. In contrast, beginning with the first flights of the Saturn V launch vehicle and the Space Shuttle, respectively, during the subsequent 4 years the Apollo program conducted 12 launches and the Space Shuttle Program conducted 17.

[66] Essentially, the region of gravitational influence that a body exerts on the space around it.

1.6.2 Stepping Stones

Between LEO and the martian surface are regions of space, essentially operational theaters, that have stepping-stone destinations that are reachable with foreseeable advances in the state of the art for key technologies. Those operational theaters include

- Cislunar space, which encompasses missions to the Earth–Moon L2 point,[67] lunar orbit, and the lunar surface (both lunar sorties with relatively short stays and lunar outposts with extended stays).
- Near-Earth asteroids (NEAs) in their native orbits.
- Mars, which encompasses a Mars flyby mission and missions to the moons of Mars, Mars orbit, and the surface of Mars.

Missions to various destinations within those operational theaters could provide the necessary challenges, adventures, and diverse patterns of activity and use to sustain a program that addresses the enduring questions. The challenges of human spaceflight beyond LEO are created in part by increased requirements for propulsive energy and by longer mission durations. For example, Figure 1.9 shows how mission duration and delta-V, which is a measure of propulsive energy requirements, vary for missions in each operational theater. The size and placement of the regions associated with each mission in Figure 1.9 are determined by a number of factors, such as mission selection, orbital constraints, and specific destination. These factors vary greatly for NEAs, which populate a wide variety of orbits. The figure uses all of the NEA objects in the NHATS database.[68] The number of known NEAs is increasing at a rate of approximately 1,000 per year, but only a small fraction of NEAs are known to have orbits that may be suitable for a human mission, and many of those are nonetheless unsuitable destinations for human exploration because of other characteristics, such as spin rate, that are harder to determine and remain unknown for most asteroids, even after their orbits are quantified.[69] The delta-V for the Mars missions is highly variable and depends on the year of the mission; an exploration system design for the lowest delta-V values shown will be capable of visiting Mars only once every 15 years or so, whereas a design for a higher delta-V capability would allow visits every other year. Similarly, improving the propulsive capabilities of an NEA orbital-transfer vehicle would increase launch opportunities and the diversity of targets.

As propulsive energy requirements or mission durations increase, so do mission cost and risk. Increasing mission duration increases the risk of component failures, increases radiation exposure of systems and crew, and exacerbates many other technical, physiological, and psychological risks. Human exploration missions to cislunar space have the advantage of variable mission durations (from as little as 2 weeks to 6 months or more, as desired) and modest delta-V requirements. Missions to most NEAs would require substantially more delta-V than cislunar missions, with typical mission durations of 6 months to a year. The delta-V requirements for missions to Mars orbit or to the moons of Mars with a stay time of 500 days at the destination are comparable with the delta-V requirements for missions to the lunar surface, but the Mars missions would last about 900 days. A Mars surface mission with a stay time of 500 days would require more delta-V than either the cislunar missions or the other 500-day Mars missions.

Missions to Mars with stay times of 30 days would be shorter than their 500-day counterparts, but they would also require more delta-V because such missions need to use a less-favorable Earth–Mars orbital alignment.

For those reasons and others, a Mars surface mission is the most difficult goal in terms of the time required to overcome all the technological and physiological factors associated with it. In particular, unlike other missions under consideration, a human mission to the Mars surface would require an entry, descent, and landing (EDL)

---

[67] Lagrangian points, also referred to as L points or libration points, are five relative positions in the co-orbital configuration of two bodies, one of them with a much smaller mass than the other. At each of those points, a third body that is much smaller than either of the first two will tend to maintain a fixed position relative to the two larger bodies. In the case of the Earth–Moon system, L1 is a position between the Moon and Earth where a spacecraft could be placed, and L2 is a position beyond the Moon that would be similarly fixed in orientation to the Earth and Moon.

[68] Near-Earth Object Human Space Flight Accessible Targets Study (NHATS). Table of Accessible Near-Earth Asteroids, http://neo.jpl.nasa.gov/cgi-bin/nhats.

[69] E.V. Ryan and W.H. Ryan, "Physical Characterization Studies of Near-Earth Object Spacecraft Mission Targets," Advanced Maui Optical and Space Surveillance Technologies Conference, Maui, Hawaii, 2012, http://www.amostech.com/TechnicalPapers/2013/POSTER/RYAN.pdf.

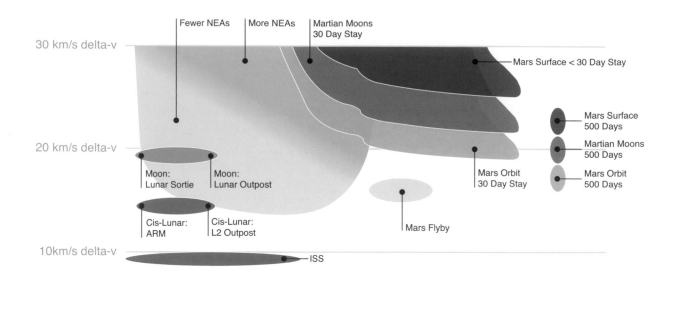

FIGURE 1.9 General comparison of human spaceflight destinations and missions in terms of mission duration and round-trip propulsive energy requirements from Earth in delta-V (velocity changes). SOURCE: Chart developed for the Committee on Human Spaceflight by the Aerospace Corporation on the basis of multiple sources, including design reference missions (DRMs) generated by NASA and the International Space Exploration Coordination Group (Human Spaceflight Exploration Framework Study, NASA, January 11, 2012, Washington, DC, http://www.nasa.gov/pdf/509813main_Human_Space_Exploration_Framework_Summary-2010-01-11.pdf; Human Space Flight Architecture Team (HAT) Technology Planning, Report to NASA Advisory Council March 6, 2012, Washington, DC, http://www.nasa.gov/pdf/629951main_ CCulbert_HAT_3_6_12=TAGGED. pdf; B.G. Drake, "Strategic Considerations of Human Exploration of near-Earth Asteroids," paper presented at the Aerospace Conference, 2012 IEEE, March 3-10, 2012; D. Mazanek et al., "Considerations for Designing a Human Mission to the Martian Moons," paper presented at the 2013 Space Challenge, California Institute of Technology, March 25-29, 2013; International Space Exploration Coordination Group, "The Global Exploration Roadmap," NASA, Washington, D.C., August 2013, https://www.globalspaceexploration.org/; D.A. Tito, G. Anderson, J.P. Carrico, J. Clark, B. Finger, G.A. Lantz, M.E. Loucks, et al. "Feasibility Analysis for a Manned Mars Free-Return Mission in 2018," paper presented at the Aerospace Conference, 2013 IEEE, March 2-9, 2013; J. Connolly, "Human Lunar Exploration Architectures," presentation to Annual Meeting of the Lunar Exploration Analysis Group, October 24, 2012, http://www.lpi.usra.edu/meetings/leag2012/presentations/). Additional information on the Design Reference Missions appears in Chapter 4.

system to land massive payloads and crew. That is a major cost, schedule, and risk item. The committee has concluded the following:

> **Given the magnitude of the technical and physiological challenges, should the nation decide to embark on a human exploration program whose horizon goal is Mars, NASA would need to begin to focus right away on the high-priority research and technology investments that would develop the capabilities required for human surface exploration of Mars. As discussed in Chapter 4, the most challenging of these will be entry, descent, and landing for Mars; in-space propulsion and power; and radiation safety (radiation health effects and amelioration).**

NASA's current proposal for the next step in deep space exploration—retrieving a small asteroid and towing it into a retrograde lunar orbit for examination by astronauts—is, with respect to the realm of human operations, the least demanding of the operating theaters. But, as argued in Chapter 4, if humans are eventually to land and operate for extended periods on Mars, the capabilities required are best developed and tested on the lunar surface as well as in cislunar space. And they are best developed in such a way that significant milestones are accomplished early and at regular intervals in the program—milestones that meaningfully and progressively enhance the capabilities of humankind for space exploration.

NASA and outside experts told the committee that the current administration regards the lunar surface as the purview of other nations' space programs and that it is not of interest to the U.S. human exploration program. That argument is made despite the barely touched scientific record of the earliest history of the solar system that lies hidden in the lunar crust,[70] despite its importance as a place to develop the capabilities required to go to Mars, and despite the fact that the technical capabilities and operational expertise of Apollo belong to our grandparents' generation. The history of exploration of our own globe carries the lesson that the ones who follow the first explorers are the ones who profit from the accomplishment. Such a lesson would suggest that the United States relook at its lack of interest in the lunar surface as a site for human operations. The pathways approach outlined below and its application, detailed in Chapter 4, support the view of the present committee that the Moon and in particular its surface have important advantages over other targets as an intermediate step on the road to the horizon goal of Mars.

The committee does not recommend either a capabilities-based or a flexible-path approach, in which no specific sequence of destinations is specified (see Chapter 4). Instead, as discussed below, the committee recommends a "pathways approach" to human space exploration: a specific sequence of intermediate accomplishments and destinations, normally of increasing difficulty and complexity, that lead to an ultimate (horizon) goal with technology feed-forward from one mission to the next.

If it is properly planned, funded, and executed, a pathways approach will enable taxpayers to see progress as missions explore significant destinations, and it will support a manageable level of development risk and an operational tempo that ensures retention of critical technical capability, proficiency of operators, and effective use of infrastructure. A pathways approach also would include technology development that addresses mission requirements and integration, placing crew and launch vehicles into a system architecture that includes such key components as radiation safety, advanced in-space propulsion and power, and EDL technologies for Mars.

Chapter 4 describes the process by which pathways to a human Mars surface mission might be developed. Each of the three example pathways could be attempted using a budget-limited approach that would not require a large bump in funding as Apollo did. However, the budget-limited scenarios feature a very low operational tempo (at best, one piloted mission every 2.4 years), which is much lower than that of any previous successful U.S. human spaceflight program. The committee has concluded as follows:

NASA can sustain a human space exploration program with meaningful milestones that simultaneously reasserts U.S. leadership in space and allows ample opportunity for substantial international collaboration when that program

- **Has elements that are built in a logical sequence.**
- **Can fund a frequency of flights sufficiently high to ensure retention of critical technical capability, proficiency of operators, and effective use of infrastructure.**

However, a NASA human spaceflight budget that increases with inflation does not permit a viable pathway to Mars (Chapter 4). **The program will require increasing the budget by more than the rate of inflation.**

---

[70] See, for example, R.M. Canup and K. Righter, eds., *Origin of the Earth and the Moon,* University of Arizona Press, Tucson Ariz., 2000; National Research Council, *The Scientific Context for Exploration of the Moon,* The National Academies Press, Washington, D.C., 2007.

1.6.3 Pathway Principles and Decision Rules

The pathways approach applies principles and decision rules that are designed to maximize efficient use of feed-forward systems and subsystems. The cost, scope, and challenges of human spaceflight beyond LEO demand that a set of carefully thought-out principles be applied before any pathway is initiated. Progress toward deep-space destinations will be measured in decades with costs measured in hundreds of billions of dollars and significant risk to human life. In what follows, the committee does not recommend one pathway over another but rather proposes principles by which national leadership might decide on pursuing a given pathway, measure progress along the pathway, move off the pathway to another, or cease the endeavor altogether. The resulting pathway principles are intended to be used in establishing a sustainable long-term course. In the environment of constrained federal budgets for the foreseeable future, the application of the principles should result in a pathway that includes *only* essential major hardware and mission elements so that it is possible to live within expected funding constraints. This approach leaves limited opportunities for major reductions in scope ("descoping") later, as described in detail in Chapter 4. Therefore, as its highest-priority recommendation, the committee recommends as follows:

NASA should adopt the following pathway principles:
  I.   **Commit to designing, maintaining, and pursuing the execution of an exploration pathway beyond low Earth orbit toward a clear horizon goal that addresses the "enduring questions" for human spaceflight.**
  II. **Engage international space agencies early in the design and development of the pathway on the basis of their ability and willingness to contribute.**
  III. **Define steps on the pathway that foster sustainability and maintain progress on achieving the pathway's long-term goal of reaching the horizon destination.**
  IV. **Seek continuously to engage new partners that can solve technical or programmatic impediments to progress.**
  V.   **Create a risk-mitigation plan to sustain the selected pathway when unforeseen technical or budgetary problems arise. Such a plan should include points at which decisions are made to move to a less ambitious pathway (referred to as an "off-ramp") or to stand down the program.**
  VI. **Establish exploration pathway characteristics that maximize the overall scientific, cultural, economic, political, and inspirational benefits without sacrificing progress toward the long-term goal, namely,**

  a. **The horizon and intermediate destinations have profound scientific, cultural, economic, inspirational, or geopolitical benefits that justify public investment.**
  b. **The sequence of missions and destinations permits stakeholders, including taxpayers, to see progress and to develop confidence in NASA's ability to execute the pathway.**
  c. **The pathway is characterized by logical feed-forward of technical capabilities.**
  d. **The pathway minimizes the use of dead-end mission elements that do not contribute to later destinations on the pathway.**
  e. **The pathway is affordable without incurring unacceptable development risk.**
  f. **The pathway supports, in the context of available budget, an operational tempo that ensures retention of critical technical capability, proficiency of operators, and effective use of infrastructure.**

Upfront pathway principles are applied with recognition that a set of operational decision rules will also be required and applied as NASA, the administration, and Congress face inevitable programmatic challenges along a selected pathway. The decision rules that this committee has developed provide operational guidance that can be applied as major technical, cost, and schedule issues arise as NASA progresses along a pathway. They have been designed to provide a framework for a sustainable program through the lifetime of the selected pathway and to allow a program to stay within the constraints accepted and developed when applying the pathway principles. The committee recommends the following:

Whereas the overall pathway scope and cost are defined by application of the pathway principles, once a program is on a pathway, technical, cost, or schedule problems that arise should be addressed by the administration, NASA, and Congress by applying the following decision rules:

A. **If the appropriated funding level and 5-year budget projection do not permit execution of a pathway within the established schedule, do not start down that pathway.**[71]

B. **If a budget profile does not permit the chosen pathway, even if NASA is well along on it, take an "off-ramp."**

C. **If the U.S. human spaceflight program receives an unexpected increase in budget for human spaceflight, NASA, the administration, and Congress should not redefine the pathway in such a way that continued budget increases are required for the pathway's sustainable execution; rather, the increase in funds should be applied to rapid retirement of important technology risks or to an increase in operational tempo in pursuit of the pathway's previously defined technical and exploration goals.**

D. **Given that limitations on funding will require difficult choices in the development of major new technologies and capabilities, give high priority to choices that solve important technological shortcomings, that reduce overall program cost, that allow an acceleration of the schedule, or that reduce developmental or operational risk.**

E. **If there are human spaceflight program elements, infrastructure, or organizations that are no longer contributing to progress along the pathway, the human spaceflight program should divest itself of them as soon as possible.**

1.6.4 Two Examples of Futures for Human Spaceflight: The Fiscal Challenge Ahead

The committee provides here examples derived from Chapter 4 to demonstrate the fiscal challenge that the United States faces in any exploration program beyond LEO. The examples are for illustrative purposes and should not be construed as recommendations regarding options. In each case, the "sand charts" depict in a linear fashion the annual estimated cost[72] of the human spaceflight program. Two budget profiles are shown for reference: a flat budget (constant then-year dollars) and a budget that increases with inflation.

1.6.4.1 Example 1—A Minimalist Program That Ends at L2

The scenario shown in Figure 1.10 provides an example of what could be accomplished with a flat human spaceflight budget. It continues the ISS to 2028, conducts the asteroid redirect mission, and establishes an intermittent human presence at the Earth–Moon L2 point. This scenario could be executed with a budget that is essentially flat from 2015 through 2045. However, no additional missions could be conducted after the L2 missions. Also, by 2045 the entire human spaceflight budget would be consumed by the fixed cost of the SLS and the Orion program and by core research, technology development, and support activities. Because this scenario does not accommodate any missions beyond cislunar space, it violates pathway principles I and III.

1.6.4.2 Example 2—A Budget-Driven Pathway Toward Mars That Does Not Satisfy the Principles

The scenario shown in Figure 1.11 was generated to conduct a technical analysis and affordability assessment of a notional pathway to Mars with a human spaceflight budget that increases at or about the rate of inflation while

---

[71] The committee recognizes that budget projections are unreliable, but they are also indispensable. One way to make the use of such projections more robust would be for NASA to conduct sensitivity analysis and evaluate plans against a range of possible 5-year budget projections that may vary by 10 percent or more. That might be done as part of the risk-mitigation plan.

[72] Because of the notional nature of the cost projections in this study, the vertical cost axes on Figure 1.10 and similar figures are not marked with dollar values. However, the committee is confident that the cost projections that are summarized in these figures provide a sound basis for making relative comparisons among the pathways and between the pathways and budget projections.

L2 OUTPOST PATHWAYS DRMs ANNUAL COST (THEN-YEAR $)

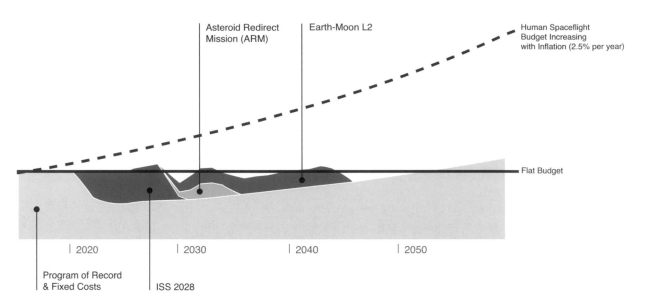

FIGURE 1.10 Pathway showing a minimalist program that ends at L2.

BUDGET-DRIVEN ENHANCED EXPLORATION ANNUAL COST (THEN-YEAR $)

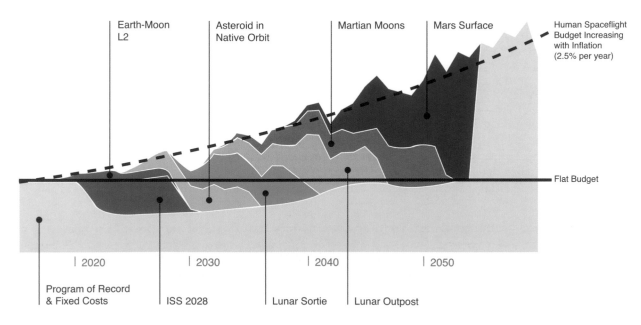

FIGURE 1.11 A budget-driven pathway toward Mars.

adhering to pathway principles VIa and VId by including targets that provide intermediate accomplishments and minimize the use of systems that do not contribute to achieving the horizon goal. Astronauts would explore new destinations at a steady pace: operation at L2 in 2024, a rendezvous with an asteroid in its native orbit in 2028, and the lunar sortie in 2033; a lunar outpost would be constructed in 2036, and the martian moons would be reached in 2043. Humans would land on Mars at the midpoint of the 21st century. This scenario violates pathway principle VIf in that the flight rate is too low to maintain proficiency (Chapter 4): on the average, one crewed mission every 2.1 years with gaps of up to 5 years in which there are no crewed missions.[73] This scenario could be modified to allow higher mission rates (see Chapter 4), but that would require funding to increase at a rate substantially higher than the rate of inflation for more than a decade, which, in the current fiscal environment, would violate pathway principle VIe.

On the basis of the lessons learned from those and other scenarios presented in Chapter 4, the committee has concluded the following:

> **As long as flat NASA human spaceflight budgets are continued, NASA will be unable to conduct any human space exploration programs beyond cislunar space. The only pathways that successfully land humans on the surface of Mars require spending to rise above inflation for an extended period.**

That conclusion could be modified in the case of robust international cost-sharing (that is, cost-sharing that greatly exceeds the level of cost-sharing with the ISS).

1.6.5 Risk Tolerance in a Sustained Program of Human Space Exploration

A sustained human exploration program beyond LEO, despite all reasonable attention paid to safety, will almost inevitably lead to multiple losses of vehicles and crews over the long term. For each step along the pathway, it will be important for NASA leadership and other stakeholders to discuss risk honestly and to establish acceptable levels of risk to missions and crews for deep-space missions. At the agency level, the risk discussion will be more detailed and will use relative or probabilistic levels to define the risk threshold, inform the design, and set priorities.

NASA should make all reasonable efforts to manage its technical risk in a way that emphasizes crew safety through the use of robust designs, failure tolerant approaches, and safe operations in adverse environments. However, a national failing to acknowledge that there are limits to the ability to mitigate the risks of human exploration inevitably undermines the ability of the program to accomplish high-risk goals and thus precludes a stable, sustainable program of exploration. A nation that chooses to extend human presence beyond the bounds of Earth affirms its commitment to that endeavor and accepts the risk to human life by continuing to pursue the program despite the inevitability of major accidents.

1.7 SUMMARY: A SUSTAINABLE U.S. HUMAN SPACE EXPLORATION PROGRAM

Human space exploration requires a long-term commitment by the nation or entity that undertakes it. Therefore, the committee has concluded the following:

> **National leadership and a sustained consensus on the vision and goals are essential to the success of a human space exploration program that extends beyond LEO. Frequent changes in the goals for U.S. human space exploration waste resources and impede progress. The instability of goals for the U.S. program of human spaceflight beyond LEO threatens our nation's appeal and suitability as an international partner.**

---

[73] The enhanced-exploration pathway includes many more SLS launches, for both crewed and uncrewed (cargo) vehicles, than the other two pathways. However, even for the enhanced-exploration pathway and even if uncrewed launches are considered, for much of the pathway the total SLS launch rate would be far lower than that of the Apollo or space shuttle programs. In particular, between 2022 and 2030, there would be an average of one SLS launch every 18 months.

The United States has had a sustained program of human spaceflight for more than a half-century, paradoxically in the face of, at best, lukewarm public support. There has not been a committed, passionate minority large and influential enough to maintain momentum for the kind of dramatic progress that was predicted by many space experts at the time of Apollo.[74] That is a problem that adds to the numerous difficulties—frequent redirection, mismatch of mission and resources, and political micromanagement—that have afflicted the U.S. human spaceflight program since Apollo ended. The committee has concluded as follows:

Simply setting a policy goal is not sufficient for a sustainable human spaceflight program, because policy goals do not change programmatic, technical, and budgetary realities. Those who are formulating policy goals need to keep the following factors in mind:

- **No defensible calculation of tangible, quantifiable benefits—spinoff technologies, attraction of talent to scientific careers, scientific knowledge, and so on—is likely ever to demonstrate a positive return on the massive investment required by human spaceflight.**
- **The arguments that triggered the Apollo investment—national defense and prestige—seem to have especially limited public salience in today's post–Cold War America.**
- **Although the public is mostly positive about NASA and its spaceflight programs, increased spending on spaceflight has low priority for most Americans. However, most Americans do not follow the issue closely, and those who pay more attention are more supportive of space exploration.**

It serves no purpose for advocates of human exploration to dismiss those realities in an era in which both the citizenry and national leaders are focused intensely on the unsustainability of the national debt, the dramatic growth of entitlement spending, and the consequent downward pressure on discretionary spending, including the NASA budget. With most projections forecasting growing national debt in the decades ahead, there is at least as great a chance that human spaceflight budgets will be below the recent flat trend line as that they will be markedly above it.

Nevertheless, the committee has concluded as follows:

If the United States decides that the intangible benefits of human spaceflight justify major new and enduring public investments in human spaceflight, it will need to craft a long-term strategy that will be robust in the face of technical and fiscal challenges.

Together with the highest-priority recommendation to adopt the pathways approach, the committee offers the following prioritized recommendations as being those most critical to the development and implementation of a sustainable human space exploration program:

NASA should

1. **Commit to design, maintain, and pursue the extension of human presence beyond low Earth orbit (LEO). This step should include**

    a. **Committing NASA's human spaceflight asset base, both physical and human, to this effort.**
    b. **Redirecting human spaceflight resources as needed to include improving program-management efficiency (including establishing and managing to appropriate levels of risk), eliminating obsolete facilities, and consolidating remaining infrastructure where possible.**

2. **Maintain long-term focus on Mars as the horizon goal for human space exploration, addressing the enduring questions for human spaceflight: How far from Earth can humans go? What can humans do and achieve when we get there?**

---

[74] *Aviation Week & Space Technology* in the 1960s predicted a landing of humans on Mars by the early 1980s.

3. **Establish and implement the pathways approach so as to maximize the overall scientific, cultural, economic, political, and inspirational benefits of individual milestones and to conduct meaningful work at each step along the pathway without sacrificing progress toward long-term goals.**

4. **Vigorously pursue opportunities for international and commercial collaboration in order to leverage financial resources and capabilities of other nations and commercial entities. International collaboration would be open to the inclusion of China and potentially other emerging space powers in addition to traditional international partners. Specifically, future collaborations in major new endeavors should seek to incorporate**

   a. **A level of overall cost-sharing that is appropriate to the true partnerships that will be necessary to pursue pathways beyond LEO.**

   b. **Shared decision-making with partners, including a detailed analysis, in concert with international partners, of the implications for human exploration of continuing the International Space Station beyond 2024.**

5. **Engage in planning that includes mission requirements and a systems architecture that target funded high-priority technology development, most critically**

   a. **Entry, descent, and landing for Mars.**

   b. **Advanced in-space propulsion and power.**

   c. **Radiation safety.**

None of those steps can replace the element of sustained commitment on the part of those who govern the nation, without which neither Apollo nor its successor programs would have occurred. Hard as the above choices may appear, they probably are less difficult or less alien for conventional political decision-makers than the recognition that human spaceflight—among the longest-term of long-term endeavors—cannot be successful if held hostage to traditional short-term decision-making and budgetary processes.

Asking future presidents to preserve rather than tinker with previously chosen pathways or asking Congresses present and future to fund human spaceflight aggressively with budgets that increase by more than the rate of inflation every year for decades may seem fanciful. But it is no less so than imagining a magic rationale that ignites and then sustains a public demand that has never existed in the first place. Americans have continued to fly into space not so much because the public strongly wants it to be so but because the counterfactual—space exploration dominated by the vehicles and astronauts of other nations—seems unthinkable after 50 years of U.S. leadership in space. In reviving a U.S. human exploration program capable of answering the enduring questions about humanity's destiny beyond our tiny blue planet, we will need to grapple with the attitudinal and fiscal realities of the nation today while staying true to a small but crucial set of fundamental principles for the conduct of exploration of the endless frontier.

2

Why Do We Go There?

2.1 INTRODUCTION

This chapter contains three separate though inter-related pieces of work, responsive to various elements of the charge to this committee. First, it reports on and presents the results of two outreach efforts from the committee to communities interested in space to gather ideas and inputs that could help inform the committee's work: a call for white papers on topics related to the committee's charge and a day of Twitter conversations (hashtag #HumansInSpace). It then elaborates on the considerations that led the committee to formulate the enduring questions presented in Chapter 1. That is followed by a discussion of the rationales that have historically been and still are advanced to justify U.S. engagement in a human spaceflight program; rationales effectively define the goals and aspirations toward which such a program can contribute. Finally, as requested in the committee's charge, the combination of the rationales is addressed in the language of a value proposition, which provides an alternative way to express rationales for such a program.

2.2 OUTREACH EFFORTS

A difficult but inescapable challenge to the committee's work was the fact that much of the rationale for human spaceflight is difficult to evaluate with solely quantitative and analytic methods. The cultural significance of human spaceflight resonates in many ways in society. The committee's examination therefore included consideration of insights from a variety of avenues, including insights provided by popular culture. (Elsewhere in this report, the committee observes, for example, the prevalence of human spaceflight themes in movies, advertisements, and other media.)

As part of its outreach effort, to cast the net as broadly as possible in examining a question that speaks as much to cultural and philosophical concerns as it does to practical benefits, the committee invited comments via social media. In addition to a call for white papers that asked respondents to provide their ideas and thoughts on ensuring a sustainable human spaceflight program, a 1-day Twitter campaign sought the public's "best ideas" along the same lines. The intention was not to develop a statistically robust sampling frame but to cast a wider net to solicit ideas, thoughts, and perspectives from individuals and groups who were engaged enough with the topic to take the time to respond. Insights gained from such sources as social media and popular culture, although they do not replace the quantitative investigations of public opinion discussed in Chapter 3, have a place in the committee's inquiry. In economics, for instance, it is commonly accepted that emotional and psychological factors can weigh as heavily in human aspirations and eventual outcomes as hard calculations of cost and benefit. John Maynard Keynes memorably wrote that

> a large portion of our positive activities depend on spontaneous optimism rather than material expectations. . . . Most probably, of our decisions to do something positive, the full consequences of which will be drawn out over many days to come, can only be taken as the result of animal spirits—a spontaneous urge to action rather than inaction, and not as the outcome of a weighted average of quantitative benefits multiplied by quantitative probabilities.[1]

Keynes was referring to the decisions of investors, but his observation could reasonably be applied to the academic and career choices of young students, the passion of scientists whose work produces technological and economic bounties of unforeseen character, and explorers venturing into a new frontier.

It is important to note that the logic of the call for white papers, tweets, and similar data inputs does not produce results that can be generalized; that is, they are not results from representative samples of any group or population. No conclusions may be drawn from them about opinions or perceptions of "the public," nor can any estimates be made about trends or percentages of people who hold one opinion or another. What can be said is that the opinions, positions, and arguments communicated to the committee through those venues are valuable sources of ideas and perspectives that might not otherwise be captured through more traditional polling and sampling methods. Thus, they were useful in the committee's deliberations and helped to ensure that it did not overlook points of view that might otherwise not be visible.

The committee reached out primarily to key influencers in the science and technology communities who maintain a moderate to high level of attention to space-related topics. Among the social structures that were contacted were science institutions, universities, professional organizations, blogs (for example, Boing Boing, io9, and Wired), and social media (for example, social media at NASA, Tim O'Reilly, and Science Friday).

Respondents to the call for white papers were asked the following questions:

- What are the important benefits provided to the United States and other countries by human spaceflight endeavors?
- What are the greatest challenges to sustaining a U.S. government program in human spaceflight?
- What are the ramifications and what would the nation and world lose if the United States terminated NASA's human spaceflight program?

Almost 200 white papers were submitted to the committee in response to the call for input.[2] Many came from people who were deeply engaged with NASA's work and contributed thoughtful analyses of specific issues that they believed the committee needed to be better informed about. All were read by two or more committee members, and the committee devoted time to discussing what they had read and to point out important papers for the full committee to read. Ideas that made a strong appearance in the white papers included substantially increasing support and recognition of commercial spaceflight efforts, exploiting space for economic benefits, increasing international partnerships, and increasing focus on technology development.

Participants in the Twitter campaign were asked to answer the following question: What are your best ideas for creating a NASA human spaceflight program that is sustainable over the next several decades? Over a period of 27 hours, tweets and retweets that used the hashtag #HumansInSpace were captured and reviewed.[3] The Twitter campaign captured 3,861 tweets and retweets, which came from 1,829 unique users who had a collective 13.75 million followers. Tweets related to the promotion of the campaign itself and all retweets were filtered out. About 1,604 original tweets from 710 unique users directly answered the call for ideas.

Many of the 1,604 tweets provided unique ideas on how to pioneer a sustainable human spaceflight program (Figure 2.1). Ideas that appeared often in the tweets included increasing the frequency of crewed missions, making

---

[1] J.M. Keynes, *The General Theory of Employment, Interest and Money,* Book 4, Palgrave Macmillan, 1936, Chapter 12, Section 7, p. 161.

[2] At the time of this writing, these can be viewed at http://www8.nationalacademies.org/aseboutreach/publicviewhumanspaceflight.aspx.

[3] Access to the public discussion via Twitter was available at the committee's website at http://sites.nationalacademies.org/DEPS/ASEB/DEPS_085240.

Matt Bellis @matt_bellis   29 Oct

Treat it like a public works project, a la the Hoover Dam or federal highway system. Weave it into the economic fabric. #HumansInSpace

James @star_avi8r   29 Oct

Personal opinion (1/2) Time we divide robotic missions into 2 categories: planetary science & precursors to human missions. #HumansInSpace

xyqbed @xyqbed   29 Oct

Every time the president changes, so does space policy. A non/bi/partisan commission would fix that.#HumansInSpace

FIGURE 2.1 Examples of tweets providing ideas for a sustainable U.S. human-spaceflight program.

laws that allow NASA to focus on consistent long-term planning, investing in more research and development (R&D) directly applicable to prolonged human spaceflight journeys, selecting artists to be astronauts, building more significant partnerships with international and commercial entities, and creating a clear storyline of how robotic and human missions are moving NASA toward the goal of human settlement.

In addition to providing ideas, the tweets and white papers were useful to the committee in reviewing the set of historical rationales presented later in this chapter. All the rationales mentioned in this report were also mentioned frequently in the tweets and white papers. Notably, the survival rationale made a strong appearance in both the white papers and tweets.

2.3 ENDURING QUESTIONS

One of the charges to the committee was to identify the enduring questions that describe the rationale for and value of human exploration in a national and international context. Implicit in that charge is the thought that identifying such enduring questions can help to ensure the continuity and sustainability of choices for the U.S. program in human spaceflight. To address that task, the committee concluded that it was necessary to examine and discuss the historical rationales that have been presented as the reasons for which such a program is needed and useful. This chapter addresses both the enduring questions and the rationales.

Implied in the committee's charge was the expectation that the committee could find questions that would both deepen the rationales for human spaceflight and provide a long-term compass for the work, as perhaps certain deep-science questions have done for some fields of science. However, the committee, having examined the historic rationales (discussed below) often given for the program, found no new or deeper rationales and no questions that would suggest them. The rationales can be divided into five that the committee calls pragmatic and two that the committee calls aspirational. The pragmatic rationales are related to benefits to economic and technological strength, to national security and defense, to national stature and international relations, to education and inspiration of students and the general public, and to scientific exploration and observation. The two aspirational rationales are human survival and shared human destiny and aspiration (for exploration). Each rationale can be evoked by one or more questions. However, in the context of the more pragmatic rationales, the questions do not lead to motivation specifically for *human* programs as opposed to motivations for spaceflight and space exploration more generally, including both robotic and human ventures. Furthermore, although the questions are important and will continue to be important, they do not rise to the level that the committee considered was intended by the term *enduring question*.

Enduring questions, in the committee's view, are ones that can serve as motivators of aspiration, scientific endeavor, debate, and critical thinking in the realm of human spaceflight. The questions endure because any answers that are available today are at best provisional and will change as more exploration takes place. Enduring questions should provide a foundation for analyzing choices that is immune to external forces and policy shifts. Enduring questions are intended not only to stand the test of time but also to continue to drive work forward in the face of technological, societal, and economic constraints. The two aspirational rationales, in contrast to the pragmatic rationales, do indeed lead us to ask such questions, which require further efforts in human spaceflight if they are to be answered and which address issues of the future of humankind. They suggest an international rather than a national effort; indeed, given the breadth of the international interest and capability in spaceflight, progress in answering these questions will not depend on the U.S. spaceflight program alone.

The committee asserts that the enduring questions motivating human spaceflight are
- **How far from Earth can humans go?**
- **What can humans discover and achieve when we get there?**

The questions are deceptively simple, but the committee was convinced that, in the context of any national or international effort in human spaceflight, asking whether a program—or even a pathway step—helps to advance us toward the ability to answer these questions can provide a useful compass in making choices.

The possibility of human spaceflight has inspired a wide variety of questions throughout time, even before it was first accomplished in 1961 with the launch of Yuri Gagarin. Indeed, the task of this committee could be construed as answering the ultimate question: Why explore? At its most fundamental level, human spaceflight is a continuation of human exploration—an extension of the human drive to investigate uncharted territory. In the early 17th century, Johannes Kepler wrote *Somnium*, a work of science-based fiction that detailed human flight to the Moon on the basis of Copernican astronomy. *Somnium* explored how humans could conduct lunar astronomy and how the motions of Earth might be studied from the viewpoint of the Moon. The achievement of human flight to the Moon was still another 3 centuries away, but Kepler declared that it was scientifically possible to go there and to ask what we might discover and do when we get there.

The two enduring questions—How far from Earth can humans go? What can humans discover and achieve when we get there?—lead to more specific questions that are more closely linked to historically stated rationales for the U.S. human spaceflight program, such as these:

- Does humanity have a long-term sustainable future beyond Earth?
- What are the limits of human adaptability to environments other than Earth?
- Can humans exploit off-Earth resources for humankind?
- What can human exploration of celestial bodies—such as the martian system, the Moon, and asteroids— uniquely teach us about the origin of the solar system and the existence of life?
- How can human spaceflight enhance national security, planetary defense, international relations, and other national and global goals?

In the world of policy-making, where day-to-day pressures and changes in leadership can cause abrupt shifts of direction, enduring questions can help to provide a compass to maintain stability over the long period needed to achieve challenging goals. Similar questions from other fields include these: What is the cure for cancer? How did the universe begin?[4] Enduring questions can create a reference frame for comparing past, present, and future policy and for analysis of contemporary applications of it. By asking How far from Earth can humans go? and What can humans discover and achieve when we get there?, the United States can address the fundamental constraints on human exploration at any given time and consider what approaches can contribute to opening up the horizon for that exploration. Continued focus on whether a project or program helps us to answer these questions, or to

---

[4] National Research Council (NRC), *Connecting Quarks to the Cosmos: Eleven Science Questions for the New Century,* The National Academies Press, Washington, D.C., 2003, p. 60.

change the current conditions that provide limits on the answers to them, can also help to set priorities among alternatives and to eliminate approaches that offer little or no path toward new or better-defined answers. With the enduring questions in mind, this chapter turns next to a discussion of historically stated rationales and some analysis of them in the current context.

2.4 RATIONALES FOR HUMAN SPACEFLIGHT

The committee searched for rationales in various ways. These included reviewing past reports, questioning invited speakers, calling for white papers, reviewing public opinion as described in Chapter 3, and soliciting ideas for a sustainable program via the more novel public input provided by a Twitter event. In essence, the committee found no truly new rationales, although rationales have been grouped and stated in somewhat different ways by speakers and writers. All the arguments that the committee heard for supporting human spaceflight can be assigned to one or more of the categories discussed below. All the rationales have been used in various forms and combinations to justify the program for many years. That such justifications typically cite more than one of these rationales leads the committee to suspect that no one finds any one of them in isolation a convincing argument. As described above, there are essentially two groups of rationales that the committee finds: the pragmatic rationales—including contributions to the economy, national security, national stature and international relations, science, or education—and the more aspirational rationales, which include contributions to the eventual survival of our species and to supporting the human destiny to explore and aspire to challenging goals. One or both of the aspirational rationales of human destiny and human survival typically are invoked in arguing for the value of the program and then supported by reference to one or more of the pragmatic rationales. For the pragmatic rationales, human spaceflight can be a contributor but not the sole contributor.

2.4.1 Economic and Technology Impacts

One of the rationales often stated for government spending on spaceflight is that a vibrant space program produces economic benefits. That rationale encompasses a number of diverse impacts at both the sectoral and economy-wide levels. Some earlier studies argue that spaceflight programs have contributed to the overall productive or technological capabilities of the U.S. economy, strengthening national economic growth and competitiveness in the global economy. At the sectoral level, NASA programs have been credited with supporting the development of new technologies and their broader adoption throughout the economy. Although few if any empirical studies have attempted to compare the effects of NASA R&D investments on innovation with the effects of other federal programs, many studies have argued that NASA programs have contributed substantially to U.S. innovation.[5]

---

[5] Representative comments on the innovation-related consequences can be found in the Paine Report, 1986; 1990 Advisory commission; and 1991 Stafford Report. The following passages are from those studies of NASA's human spaceflight programs:

We are confident . . . that leadership in pioneering the space frontier will "pull through" technologies critical to future U.S. economic growth, as World War II military developments set the stage for major postwar growth industries. (National Commission on Space, *Pioneering the Space Frontier: An Exciting Vision of our Next Fifty Years in Space,* Bantam Books, New York, 1986, p. 189.)

The space program produces technology that enhances competitiveness; the largest rise and subsequent decline in the nation's output of much needed science and engineering talent in recent decades coincided with, and some say may have been motivated by, the build-up and subsequent phase-down in the civil space program. (Advisory Committee on the Future of the U.S. Human Spaceflight Program, *Report of the Advisory Committee on the Future of the U.S. Human Spaceflight Program,* Executive Summary, NASA, Washington, DC, 1990.)

America's recent history has demonstrated that our space program stimulates a wide range of technological innovations that find abundant applications in the consumer marketplace. Space technology has revolutionized and improved our daily lives in countless ways, and continues to do so. (Synthesis Group on America's Space Exploration Initiative, *America at the Threshold: America's Space Exploration Initiative,* commonly known as the Stafford Report, U.S. Government Printing Office, Washington, D.C., 1991, available at http://history.nasa.gov/staffordrep/exec_sum.pdf, p. 2.)

An investment in the high technology needed for space exploration maintains and improves America's share of the global market and enhances our competitiveness and balance of trade. It also directly stimulates the scientific and technical employment bases in our country, sectors whose health is vital to our nation's economic security. (The Synthesis Group, *America at the Threshold,* 1991, pp. 2-3.)

There are now a substantial space-based element of the communication industry and multiple commercial uses of space-based Earth-observing capability. Clearly, those industries would not exist without the original NASA and Department of Defense (DOD) satellite and rocket development work, but they benefited only modestly, if at all, from human spaceflight programs. NASA's more recent efforts to foster new transportation systems for delivering cargo and crew to the International Space Station may support a stronger U.S. share in the industries, for example, by reducing the costs of launch services and increasing competition across the sector relative to foreign launch providers.[6,7] Some of the companies offering such services are developing diversified business portfolios—including NASA, military, and private customers—but the evidence to test this hope is not yet in.[8,9]

Other manifestations of NASA's contribution to the national economy are known as spinoffs (specific technologies) as opposed to macroeconomic indicators that are more difficult to measure. The abundance of discrete spinoffs in which NASA can justifiably claim an important role has been large enough for NASA to produce a substantial publication every year that describes the benefits accrued to several consumer fields from space activities.[10] They have included the development of scratch-resistant lenses, water-purification systems, aircraft anti-icing systems, freeze-drying processes, and cryogenic insulation. It is worth noting that NASA's annual *Spinoff* publication does not single out benefits of human spaceflight but rather describes benefits of all NASA activities, including aeronautics.[11] The committee has not found any reliable analysis that separates those elements, but many of the entries appear to be related to the development of technologies to support human spaceflight.

Although examples of technology transfer can be compelling in fostering public interest in and support for space activities, they do not provide the foundation for a systematic understanding of the relationship between human spaceflight and economic benefit. Over the past few decades, a number of studies have attempted to establish correlation (if not causation) between space activities (both human and robotic) and economy-wide benefits, including the effects on U.S. industrial or technological capabilities. The studies have usually adopted one of three methods: using a macroeconomic production function model to estimate the impact of technological change resulting from R&D spending expressed as a rate of return on a given investment, assessing the returns on specific technologies through cost–benefit ratios, and evaluating the evidence of the direct transfer of technology from federal space programs to the private sector.[12] A brief summary of some of the key studies illustrates the kinds of evidence that have been used to argue that a vibrant space program leads to substantial economic benefits at both the micro and macro levels.

A number of studies of the impact of NASA on specific or broad economic indicators were undertaken in the late 1960s. Not all were commissioned by NASA, and almost all focused on local impacts.[13] A Stanford Research Institute study in 1968 found that "NASA activities have had a positive and consequential influence on the locali-

[6] J. Oberg, "Russians face their space crisis: Agency chief worries that country's aerospace industry is becoming uncompetitive," *NBC News,* September 28, 2012, http://www.nbcnews.com/id/49217472/ns/ technology_and_science-space/t/russians-face-their-space-crisis/#.Ue1OtRaOXGI.

[7] Ilya Kramnik, The new war for space is a public-private one, *Russia and India Report,* April 30, 2012, http://indrus.in/articles/2012/04/27/the_new_war_for_space_is_a_public-private_one_15428.html.

[8] A.J. Aldrin, "Space Economics and Commerce," pp. 179-200 in *Space Strategy in the 21st Century* (E. Sadeh, ed.), Routledge (Taylor & Francis), London, U.K., 2013.

[9] See, for example, "Low-Cost SpaceX Delays 1st Commercial Launch," which details a backlog of 50 launches reflecting $4 billion in orders, http://www.reuters.com/article/2013/11/26/space-spacex-launch-idUSL2N0JA1XL20131126.

[10] NASA, Office of the Chief Technologist, "NASA Spinoff," http://spinoff.nasa.gov/, last updated August 10, 2011.

[11] A recent paper, partly written by NASA representatives, focused on spinoffs from all NASA activities and proposed a set of discrete and quantitative measures for assessing the agency's impact. The measures included jobs created, revenue generated, productivity and efficiency improvements, lives saved, and lives improved. The early results of a survey suggested favorable numbers for some of the categories. See D. Comstock, D. Lockney, and C. Glass, "A Structure for Capturing Quantitative Benefits from the Transfer of Space and Aeronautics Technology," (pp. 7-10) paper presented at the International Astronautical Congress, Cape Town, South Africa, October 3-7, 2010, International Astronautical Federation, Paris, France.

[12] These approaches are summarized in H.R. Hertzfeld, "Space as an Investment in Economic Growth," pp. 385-400 in *Exploring the Unknown: Selected Documents in the History of the U.S. Civil Space Program, Volume III: Using Space* (J.M. Logsdon with R.D. Launius, D.H. Onkst, and S.J. Garber, eds.), NASA, Washington, D.C., 1998.

[13] See, for example, W. Isard, *Regional Input-Output Study: Recollections, Reflections, and Diverse Notes on the Philadelphia Experience,* MIT Press, Cambridge, Mass., 1971; W.H. Miernyk, *Impact of the Space Program on a Local Economy: An Input-Output Analysis,* West Virginia University Press, Morgantown, West Va., 1967.

ties in the South in which it has established research and development centers and production, testing, and launch facilities."[14]

More substantial studies were performed in the 1970s, some of which were directly commissioned by NASA at a time when the agency's budget was shrinking and the agency was seeking to understand arguments in support of increased investment. In 1971, the Midwest Research Institute issued the results of its analysis of a comprehensive national estimate of the returns on federal R&D expenditures, including those of NASA. It is important to note that the study estimated the national returns from all federal R&D spending on the basis of an analytic framework pioneered by Moses Abramovitz (1956) and Robert Solow (1957) and applied the estimates to NASA R&D spending. In other words, the study assumed that the rate of return on NASA R&D was similar to that on other federal R&D programs rather than much higher or lower. The Midwest study estimated a 7-to-1 return on NASA expenditures and projected a 33 percent discounted rate of return that began with the establishment of NASA in 1958 and was projected through to 1987.[15] Although the methodology and results of the study were criticized by many,[16,17] its conclusions were cited by NASA and NASA supporters for many years in arguing for the beneficial effects of NASA R&D.[18]

Perhaps the most important studies that NASA commissioned in the 1970s were those undertaken by Chase Econometrics (in 1976 and 1980). The studies were early attempts to measure overall returns to NASA in terms of measures at the level of the national economy—gross national product (GNP), employment, and productivity. The 1976 study showed that "the historical rate of return from NASA R&D spending is 43 percent." Chase found that "a sustained increase in NASA spending of $1 billion (1958 dollars) for the 1975–1984 period would" increase the GNP by $23 billion by 1984 ("a 2% increase over the 'baseline'").[19] Yet both the Chase Econometrics studies and a follow-on study by the Midwest Research Institute in 1988 that came to somewhat similar conclusions were again criticized by the General Accounting Office (now the Government Accountability Office).[20]

Further studies in the 1970s focused on NASA's contributions to specific fields. The best known of these was the study performed by Mathematica in 1976 that drew attention to NASA's contribution to four specific technologies: gas turbines, cryogenics, integrated circuits, and a software program widely used for modeling of physical structures (NASTRAN).[21] Three of those four benefited from NASA R&D associated with human spaceflight. The Mathematica study defined the economy-wide benefit stream that is attributable to NASA as benefits associated with reductions in the cost of the technologies and acceleration in their development: for example, What is the economic value of being able to use the integrated circuit in non-NASA applications 1 or 2 years earlier than might otherwise have been the case? The study's analytic approach thus attempted to address one of the most difficult questions in any evaluation of the benefits associated with government R&D investment: What would have happened if the investment had not been made?

The Mathematica study concluded that the economy-wide benefits attributable to NASA's investment in the development of the four innovations amounted to $7 billion, $5 billion of which was associated with the development of integrated circuits. Inasmuch as that estimated total benefit exceeded NASA's total 1974 budget, the study

---

[14] R.W. Hough, "Some Major Impacts of the National Space Program," Stanford Research Institute, Contract NASW-1722, June 1968, reproduced in *Exploring the Unknown, Volume III*, 1998, pp. 402-407.

[15] Midwest Research Institute, "Economic Impact of Stimulated Technological Activity," Final Report, Contract NASW-2030, October 15, 1971, reproduced in *Exploring the Unknown, Vol. III*, 408-414.

[16] A summary of the challenges in using the methodology of the Midwest Research Institute study that includes a critique of other studies of the economic benefits of NASA spaceflight programs can be found in H. Hertzfeld, Measuring returns to space research and development, pp. 155-170 in *Space Economics* (J. Greenberg and H. Hertzfeld, eds.), Progress in Astronautics and Aeronautics, AIAA, Washington, D.C., 1992.

[17] Congressional Budget Office criticisms of the Midwest Research Institute study can be found at Reinventing NASA, http://www.cbo.gov/sites/default/files/cbofiles/ftpdocs/48xx/doc4893/doc20.pdf, p. 4.

[18] An example of NASA advocates that cited the study uncritically may be found at http://www.penny4nasa.org/category/fight-for-space/.

[19] M.K. Evans, "The Economic Impact of NASA R&D Spending," Executive Summary, Chase Econometric Associates, Inc., Bala Cynwyd, Penn., Contract NASW-2741, April 1976, reproduced in *Exploring the Unknown, Volume III*, 1998, pp. 414-426.

[20] U.S. General Accounting Office, "NASA Report May Overstate the Economic Benefits of Research and Development Spending," Washington, D.C., 1977, pp. 6-11.

[21] Mathematica, Inc., "Quantifying the Benefits to the National Economy from Secondary Applications of NASA Technology—Executive Summary," NASA CR-2674, March 1976, reproduced in *Exploring the Unknown, Volume III*, 1998, pp. 445-449.

concluded that the benefits from NASA were primarily in accelerating the process of bringing technologies into the market place, not necessarily in developing the technologies themselves. Henry R. Hertzfield notes that "because this was a study of four cases and used the more traditional consumer surplus theory of microeconomics, the results were more readily accepted by the economics community than the results of the macroeconomic studies of that era."[22]

Further studies in the 1980s generally were geared to justifying large programs, such as a space-station program, rather than a broader approach that focused on R&D as a whole. The Midwest Research Institute issued a study commissioned by the National Academy of Public Administration in 1988, which repeated the study originally done in 1971.[23] It estimated a 9-to-1 return on the space program, but the study was subjected to the same types of critiques (methodology and problems with data) that plagued earlier studies, such as the Chase Econometrics study in 1976[24] and the earlier Midwest study.[25]

Besides studies at the macroeconomic level, a number of studies in the 1970s and 1980s looked at technology transfer. Commissioned by NASA, the studies examined arguments that the favorable impacts of NASA activities went beyond abstract scientific benefits to the everyday life of the average American. In a similar vein, NASA expanded a program to showcase particular technologies with its annual *Spinoff* publication (first published in 1976), which highlighted the many different areas of life that have been affected by NASA-related innovations. One common criticism of NASA's attempts to showcase spinoffs as important to the average American was that "most of the reported technological successes in *Spinoff* [were] either demonstration projects (that is, not fully commercialized) or public-sector uses of space technology."[26] Although that may lower their value as a public-relations tool, it does not diminish their economic impacts if these were calculated on the basis of actual rather than projected uses.

A more recent study by Henry R. Hertzfeld examined the economic benefits associated with 15 private companies' successful commercialization of innovations derived from NASA life-sciences R&D programs, most of which are associated with NASA's human spaceflight activities.[27] According to the study, NASA invested roughly $3.7 billion in life-sciences R&D during 1958-1998. The NASA-related R&D investment in the 15 technologies that were the subjects of the study amounted to $64 million, and the companies spent about $200 million in private funds in further development and commercialization activities. The results show that the 15 companies contributed $1.5 billion in value added to the U.S. economy during the 1975-1998 period. Thus, the study highlights the economic benefits associated with successful commercialization of technologies that are based on NASA life-sciences R&D and suggests further that the benefits associated with commercial success are substantial relative to the magnitude of the total NASA R&D investment in the fields in question. A comparison of the benefits of NASA life-sciences R&D associated with human spaceflight and those of other federal biomedical R&D programs is, however, beyond the scope of Hertzfeld's study.

2.4.1.1 Evaluation of Economic and Technological Rationales

Most economic studies conclude that the federal investment in NASA's space activities has benefited the U.S. economy, but they also agree that the benefits are difficult or impossible to measure or quantify. The results are particularly inconclusive regarding the degree to which NASA's *human* spaceflight programs contribute to economic growth. No systematic attempt has been made by NASA or other analysts to compare the economic benefits of NASA human spaceflight programs with those of other federal R&D programs.[28]

---

[22] H.R. Hertzfeld, "Space as an Investment in Economic Growth," p. 391.

[23] Midwest Research Institute, "Economic Impact and Technological Progress of NASA Research and Development Expenditures," Executive Summary, for the National Academy of Public Administration, September 20, 1988, reproduced in *Exploring the Unknown, Volume III*, 1998, pp. 427-430.

[24] M.K. Evans, "The Economic Impact of NASA R&D Spending," 1976.

[25] Midwest Research Institute, "Economic Impact of Stimulated Technological Activity," 1971.

[26] H.R. Hertzfeld, "Space as an Investment in Economic Growth," p. 391.

[27] H.R. Hertzfeld, Measuring the economic returns from successful NASA life sciences technology transfers, *Journal of Technology Transfer* 27.4:311-320, 2002.

[28] An early study that focused on human spaceflight was M.A. Holman, *The Political Economy of the Space Program,* Pacific Books, Palo Alto, Calif., 1974. See also M.A. Holman and R.M. Konkel, Manned space flight and employment, *Monthly Labor Review* 91(3):30, 1968; R.M. Konkel and M. Holman, *Economic Impact of the Manned Space Flight Program,* NASA, Washington, D.C., January 1967.

At heart here is a counterfactual issue: even if NASA's human spaceflight activities have had a substantial favorable effect on U.S. technical, industrial, and innovative capabilities, it is difficult or impossible to ascertain whether similar effects could have resulted from similarly large R&D investment by other federal agencies. Besides that analytical problem, most of the older substantive studies have limited use for the present study for two other reasons: they do not distinguish between robotic and human spaceflight, and they tend to focus on data from the 1960s.[29] The latter is an important issue because of the vast disparity between the high levels of funding in the 1960s and those in the 21st century. For example, in fiscal year (FY) 1967, NASA accounted for nearly 30 percent of total federal R&D spending and almost 35 percent of all federally funded development spending.[30] By 2009, those shares stood at roughly 4.5 percent and 6 percent, respectively. In addition, reported NASA R&D spending in FY 2009 included a much larger share devoted to robotic space exploration, suggesting that NASA human spaceflight R&D may account for as little as 3-4 percent of total federal R&D spending, and perhaps even less. Such data indicate that it is at best hazardous to use even the imperfect and often inconclusive results of studies that are based on NASA data from the 1960s to project the effects on U.S. innovative performance of the far smaller NASA human spaceflight budgets that are likely to characterize the program for the near future. A more recent review could substantially revise the current understanding of likely economic benefits given the different budgetary footprint of the human spaceflight program of the past 20 years relative to its peak of the 1960s and the high probability that future human spaceflight budgets will resemble those of the post-2000 period more closely than those of the 1960-1970 period. As this report notes below, the complexities of the channels through which the economic benefits of federal R&D investment are realized means that analytically defensible evaluations are rare. Nevertheless, a more recent evaluation could be of value to policy-makers who are seeking to understand the economic impact of NASA programs.

In summary, the economic rationale for a sustainable human spaceflight program is an oft-repeated one that rests on a generally accepted notion that such programs have generated substantial benefits to the U.S. economy. There are many individual examples of technological spinoffs of space activities, especially those involving robotic spaceflight.[31] Nonetheless, although the economic and technological benefits of human spaceflight are anecdotally impressive, they are extremely difficult to measure, and the uncertainty in them makes it impossible to compare them with the benefits of other federal investment in R&D that might have achieved the same or better economic results. Moreover, there is little basis on which to predict whether future NASA human spaceflight programs will have anything like the influence on U.S. technological innovation or on the education of scientists and engineers that the programs arguably had during the 1960s, simply because future NASA programs will account for a much smaller share of overall federal R&D and procurement spending. The absence of evidence suggesting that the economic return on investment in NASA human spaceflight is either more or less than the return on other R&D investments made by the federal government, all of which are generally thought to affect the economy favorably, should, of course, not be taken to imply that there is no economic benefit from such investment.

More recently, there has been an emphasis on commercial exploitation of space, either for low Earth orbit (LEO) travel or for going beyond LEO, with ideas of eventual commercial exploitation of space resources. NASA has supported the entry of private companies into the market as cargo or crew transportation providers in early-stage development roles that were previously the territory of NASA centers, for example, designing and developing vehicles to transport goods, and eventually astronauts, to the ISS. Such investment has encouraged a small number of individuals and companies to invest their own resources in addition to the NASA funding that they have received. The hope is that by transferring development risk to a commercial sector, which expects to have a broader customer base and broader income stream than NASA alone, NASA will eventually realize significant cost savings and at the same time stimulate new industries. It remains to be seen whether any of these ventures

---

[29] R.A. Bauer, *Second-order Consequences: A Methodological Essay on the Impact of Technology,* MIT Press, Cambridge, Mass., 1969; F.I. Ordway III, C.C. Adams, and M.R. Sharpe, *Dividends from Space,* Crowell, New York, 1971.

[30] See National Science Foundation, *Federal Funds for Research and Development, Fiscal Years 1951-2002,* NSF 03-325, August 14, 2003, http://www.nsf.gov/statistics/fedfunds/; NSF, "Federal Funds for Research and Development, Fiscal Years 2009-2011," http://www.nsf.gov/statistics/nsf12318/content.cfm?pub_id=4177&id=2; accessed January 21, 2014.

[31] For a useful summary of such itemized benefits, see M. Bijlefeld, *It Came from Outer Space: Everyday Products and Ideas from the Space Program,* Greenwood Press, Westport, Conn., 2003.

will become self-sustaining and be profitable without NASA, although at least one has a significant backlog of non-NASA orders.

A number of investors have placed large bets that some of those efforts will eventually present continuing business opportunities, and investor interest in the sector is not limited to orbital flights. Wealthy individuals also have invested in suborbital flight opportunities either by starting new companies or by booking seats for planned future suborbital trips and beyond. The prospect of economic return from mining space resources is even more remote; it may exist at some future time, but with current costs and values of resources, back-of-the-envelope calculations suggest that it is highly unlikely to become viable within the term of this study, even if the uncertain legal status of off-Earth mining claims is resolved.

Public–private partnerships similar to those involved in development of new transportation systems for the ISS are also under consideration by NASA for broader commercial use of the ISS itself and to encourage other private activity in LEO—for example, to develop commercial space platforms that could lease services or facilities to the government and other users.[32] At the same time, NASA is soliciting inputs from commercial entities that are interested in developing capabilities for beyond-LEO exploration efforts. Increased activity in cislunar space may create new opportunities for private interests to enter into partnerships with the agency. NASA recently entered into agreements with three firms to develop commercially sourced capabilities for a robotic lunar lander, and at least one company has been founded with the goal of commercial exploitation of the Moon, including the development of tourism. Other companies have expressed interest in the development of "infrastructure" to support cislunar activity, including development of communication networks and habitats.

A subject of recent commercial interest is the opportunity for exploitation of space resources beyond Earth orbit, which can be achieved robotically. The committee considers even robotic exploitation of space resources for on-Earth use to be highly speculative in that the cost–benefit ratio would need to change substantially for such exploitation to be commercially viable. Exploitation that requires human spaceflight as an element of the work would be much more expensive and hence even less commercially viable.

It is currently impossible to assess whether commercial capabilities will develop to the point where they can create substantial cost savings (on the order of tens of billions of dollars) for NASA human space exploration efforts beyond LEO. In addition, investments to foster new commercial partners may create a tension in NASA in that the goal of facilitation of new commercial ventures can compete with that of exploration (that is, the goal of answering the enduring questions) in making decisions about program priorities.

There is no widely accepted, robust quantitative methodology to support comparative assessments of the returns on investment in federal R&D programs in different economic sectors and fields of research. Nevertheless, it is clear that the NASA human spaceflight program, like other government R&D programs, has stimulated economic activity and has advanced development of new products and technologies that have had or may in the future generate significant economic impacts. It is impossible, however, to develop a reliable comparison of the returns on spaceflight versus other government R&D investment.

2.4.2 National Security and Defense

2.4.2.1 Space and National Security

A second commonly stated rationale is that investment in human spaceflight contributes to national security. DOD and other national security agencies have long recognized the potential advantages of performing various operations from the "high ground" of space. Over the past 3 decades, the number and scope of space-related applications that directly support military operations conducted on or near Earth's surface have expanded dramatically. However, as discussed below, the role of human spaceflight in such efforts is limited.

---

[32] NASA, "Evolving ISS into a LEO Commercial Market," released April 28, 2014, https://prod.nais.nasa.gov/cgibin/eps/synopsis.cgi?acqid=160471.

The original impetus for developing satellites during the Eisenhower administration in the 1950s was to gather intelligence over the Soviet Union and thereby avoid the inherent risks associated with using piloted aircraft for this purpose. Although the Soviets initially objected to the notion of observing sovereign national territory (particularly its own) from space, they eventually followed suit.[33] Both superpowers also subsequently deployed early-warning satellites to detect the launch of ballistic missiles and communication satellites to ensure connectivity between national leaders and their own military forces. The U.S. military operated a small fleet of weather satellites to facilitate strategic planning and targeting. Although those military satellites were occasionally employed in support of routine peacetime activities as well as crises, the underlying rationale for their existence during the Cold War was to support nuclear forces.[34]

Beginning with the 1991 Gulf War and still continuing, U.S. military services have systematically explored ways in which space systems can also better support conventional military operations. As the technical capabilities of on-orbit systems and related ground equipment have improved, the national security community has relied increasingly on space-based information and services, as the examples below illustrate.

- Intelligence, surveillance, and reconnaissance (ISR) systems can assist in locating, identifying, and targeting enemy forces and in assessing the effects of air, ground, and maritime operations. In 2013, describing its more recent contributions to the "warfighter," the director of the National Reconnaissance Office noted that "we've brought dozens of innovative ISR solutions to the fight. These services, products and tools directly contribute to the highest priority missions, to include: counter-Improvised Explosive Device (IED) efforts; identifying and tracking High-Value Targets; countering narcotics trafficking; and special communications."[35]

- Early-warning satellites can detect the launch of enemy missiles and alert missile-defense units. They can also play an important role in warning civilian populations of impending attack, for example, during the Iraqi Scud missile attacks on Israel and Saudi Arabia in 1991.[36]

- The vast majority of all the long-distance communication used by U.S. and allied forces in recent conflicts has been routed through space, including both commercial and military communication satellites. With ever-increasing bandwidth, the most up-to-date intelligence information can now be sent to troops on the ground and to pilots in their cockpits for nearly real-time strikes on sensitive, fleeting targets. The safe and effective operation of remotely piloted aircraft in many cases depends on secure and reliable data links provided by communication satellites.

- Weather satellites help to forecast conditions that could affect military operations by obscuring targets from aircraft or adversely affecting ground and maritime movements.

- Originally developed as an aid to long-range navigation across the oceans or featureless terrain, the Air Force's Global Positioning System (GPS) has proved so accurate and so reliable that older forms of navigation have largely fallen into disuse and become a lost art. In addition, the military has applied GPS to an increasingly wide range of combat and logistical activities, including the development of highly precise munitions that can destroy targets more effectively while producing less collateral damage to the surrounding areas. GPS has also been used with steerable parachutes that can deliver supplies quickly and safely to troops in remote, inaccessible areas.

---

[33] Excellent accounts of the Eisenhower administration's space policy are provided in D.A. Day, J.M. Logsdon, and B. Latell, *Eye in the Sky: The Story of the Corona Spy Satellites,* Smithsonian Institution Press, Washington, D.C., 1998; and W.A. McDougall, *The Heavens and the Earth: A Political History of the Space Age,* Johns Hopkins University Press, Baltimore, Md., 1985, pp. 112-209.

[34] For more on the development of the American military's interest in space, see D.N. Spires, *Beyond Horizons: A History of the Air Force in Space, 1947-2007,* 2nd ed., Air Force Space Command, Peterson AFB, Colo., 2007; M. Erickson, *Into the Unknown Together: The DoD, NASA, and Early Spaceflight,* Air University Press, Maxwell AFB, Ala., 2005; and F.G. Klotz, *Space Commerce, and National Security,* Council on Foreign Relations, New York, 1999, pp. 7-10, http://www.cfr.org/world/space-commerce-national-security-cfr-paper/p8617.

[35] Betty Sapp, Director, National Reconnaissance Office, Statement for the Record Before the House Armed Services Committee, Subcommittee on Strategic Forces, U.S. House of Representatives, April 25, 2013, http://docs.house.gov/meetings/AS/AS29/20130425/100708/HHRG-113-AS29-Wstate-SappB-20130425.pdf.

[36] U.S. Department of Defense, *Conduct of the Persian Gulf War: Final Report to Congress,* Washington, D.C., April 1992, p. 177.

In light of those examples, it is virtually impossible to imagine how a modern military force could conduct operations successfully in crisis and conflict, or provide disaster relief and humanitarian assistance, without access to the data delivered from and through space systems. For that reason, the national security community has devoted increasing attention in recent years to potential threats to its space systems, whether natural or artificial, including improved capabilities to detect, track, identify, and characterize objects that are orbiting Earth ("space situational awareness"). It is also currently in the process of "recapitalizing" existing satellite constellations, including new generations of even more capable navigation and timing, communication, missile-warning, and meteorological systems.[37]

2.4.2.2 The Role of Human Spaceflight in National Security

Is it noteworthy that none of the capabilities discussed above depends on human activities in space. The military, in fact, studied and pursued several projects that would involve human spaceflight in the 1960s, including the *Dyna-Soar* delta-winged orbital vehicle and the Manned Orbiting Laboratory. According to historian David Spires, the latter was justified in the Air Force's 1961 draft Space Plan as a platform for evaluating potential military missions in space, including "space command posts, permanent space surveillance stations, space resupply bases, permanent orbiting weapon delivery platforms, subsystems, and components."[38] Both programs, however, were eventually terminated because of cost, technical issues, and, in the face of increasing robotic capability, lack of a compelling military requirement for the human element.

Although essentially a NASA program, the U.S. government at one point directed that all U.S. payloads, including national security payloads, be launched with the space shuttle and that the existing fleet of expendable launch vehicles be retired.[39] However, after the 1986 *Challenger* disaster, Congress forbade the shuttle to carry commercial payloads, whereas the intelligence establishment moved its national security payloads to conventional boosters—a move anticipated even before *Challenger* given the shuttle's inability to match the originally projected high launch rates. The Air Force opted to concentrate on uncrewed boosters to meet its launch needs even though this placed some restrictions on the size of satellites that could be launched. Termination of the Space Shuttle Program at Vandenberg Air Force Base in December 1989 symbolized the end of Air Force interest in direct participation in human spaceflight. While the Clinton administration's 1999 DOD space policy did mention that "humans in space may be utilized to the extent feasible and practical to perform in-space research, development, testing, and evaluation as well as enhance existing and future national security space missions," those missions were not defined.[40]

National space-policy documents of subsequent administrations did not directly link human spaceflight to national security missions. Rather, the association between human spaceflight and broader policy and goals has been generally described in terms of enhancing national stature and international cooperation, both of which are viewed as important although indirect contributors to overall national security.

During the George W. Bush administration, the 2004 *Vision for Space Exploration* contained several references to the importance of human spaceflight for broader national security and policy goals, including accelerating the development of critical technologies that underpin and advance the U.S. economy and help to ensure national security, serving as "a particularly potent symbol of American democracy," and contributing to change and growth in the United States.[41] The 2006 *National Space Policy* did not connect human spaceflight to broader national

---

[37] General William L. Shelton, Commander, Air Force Space Command, Statement to the Subcommittee on Strategic Forces, House Armed Services Committee, U.S. House of Representatives, April 25, 2013, http://docs.house.gov/meetings/AS/AS29/20130425/100708/HHRG-113-AS29-Wstate-SheltonUSAFG-20130425.pdf.

[38] D.N. Spires, *Beyond Horizons: A History of the Air Force in Space, 1947-2007,* 2007, p. 121.

[39] RAND National Defense Research Institute, *National Security Space Launch Report,* Santa Monica, Calif., 2006, p. ix, http://www.rand.org/pubs/monographs/MG503.html.

[40] U.S. Department of Defense, "Department of Defense Space Policy," Memorandum for Secretaries of the Military Departments, et al., July 9, 1999, section 4.11.4, http://www.fas.org/spp/military/docops/defense/ d310010p.htm.

[41] NASA, *The Vision for Space Exploration,* February 2004, http://www.nasa.gov/pdf/55583main_vision_space_exploration2.pdf.

policy or security concerns.[42] And the 2006 *National Security Strategy* did not mention space other than in the context of technologies that state and nonstate actors might use "in new ways to counter military advantages the United States currently enjoys."[43]

In his 2010 speech at the Kennedy Space Center, President Obama explicitly linked the "capacity for people to work and learn and operate and live safely beyond the Earth for extended periods of time" to strengthening "America's leadership here on Earth."[44] Although the 2010 *National Security Strategy* explicitly notes that U.S. space capabilities bolster "our national security strengths and those of our allies and partners," it does not mention human spaceflight in this regard.[45] Likewise, the 2010 *National Space Policy* lays out the Obama administration's revised approach to human spaceflight but does not link it to broader national policy or security goals.[46]

2.4.2.3 Evaluation of National Security and Defense Rationales

Both today and for the foreseeable future, direct U.S. national security requirements in space are likely to be met entirely by uncrewed systems. At the moment, there is no compelling rationale for humans in space to carry out national security missions. That said, human spaceflight is not totally irrelevant to national security considerations. The intellectual capital, technical skills, and industrial infrastructure required to design, develop, launch, and operate human-rated spacecraft clearly overlap those involved with robotic spacecraft, including national security payloads, although the size of the NASA human spaceflight program now is so small that it exerts only a modest influence on the portions of the U.S. industrial base that are relevant to national defense.[47] In addition, new options may arise, such as rapid suborbital transport of small troop units for special operations, which is under consideration by the Defense Advanced Research Projects Agency and the Air Force Research Laboratory (the SUSTAIN 2002 and Hot Eagle 2008 concepts).

"Soft power," or "getting what you want [in international relations] through attraction rather than coercion," is a benefit of NASA's human spaceflight programs.[48,49] A previous National Research Council (NRC) report recommended that the United States consider "using human spaceflight to enhance the U.S. soft power leadership by inviting emerging economic powers to join with us in human spaceflight adventure."[50] In addition to the economic and industrial benefits that may be generated by those international efforts,[51] the soft-power value of human space exploration is rooted in the demonstrated technical achievements required to execute such programs. Using technology to enhance national stature and build a perception of power has useful geopolitical consequences, as both the U.S. Apollo Program and, more recently, the Chinese human spaceflight program have demonstrated. Finally, the international standing needed to have a strong voice in future international agreements about space use and settlement has important security implications. Such a voice will be stronger if the United States has an active presence among the nations that are sending humans into space and supporting space developments.

---

[42] Executive Office of the President, Office of Science and Technology Policy, *U.S. National Space Policy,* August 31, 2006, http://www.whitehouse.gov/sites/default/files/microsites/ostp/national-space-policy-2006.pdf.

[43] Executive Office of the President, *The National Security Strategy of the United States of America,* March 2006, p. 44, http://georgewbush-whitehouse.archives.gov/nsc/nss/2006/index.html.

[44] Executive Office of the President, "Remarks by the President on Space Exploration in the 21st Century," Kennedy Space Center, Fla., April 15, 2010, http://www.whitehouse.gov/the-press-office/remarks-president-space-exploration-21st-century.

[45] Executive Office of the President, *National Security Strategy,* May 2010, p. 31, http://www.whitehouse.gov/sites/default/files/rss_viewer/national_security_strategy.pdf.

[46] Executive Office of the President, *National Space Policy of the United States,* June 28, 2010, http://www.whitehouse.gov/sites/default/files/national_space_policy_6-28-10.pdf.

[47] NASA's total procurement spending in FY 2010 (including uncrewed exploration and aeronautics programs and human spaceflight) was smaller than that of the U.S. Postal Service, the Department of Veterans Affairs, or the Department of Energy (U.S. Census Bureau, *Consolidated Federal Funds Report for Fiscal Year 2010,* USGPO, 2011).

[48] J.S. Nye, *Soft Power: The Means to Success in World Politics,* Public Affairs Press, New York, 2004, p. 2.

[49] J. Johnson-Freese, *Space as a Strategic Asset,* Columbia University Press, New York, 2007, pp. 51-81.

[50] See NRC, *America's Future in Space: Aligning the Civil Space Program with National Needs,* 2009, p. 4.

[51] One example is the infusion of cash into the Russian space program in the 1990s as the Russians entered the partnership that was developing the International Space Station. That was part of a broader initiative to stabilize the Russian aerospace sector and encourage collaboration after the end of the Soviet system.

Space-based assets and programs are an important element of national security, but the direct contribution of human spaceflight in this realm has been and is likely to remain limited. An active U.S. human spaceflight program gives the United States a stronger voice in an international code of conduct for space, enhances U.S. soft power, and supports collaborations with other nations; thus, it contributes to our national interests, including security.

2.4.3 National Stature and International Relations

A third rationale for human spaceflight is that it contributes to both national stature (viewed both internally and externally) and to the promotion of peaceful international relations.

From an internal perspective, there is little doubt that the U.S. space program has contributed favorably to the national self-image. There are few moments in our national narrative when as a nation of individuals we focus on a single event in real time. It is remarkable that the vast majority of Americans alive at the time—from those who were children to the oldest citizen—can remember where they were when John F. Kennedy was shot, when Neil Armstrong first set foot on the Moon, when the space shuttle *Challenger* broke apart, and when the World Trade Center buildings crumbled in the terrorist attack of 9/11. Space exploration has provided such moments through its successes and, equally, through its public disasters, which serve as a reminder of the risks posed by exploration. Thus, in the broadest terms, space exploration makes unique contributions to U.S. political and social culture. It plays a role in defining what it means to be "an American" and reinforces the identity as explorers who take the risk of challenging new frontiers that has long been a part of the national culture and history. The desire to be "the best" and to maintain this identity as a country that undertakes bold ventures serves for some as a rationale for continued space exploration efforts. As will be discussed further in Chapter 3, space exploration is not a primary concern for most Americans but remains a source of pride.

The U.S. space program has since its inception been viewed as a contributor to U.S. international prestige and stature. Initial robotic efforts ramped up rapidly and immediately after the success of Sputnik as a direct response to a perceived threat of Soviet dominance of space and the implied possibility of military use. The large increases in funding to undertake human spaceflight during the Apollo program were supported by President Kennedy as a way to take the lead in space and demonstrate U.S. technological superiority to the world. Joint efforts with other nations have since become a larger part of the human-spaceflight element of NASA's programs, and these cooperative ventures have played a role in reinforcing international relationships at the same time as they serve other national goals, including the enhancement of U.S. soft power, as discussed in the previous subsection. As more nations venture into space and engage in human spaceflight, one rationale frequently articulated for continued U.S. human spaceflight is the need to be engaged in this domain in order to have a strong and ongoing voice in any international agreements that are needed so that the United States can guard against aggression from space and participate in setting policies regarding exploitation of possible space resources or possible space settlement. A number of individuals who presented before the committee drew an analogy with international agreements about exploration and use of Antarctica.[52]

2.4.3.1 Evaluation of National Stature and International Relations Rationales

There is a potential tension between engagement in human spaceflight as a demonstration of national capability and stature and engagement in cooperative projects in human spaceflight as an element of international cooperation that contributes to relationships that foster international stability. National stature is a subjective judgment that is made differently by different communities and nations; however, it is fair to say that human spaceflight is a peaceful activity that, in most eyes, increases the stature of those who achieve it (both the individuals and the nations that support them). Furthermore, when such achievements occur in internationally cooperative projects, they involve a level of negotiation and international trust that produces positive international relationships provided

---

[52] P. Ehrenfreund, M. Race, D. Labdon, Responsible space exploration and use: Balancing stakeholder interests, *New Space* 1/2(201):60-72, http://www2.gwu.edu/~spi/NEW_Space.2013.0007.pdf, 2013.

that all sides meet their obligations under the agreements. The international cooperation required to agree on goals and to maintain and revise programs through technical milestones can have broader benefits in developing or helping to maintain strong international relationships that go beyond the human-spaceflight endeavor and affect the overall relationships of the nations involved. As U.S. human spaceflight programs transitioned from the competitive stance of the 1960s to the cooperative stance of the ISS program, the United States continued to maintain a leadership position in funding and carrying out the work and thereby served the goals of both national stature and international relationships. Choosing to pursue an international cooperative path to a science or exploration goal may on the one hand lead to cost-sharing that potentially reduces U.S. costs and on the other bind the United States to continue supporting the project or face adverse consequences for its international reputation. Cooperative agreements can lend stability to international efforts that are less likely to be subject to year-to-year funding fluctuations than national projects that lack such agreements to maintain them.

International cooperation on the ISS served to support and employ Russian rocket scientists at a time of instability when the breakup of the Soviet Union put its nuclear and rocket capabilities at risk of contributing to nuclear proliferation, and it arguably helped to prevent such an outcome. In a letter to Representative Sensenbrenner, Assistant Secretary of State for Legislative Affairs Barbara Larkin described the Department of State's position as "seeking to keep Russia constructively engaged in the international arena and perhaps more importantly, Russian participation in the ISS plays a vital role in our non-proliferation program."[53] Secondary considerations included access to Russian technology and a desire to help Russia to maintain a peaceful emblem of superpower status—its human spaceflight program.[54] This continuing space partnership has been a favorable element of the U.S.-Russia relationship. However, the extent of the effect on the bilateral international relationships is hard to quantify. U.S.-Russia relationships were changing for many reasons at the time and are doing so as this report goes to press.

To achieve such highly ambitious goals as sending humans to the Moon, near-Earth asteroids, and Mars, international cooperation is probably essential to exploit worldwide expertise, share costs and eliminate duplication of efforts, and maintain the course over the long term needed for success. In this context, the technological roadmaps of the International Space Exploration Coordination Group (ISECG), which represent the work of representatives of 14 space agencies, are particularly notable: In 2007, a report titled *Global Exploration Strategy: The Framework for Cooperation*[55] was released as the first product of an international coordination process among the agencies. Two global exploration roadmaps that build on the discussion of ISECG members have followed.[56] The cooperative work that was required to develop those plans has built a network among the space agencies of the nations involved that contributes, with many other such networks, to peaceful relationships among the nations. True international cooperation on such projects as human spaceflight needs to be built on the basis of shared planning and requires long-term agreements, which in turn can contribute to well-chosen pathways and goals for the program and to the stability of program goals through changes in administrations in any one of the partner countries.

Being a leader in human space exploration enhances international stature and national pride. Because the work is complex and expensive, it can benefit from international cooperative efforts. Such cooperation has important geopolitical benefits.

2.4.4 Education and Inspiration

Many have argued that the space program has an important role in inspiring the next generation of scientists and engineers and in educating them. In prior documents reviewed by the committee and in presentations to the committee, that role has been advanced as one of the rationales for supporting human spaceflight.

---

[53] Barbara Larkin (Assistant Secretary for Legislative Affairs, Department of State) to Representative James Sensenbrenner, December 22, 1998. Cited in J. Johnson-Freese, *Space as a Strategic Asset,* 2007, p. 67.

[54] J.M. Logsdon and J.R. Millar, U.S.-Russia cooperation in human spaceflight: Assessing the impacts, *Space Policy* 3:171-178, 2001.

[55] International Space Exploration Coordination Group, *Global Exploration Strategy: The Framework for Cooperation,* 2007, http://www.globalspaceexploration.org/documents.

[56] ISECG, http://www.globalspaceexploration.org, accessed January 6, 2014.

Indeed, the U.S. government made spending on science education at all levels a fundamental component of the massive investment in the aftermath of the Soviet Sputnik satellite in 1957 and passage of the National Defense Education Act in 1958 (Public Law 85-164). Many who later became NASA scientists, engineers, and astronauts point to Sputnik and the Apollo program as their childhood inspiration and to the opportunities to study and make careers in NASA-related fields that that funding offered as critical to their educational experience.

The drive to use space to inspire students to pursue science and engineering careers motivated Congress to authorize and NASA programs to attach high priority to the expenditure of a small fraction of every NASA effort for programs labeled Education and Public Outreach (EPO). The programs have included those directed to K-12–age students in and out of school, programs in science museums, opportunities for students and the public to contribute to particular science projects by engaging in "participatory exploration" or "citizen science," and programs that support graduate study in NASA-needs areas. Different NASA directorates have emphasized different elements of this spectrum of programs. The human spaceflight program has recognized and used the inspirational role that astronauts can have for young students as an element of its outreach. Astronauts interact with elementary-school and middle-school students on the ground and from the ISS. The nation remembers that one of those who died in the *Challenger* disaster was a teacher, Christa McAuliffe, whose job aboard the space shuttle was to have been to conduct demonstrations of basic science experiments and talks for students. Such EPO efforts no longer garner a legislatively mandated fraction of NASA project spending, but they continue to have high priority in the agency.

The human space-exploration program is still often cited in NASA-related reports and in presentations to this committee as an inspiration to students to pursue science, technology, engineering, and mathematics (STEM) careers. It would take a detailed survey beyond the scope of this study to separate the role of human and robotic science projects in this respect or to compare NASA-related inspiration with inspiration from other sources.

It has been widely stated that attracting enough students to STEM careers[57,58] is critical to the nation's competitiveness, economic health, and development. Research has documented a decline in U.S. student interest in those fields, most notably the landmark research that produced the 1983 report *A Nation at Work: Education and the Private Sector.*[59] The 2012 National Science Foundation *Science and Engineering Indicators* report on science and technology[60] showed little to no improvement over the years on a number of indicators, whereas other countries have greatly increased production of STEM students, and the American Institute of Aeronautics and Astronautics reports industry CEOs' concern about the lack of incoming science and technology students, noting that the majority of them are of foreign origin.[61] The 2008 General Social Survey[62] found that the majority of Americans in all demographic groups believe that the quality of science and mathematics education in U.S. schools is inadequate; a 2007 Gallup poll[63] reported that about half of all Americans believed that their local schools did not put enough emphasis on teaching science and mathematics, and just 2 percent said that there was "too much" emphasis on these subjects. The President's Commission on Implementation of United States Space Exploration Policy in 2004 noted that long-term competitiveness requires a skilled workforce and that the ability of American children to compete in the 21st century was in decline compared with those in other countries.[64] One of its conclusions was that space exploration "can be a catalyst for a much-needed renaissance in math and science education in the United States. . . [and] offers an extraordinary opportunity to stimulate math, science and engineering

---

[57] NASA, *Societal Impact of Spaceflight,* NASA SP-2007-4801, Washington, D.C., 2007.

[58] Institute of Medicine, National Academy of Sciences, and National Academy of Engineering, *Rising Above the Gathering Storm: Energizing and Employing America for a Brighter Economic Future,* The National Academies Press, Washington, D.C., 2007, pp. 9-10.

[59] National Advisory Council on Vocational Education, *A Nation at Work: Education and the Private Sector,* National Alliance of Business, Washington, D.C., 1983.

[60] National Science Board, *Science and Engineering Indicators 2012,* NSB 12-01, National Science Foundation, Arlington, Va., 2012.

[61] *A Journey to Inspire, Innovate, and Discover: Report of the President's Commission on Implementation of United States Space Exploration Policy,* ISBN 0-16-073075-9, U.S. Government Printing Office, Washington, D.C., 2004, p. 41.

[62] The 2008 General Social Survey conducted by the National Opinion Research Center at the University of Chicago, data collected between April 17, 2008, and September 13, 2008, a new cross-sectional survey of 2,023 cases.

[63] L.C. Rose and A. Gallup, The 39th Annual Phi Delta Kappa/Gallup Poll of the Public's Attitudes Toward the Public Schools, *Phi Delta Kappan* 89(1):33-48, 2007.

[64] *A Journey to Inspire, Innovate, and Discover,* 2004, p. 12.

FIGURE 2.2 Earthrise as seen from the Moon and captured by Apollo 8 astronaut William Anders. SOURCE: Courtesy of NASA.

excellence for America's students and teachers."[65] In contrast, multiple economic manpower studies dispute the claim of a shortage of trained scientists and engineers, and indeed demand for particular professional skills can fluctuate with economic shifts. However, almost all analysts agree that a rising fraction of jobs, particularly those in manufacturing, require a higher level of technical capabilities and science-related skills than in the past, and they do not dispute the need for basic education in STEM fields even for those who will not become scientists and engineers. NASA education programs have often been geared as much to that broader goal as to the more specific one of increasing output of scientists and engineers, and one cannot pursue the latter without at the same time considering the former.[66]

Not only students but also a broader public has been inspired and affected by human space exploration. *Time Magazine*'s "Photo of the 20th Century," Earthrise as seen from the Moon captured by Apollo 8 astronaut William Anders (Figure 2.2), and other iconic images of the view of Earth from space have given humankind a new perspective from which to view our planet.

---

[65] *A Journey to Inspire, Innovate, and Discover,* 2004, p. 12.

[66] NRC, *NASA's Elementary and Secondary Education Program: Review and Critique,* The National Academies Press, Washington, D.C., 2008, p. 21-43.

2.4.4.1 Evaluation of Education and Inspiration Rationales

Although education effects are, like economic effects, difficult to quantify, there is no doubt that for some the vision of the frontier with astronauts as explorers provides an added inspiration and impetus for study of challenging STEM subjects.[67] It motivates, and in the view of this committee justifies, NASA's efforts to engage in public outreach efforts and to communicate its challenges and its achievements to the public.

A 2008 NRC study of NASA education programs[68] concluded that NASA's education efforts should capitalize on its science and exploration activity and its scientists and engineers to engage and inspire young students to become interested in science. The report also suggested that to have the strongest effect NASA, in its education and outreach efforts, should partner with institutions that have more direct understanding of formal or informal science education. The study viewed NASA's expenditures on outreach and education programs as worthwhile for education but assumed that the expenditures on spaceflight programs were justified by other rationales.

To state those conclusions more strongly: Although the effect on students can be used as a rationale for some spending on EPO efforts within the human spaceflight program, few in the education realm would view this effect itself as a rationale for funding the human spaceflight program. The sources of inspiration and engagement of students in science today are many, and whereas NASA human spaceflight clearly appeals to and inspires some students, it is not unique in this role.[69] Furthermore, whereas initial engagement or inspiration can be important and memorable, the path to becoming a scientist or engineer requires more; it depends on continuing educational experiences that build on and support a student's interest and converts it from a casual interest to an identity that causes the student to persevere as the path becomes more challenging.[70,71]

The United States needs scientists and engineers and a public that has a strong understanding of science. The challenge and excitement of space missions can serve as an inspiration for students and citizens to engage with science and engineering although it is difficult to measure this. The path to becoming a scientist or engineer requires much more than the initial inspiration. Many who work in space fields, however, report the importance of such inspiration, although it is difficult to separate the contributions of human and robotic spaceflight.

2.4.5 Scientific Exploration and Observation

Exploration of our solar system and objects within it for the purpose of gaining a better scientific understanding of them is a major goal of science. Exploring physics, chemistry, and biology in low-gravity environments offers additional scientific horizons. In both of those endeavors, the U.S. space program has been reviewed and projected in major NRC studies, and studies have discussed which parts of the program can be achieved through human spaceflight.[72,73,74] A similar review of the European Space Agency (ESA) science programs also discussed both robotic and human elements.[75] In specific contexts, human spaceflight and robotic missions can play complementary roles, and achieving certain science goals has been cited as another rationale for human spaceflight. All those studies have discussed the importance of including science goals in planning human spaceflight and considering

---

[67] R. Monastersky, Shooting for the Moon, *Nature* 460(7253):314-315, 2009.

[68] NRC, *NASA's Elementary and Secondary Education Program,* 2008, p. 113-118.

[69] NRC, *Learning Science in Informal Environments: People, Places, and Pursuits,* The National Academies Press, Washington, D.C., 2009, p. 100-102.

[70] K.A. Renninger and K.R. Riley, Interest, cognition, and the case of L- in science, pp. 325-382 in *Cognition and Motivation: Forging an Interdisciplinary Perspective* (S. Kreitler, ed.), Cambridge University Press, New York, 2013.

[71] K.A. Renninger and S. Su, Interest and its development, pp. 167-187 in *Oxford Handbook of Motivation* (R. Ryan, ed.), Oxford University Press, New York, 2012.

[72] NRC, *Vision and Voyages for Planetary Science in the Decade 2013-2022,* The National Academies Press, Washington, D.C., 2011.

[73] NRC, *Recapturing a Future for Space Exploration: Life and Physical Sciences Research for a New Era,* The National Academies Press, Washington, D.C., 2011.

[74] NRC, *The Scientific Context for Exploration of the Moon: Final Report,* The National Academies Press, Washington, D.C., 2007.

[75] European Science Foundation, *Independent Evaluation of ESA's Programme for Life and Physical Sciences in Space (ELIPS): Final Report,* Strasbourg, France, December 2012.

it as an element of a science program that also includes robotic missions. The particular skill of humans in noticing anomalous or emergent features and events and rapidly scanning an environment for sought features is what continues to give humans an edge over robots in the context of exploratory science even though robotic programs are typically easier (although still extremely complex and challenging) and, on a per-mission basis, much less expensive. For example, the ISECG roadmaps, produced through the cooperation of many space agencies, suggest that a mix of robotic and human missions will best enable the science and at the same serve the more aspirational goal of extending human presence further into space.

After more than 4 decades of planetary exploration by the space-faring nations, a transition point has come. The initial phase of planetary reconnaissance is closing as Voyager's two spacecraft pass through the heliopause and leave the solar system and as the New Horizons spacecraft proceeds on its way to Pluto. The intensive robotic observation phase is now well under way. A number of missions are intensively studying many of the bodies of our planetary system. MESSENGER is still sending back observations from Mercury; DAWN has just visited the asteroid Vesta and is on its way to Ceres; Cassini is still making important discoveries regarding Saturn, its rings, and its moons; and the Juno mission will encounter Jupiter in 2016. The rationale for those missions and their predecessors followed the imperative from the founding National Aeronautics and Space Act of 1958 to carry out the exploration of space.[76] Missions were chosen through extensive consultation with the science community with due recognition of the engineering and budgetary realities. Starting a decade ago, the planetary-exploration missions have followed the successful practice of the astronomy and astrophysics community by carrying out decadal surveys, which were combined with roadmap exercises.

The course of human in situ studies of the bodies of our solar system has had a more complex history. The Apollo program, which successfully landed astronauts on the Moon, was engendered by political considerations during the Cold War era. As it developed, however, it took on an increasingly scientific cast, and astronauts undertook scientist-guided experiments that have led to a dramatic revision of knowledge of the lunar surface and of understanding of lunar genesis and history. With the termination of the Apollo program, however, progress toward further human in situ investigations has been constrained.

The participation of humans in landing on and exploring the surfaces of the Moon and Mars is now being studied, but no new missions have been undertaken by the United States or its international partners. An essential motive for reviving human exploration is the recognition that the alert examination of nature has been the route to many of the most dramatic discoveries in science and that further such discoveries are likely when humans can investigate the surfaces of the Moon and Mars directly. In this context, major spacefaring nations are currently engaging in robotic space missions that target the environment "where humans can go," namely, the Moon, the martian system, and near-Earth objects. Many of these missions have the dual goals of achieving steps toward eventual human missions and doing scientific research.

The Moon represents a window through which the origin and evolution of our solar system, as well as the dynamics of the Earth-Moon system, can be explored.[77] The rich science that can be conducted on and from the Moon can add value to those pathways to Mars that include intermediate missions to the lunar surface, provided that the missions are also designed to address high-priority elements of this science. Several orbiters from the United States, Europe, Japan, India, and China have obtained data with unprecedented resolution in the past decade, leading to new discoveries. The first soft-landing mission from the Chinese Space Agency, Chang'e 3, arrived in early December 2013. China is the third nation to have achieved that goal, and further missions are projected, perhaps including an eventual human landing and return.

NASA's successful multidecadal Mars program of orbiters and rovers and Europe's MarsExpress spacecraft have contributed much to the reconnaissance of our sister planet. Through NASA's strategy to "follow the water" and the detailed mapping of the surface mineralogy from orbit and at specific landing sites, excellent data on the evolutionary history and habitability of Mars have been collected. The NASA rovers Opportunity and Curiosity continue operating on the surface, and their science teams continue to generate reports of new scientific findings.

---

[76] National Aeronautics and Space Act of 1958, Public Law 85-568, 72 Stat., 426, signed on July 29, 1958.

[77] NRC, *The Scientific Context for Exploration of the Moon,* 2007.

Near-Earth asteroids (NEAs) constitute a threat to humankind and life on Earth. The recent Chelyabinsk meteor event, which affected more than 1,000 people, contributed much to public awareness about NEAs. The investigation of NEAs has a dual purpose: to gather data for hazard mitigation and to add to understanding of the early solar system and the impact history of early Earth. The proximity of NEAs also makes them ideal targets for the exploration of raw materials that could eventually be subject to exploitation for commercial use or to support interplanetary journeys. The Japanese asteroid-sample return mission Hayabusa achieved the first sample return from the asteroid Itokawa in 2010.

In recent decades, international space-exploration working groups have defined key drivers in science and technology for exploring the Moon, Mars, and NEAs. In particular, the Lunar Exploration Analysis Group (LEAG), the International Lunar Exploration Working Group, the Mars Exploration Program and Analysis Group (MEPAG),[78] and the International Primitive Exploration Working Group have provided extensive roadmaps with strong support from the science community. They are continually updated and provide constructive and strong rationales that every space agency can use to develop space-exploration plans and architectures. They have included both robotic and human missions in their considerations.

Key scientific drivers of lunar exploration, as discussed in recent roadmaps, include investigating the bombardment history of the inner solar system that is uniquely revealed on the Moon, the structure and composition of the lunar interior and lunar crustal rocks that lead to understanding of evolutionary planetary processes, and lunar poles that may harbor important volatiles. Many of these goals are or will be addressed robotically in the coming years by an international fleet of spacecraft; however, many spacefaring nations also have plans for human exploration missions, including outposts. The added value of a human return to the Moon includes efficient sample identification and collection, enhanced sample-return capacity, and increased opportunities for serendipitous discoveries. Furthermore, synergistic robotic-human exploration of the Moon would facilitate large-scale exploratory activities (such as drilling), deployment, and maintenance of complex equipment and would provide operational experience for future Mars exploration.[79]

Key scientific drivers of the exploration of Mars as defined by MEPAG[80] are a determination of whether life ever arose on Mars, an understanding of the processes and history of climate on Mars, an understanding of the evolution of the surface and interior of Mars, and an understanding of what is needed to prepare for human exploration.[81] Similar goals are discussed in the 2011 NRC decadal report on planetary science.[82] To prepare for human exploration of Mars, research on many topics is needed, including atmospheric measurements, biohazard and planetary protection, in situ resource utilization (ISRU), radiation, toxic effects of martian dust on humans, atmospheric electricity, effects of dust on surface systems, and trafficability.[83] Some of those technological challenges are discussed in Chapter 4.

In summary, the scientific drivers of past planetary exploration have had a common theme. The overall scientific driver is to understand how our solar system came into being and how it has evolved. A different kind of historical question has also served as a powerful driver for planetary exploration: Has life existed on other bodies of our solar system, and does it exist today? Mars is one of several promising sites in the solar system to search for life,[84] and although it is an unlikely locale for life today, there has been liquid water on the martian surface in the past, and some form of life might well have existed there. Both questions have relevance to Earth. As noted in a recent review, "Human space exploration can eventually help answer some of the main questions of our existence, namely how our solar system formed, whether life exists beyond Earth and what our future may be."[85]

---

[78] See the MEPAG website, http://mepag.jpl.nasa.gov/.

[79] I. Crawford, M. Anand, M. Burchell, et al., "The Scientific Rationale for Renewed Human Exploration of the Moon," white paper submitted to the National Research Council Planetary Science Decadal Survey, http://www8.nationalacademies.org/ssbsurvey/publicview.aspx, 2009.

[80] It should be noted that MEPAG goals are not limited to ones that would be pursued through human exploration.

[81] Mars Exploration Program Analysis Group (MEPAG), "Mars Scientific Goals, Objectives, Investigations, and Priorities" (J.R. Johnson, ed.), white paper posted September 2010, http://mepag.jpl.nasa.gov/reports/index.html.

[82] NRC, *Vision and Voyages for Planetary Science in the Decade 2013-2022,* 2011.

[83] MEPAG, "Mars Scientific Goals, Objectives, Investigations, and Priorities," 2010.

[84] Other bodies cited include Enceladus and Europa for their extant liquid water and Titan for its extensive organics and Earth-like processes.

[85] P. Ehrenfreund, C. McKay, J.D. Rummel, B.H. Foing, C.R. Neal, T. Masson-Zwaan, M. Ansdell, et al., Toward a global space exploration program: A stepping stone approach, *Advances in Space Research* 49:2-48, 2012.

2.4.5.1 Human Spaceflight in Low Earth Orbit: The International Space Station

The ability to conduct laboratory science in LEO was a major rationale for and goal of the International Space Station (ISS) program. The science can be divided into two major categories: the study of human factors and physical phenomena in space in order to enable longer-range exploration and studies that benefit from the low gravity and the human-staffed laboratory provided by the ISS that are not necessarily related to enabling human spaceflight. Both are discussed below; however, it should be kept in mind that the second category is the only one that is conducted primarily to benefit science rather than to further human spaceflight. The impacts of human-factors science on Earth-based medical treatments make up the type of spinoff that is included in the discussion of economic rationales.

In the past decade, scientists and astronauts have performed research to prepare for human exploration—research related to lack of gravity, altered circadian rhythms, and increased exposure to cosmic radiation. The investigation of adaptations of the human body to extraterrestrial conditions is of utmost importance for safeguarding human health during exploratory missions.[86] NASA ISS science is focused on the Human Research Program (HRP), which is aimed at research and technology that will enable productive and healthy human work and life in space (Figure 2.3). NASA has identified a set of risks to humans, including radiation, bone and muscle loss, increased intracranial pressure, and other physiologic responses to the microgravity environment. The role of the HRP is to characterize these risks and explore mitigating factors—or to determine which risks cannot be mitigated. The HRP's role in enabling an extended human presence in cislunar space and beyond is therefore critical.

ESA, through its European Programme for Life and Physical Science in Space (ELIPS), makes Europe the largest scientific user of the ISS at present.[87] ELIPS conducts studies in support of exploration that involve radiation biology and physiology and health-care, life-support, and contamination studies. (See recent evaluation by the European Science Foundation.[88]) An example of the benefits of human-tended laboratory research onboard the ISS may emerge from a suite of plasma-physics experiments called PK-3. Conducted by Russian and European researchers in multidisciplinary teams, the project has produced a new type of matter called cold atmospheric plasma (CAPs) that has antibacterial properties against dozens of organisms.[89,90] In laboratory and clinical trials, applications of CAPs show promise for clinical applications, particularly wound healing.[91,92,93] Cold plasmas do not occur on Earth and could not have been discovered without human-tended experiments conducted in microgravity.[94]

In the most recent addition to the ISS, the Japanese experiment module Kibo, scientists perform experiments on space medicine, biology, and biotechnology in addition to Earth observations. Through Kibo's airlock, experiments can be transferred and exposed to the external environment of space and manipulated with robotic arms. Both the U.S. and Russian ISS research programs also have a major emphasis on studies of human performance and endurance in space and on ways in which they can be supported and extended. Other ISS science experiments explore fluid dynamics in microgravity and other physical-science phenomena in the environment of space.

---

[86] C.A. Evans, J.A. Robinson, J. Tate-Brown, et al. *International Space Station Science Research: Accomplishments During the Assembly Years: An Analysis of Results from 2000-2008,* 2008, http://www.nasa.gov/pdf/389388main_ISS%20Science%20Report_20090030907.pdf.

[87] Since 2002, 15 European countries have invested in the ELIPS program, which is running in its fourth term (ELIPS-4) and involves more than 1,500 scientists.

[88] European Science Foundation, *Independent Evaluation of ESA's Programme for Life and Physical Sciences in Space (ELIPS): Final Report,* Strasbourg, France, December 2012.

[89] H. Thomas, "Complex Plasma Applications for Wound Healing," presentation at the 2nd Annual International Space Station Research and Development Conference, Denver, Colo., July 13, 2013, http://www.youtube.com/watch?v=sbhlA0OON4s.

[90] T. Miasch, T. Shimizu, Y.-F. Li, J. Heinlin, S. Karrer, G. Morfill, and J. Zimmermann, Decolonization of MRSA, S aureus and E. coli by cold-atmospheric plasma using a porcine skin model in vitro, *PLoS ONE* 7(4):e34610, 2012.

[91] Plasma Technologies, http://www.ptimed.com/index.html, accessed October 12, 2013.

[92] G. Isbary, G. Morfill, H.U. Schmidt, M. Georgi, K. Ramrath, J. Heinlin, S. Karrer, et al., A first prospective randomized controlled trial to decrease bacterial load using cold atmospheric argon plasma on chronic wounds in patients, *British Journal of Dermatology* 163:78-82, 2010, doi:10.1111/j.1365-2133.2010.09744.x.

[93] G. Isbary, T. Shimizu, J.L. Zimmermann, H.M. Thomas, G.E. Morfill, and W. Stolz, Cold atmospheric plasma for local infection control and subsequent pain reduction in a patient with chronic post-operative ear infection, *New Microbes and New Infections* 1:41-43, 2013, doi:10.1002/2052-29775.19/full.

[94] Thomas, "Complex Plasma Applications for Wound Healing," 2013.

FIGURE 2.3 Expedition 36/37 Flight Engineer Karen Nyberg of NASA uses a fundoscope to take still and video images of her eye while in orbit. This was the first use of the hardware and new vision-testing software. SOURCE: Courtesy of NASA, http://www.nasa.gov/content/it-s-all-in-your-head-nasa-investigates-techniques-for-measuring-intracranial-pressure/.

In 2010, Congress directed NASA to contract with a nongovernment organization to manage the ISS National Laboratory, which accounts for half the volume and resources allocated to utilization of the U.S. segment of the ISS.[95] The mission of the Center for the Advancement of Science in Space (CASIS) is to identify and develop science and applications for the benefit of human activities and life on Earth "and for the public good."[96] Microgravity research conducted during the Space Shuttle era produced an analogue of muscle-wasting and bone-loss syndromes on Earth that can be studied in an accelerated timeframe because of rapid changes on orbit relative to slower clinical progression on Earth.[97] Other research identified changes in genetic expression obtained only in microgravity. Those findings and others have led Fortune 500 companies—including Procter & Gamble, Merck, and Novartis—to invest in flight projects now destined for the ISS National Laboratory with its longer-duration flight opportunities in the hope of using microgravity research to facilitate commercial product development.[98]

[95] 2010 NASA Authorization Act.

[96] CASIS Strategic Plan.

[97] L. Stodieck, AMGEN countermeasures for bone and muscle loss in space and on Earth in Proceedings of the 2nd Annual International Space Station Research and Development Conference, Denver, Colo., July 2013, American Astronautical Society, Washington, D.C., http://www.astronautical.org/sites/default/files/issrdc/2013/issrdc_2013-07-17-0800_stodieck.pdf.

[98] Center for the Advancement of Science in Space (CASIS), "Supporting Entrepreneurs in Space," October 1, 2013, http://www.iss-casis.org/NewsEvents/NewsDetail/tabid/122/ArticleID/ 87/ArtMID/581/Supporting-Entrepreneurs-in-Space.aspx.

The life and microgravity science decadal survey conducted by the NRC, which resulted in the 2011 report *Recapturing a Future for Space Exploration: Life and Physical Sciences Research for a New Era*,[99] investigated objectives for life-sciences and physical-sciences research to support future exploration missions. The research portfolio recommended ground-based and space-based experiments that included investigations of "the effects of the space environment on life support components, the management of the risk of infections to humans, and fundamental physical challenges."[100] Since then, the life-sciences research portfolio has focused on an integrated pursuit to manage health risks to space explorers while at the same time advancing fundamental science discoveries. A key point is that the ISS affords the opportunity to conduct laboratory science that explores the use of microgravity as a tool for examining a number of physical and biological processes. For the foreseeable future, investigations of this sort will require human beings to configure experiments and perform multiple runs. Those activities cannot be performed by robots. Thus, the opportunities for science returns that could yield revolutionary systems and input to exploration architectures depend on human-tended research onboard the orbiting ISS laboratory.

2.4.5.2 Evaluation of Scientific Exploration and Observation Rationales

As various studies have argued,[101] human spaceflight, planned in coordination with robotic missions, can help to address important science goals. Answering the enduring question of how far humans can go into space requires further scientific research on the effects of long-term stays in space on human physiology and psychology. Human spaceflight can certainly benefit from further investment in that science. However, the inverse—that science benefits from human spaceflight—must be examined on a case-by-case basis. For planetary science, a cost/risk/benefit analysis is required at each stage to decide whether the path to the science is best served by human or by robotic missions. Science done in LEO, other than that directed at furthering human exploration, provides a rationale only for LEO. In addition, in order for the human spaceflight program to serve science well, it is important that scientists and science goals play a role in mission planning.

The relative benefits of robotic versus human efforts in space science are constantly shifting as a result of changes in technology, cost, and risk. The current capabilities of robotic planetary explorers, such as Curiosity and Cassini, are such that although they can go farther, go sooner, and be much less expensive than human missions to the same locations, they cannot match the flexibility of humans to function in complex environments, to improvise, and to respond quickly to new discoveries. Such constraints may change some day.

2.4.6 Survival

Long-term survival of the human species is often cited in the space community,[102] by futurists,[103] and by space enthusiasts[104] as a rationale for human spaceflight. Robotic spacecraft have given scientists insights into some of our potential futures by probing the history of other planets in our solar system. Through space exploration, we have discovered the runaway greenhouse effect on Venus, recognized that Mars effectively "dried up" 3 billion years ago, and monitored the decline of Earth's ozone layer. It is often said that it is difficult to know one's own country until one visits other countries; in the same sense, space exploration has given us the ability to know our own planet better as we contrast it with others. By continuing to obtain scientific knowledge of other planets and our own, we become more aware of Earth's fragile nature.

---

[99] NRC, *Recapturing a Future for Space Exploration,* 2011.

[100] Ibid.

[101] See NRC, *Vision and Voyages for Planetary Science in the Decade 2013-2022,* 2011, and *Recapturing a Future for Space Exploration,* 2011, both published by The National Academies Press, Washington, D.C.

[102] M. Huang, Sagan's rationale for human spaceflight, *The Space Review,* November 8, 2004, http://www.thespacereview.com/article/261/1.

[103] S. Brand, *Space Colonies,* Penguin Books, New York, 1977.

[104] G. Anderson, A rationale for human spaceflight, *The Space Review,* 2011, http://www.thespacereview.com/article/1920/1.

Proponents of the survival rationale point to various types of future events as cause for concern: the depletion of global resources to such an extent that civilization can no longer be sustained on Earth; the likelihood—always from a probabilistic standpoint—that Earth will suffer a collision with an asteroid or comet of sufficient size to disrupt civilization or even cause the extinction of humankind; and extreme excursions in solar activity (including coronal mass ejections) or other far-future developments in the evolution of the Earth-Sun system that would make Earth uninhabitable.[105] Discussion of this rationale requires consideration of how far in the future the events are likely to occur and of whether amelioration of their effects is more feasible than moving a self-sustaining population off Earth. Amelioration seems on its face to be far more plausible for all but planet-clearing events.[106] However, it is equally clear that without continuing investment in human exploration, questions of possibility, timeline, and cost associated with an off-Earth settlement cannot be answered.

The survival rationale can be directly paired with the "ultimate goal" of human settlement on another celestial body. The most recent Augustine report from 2009, *Seeking a Human Spaceflight Program Worthy of a Great Nation*, concluded that "there was a strong consensus within the Committee that human exploration also should advance us as a civilization towards our ultimate goal: charting a path for human expansion into the solar system. It is too early to know how and when humans will first learn to live on another planet, but we should be guided by that long-term goal."[107] Indeed, in the present committee's recently collected Twitter-based commentary, survival of the human species by way of an off-Earth settlement was repeatedly mentioned as a strong rationale for continuing to advance the frontier of human spaceflight beyond LEO.

Some scientists who are both distinguished and well-known popularizers of science have helped to fuel the enthusiasm for considering space exploration as a means of species survival. Carl Sagan wrote that "every surviving civilization is obliged to become spacefaring—not because of exploratory or romantic zeal, but for the most practical reason imaginable: staying alive. . . . The more of us beyond the Earth, the greater the diversity of worlds we inhabit . . . the safer the human species will be."[108] Stephen Hawking spoke in 2013 about the need for humanity to populate itself beyond Earth to survive: "We must continue to go into space for humanity. If you understand how the universe operates, you control it in a way. We won't survive another 1,000 years without escaping our fragile planet."[109] Although a viable off-Earth settlement would by its very existence increase the odds of long-term human survival, it is not currently known whether an independently surviving space settlement could be developed. There are many technical challenges along the path from current capabilities to such a development, so this rationale speaks to a far-future aspirational goal. However, any progress in addressing the challenges requires a continuing human spaceflight program.

It is not possible to say whether human off-Earth settlements could eventually be developed that would outlive human presence on Earth and lengthen the survival of our species. That question can be answered only by pushing the human frontier in space.

2.4.7 Shared Human Destiny and Aspiration

That space exploration is a shared human aspiration and an aspect of human destiny is a broad rationale for its pursuit that has been espoused by practitioners in fields as diverse as science fiction and international policy. The rationale can be defined as the conviction that human space exploration is transpersonal in nature and that space is a frontier for humanity's collective aspiration.[110] In this context, human spaceflight aims to study humanity's future—to dare discover how far humans can go and to investigate what they have a chance to become. From

---

[105] In listing these concerns, the committee is recognizing—not endorsing—the link in popular media between each of them and the question of human expansion beyond Earth.

[106] N. Tyson, *Space Chronicles: Facing the Ultimate Frontier,* W.W. Norton, New York, 2012.

[107] Review of U.S. Human Space Flight Plans Committee, *Seeking a Human Spaceflight Program Worthy of a Great Nation,* 2009.

[108] C. Sagan, *Pale Blue Dot,* Random House, New York, 1994, p. 371.

[109] *Reuters,* Hawking: Mankind has 1,000 years to escape Earth, 2013, http://rt.com/news/earth-hawking-mankind-escape-702/.

[110] Nikki Griffin, "A science officer speaks—An interview with JPL Flight Director Bobak Ferdowsi," *Geek Exchange,* May 28, 21013, http://www.geekexchange.com/a-science-officer-speaks-an-interview-with-jpl-flight-director-bobak-ferdowsi-62643.html.

space stations and starships to planetary outposts and terraforming, human imagination acts as a forecaster of a potential future to be reached only via continued development of humankind's capabilities for human spaceflight.

Most countries do not have human spaceflight capabilities, and many countries cannot afford to contribute to human spaceflight systems. In 50 years of human spaceflight, only a few more than 500 people have ever been in space, and yet human spaceflight has become part of the world's culture. A sense of shared human destiny and common aspiration does not necessarily depend on wider accessibility (although that is a large part of it). In this sense, the world relies on countries and organizations that have human spaceflight capabilities to adopt an inclusive approach to partnering in service of this goal. The Outer Space Treaty of 1967 is deeply rooted in this rationale:

> Recognizing the common interest of all mankind in the progress of the exploration and use of outer space for peaceful purposes, Believing that the exploration and use of outer space should be carried on for the benefit of all peoples irrespective of the degree of their economic or scientific development, . . . The exploration and use of outer space, including the moon and other celestial bodies, shall be carried out for the benefit and in the interests of all countries, irrespective of their degree of economic or scientific development, and shall be the province of all mankind.[111]

In 1958, President Eisenhower wrote a preface to a White House pamphlet titled "Introduction to Outer Space,"[112] in which he stated that "we and other nations have a great responsibility to promote the peaceful use of space and to utilize the new knowledge obtainable from space science and technology for the benefit of all mankind." Although that statement applied to space programs in general, not specifically to human spaceflight, the pamphlet is said to have directly influenced *Star Trek*,[113] which, beginning in the 1960s, became a franchise of such longevity that it is now solidly embedded in popular American culture. Many space scientists, space enthusiasts, and the general public often cite aspects of the *Star Trek* franchise as their aspirational vision of the kind of future that should be pursued through human spaceflight.[114]

Since its inception, human spaceflight has been a planetwide experience, thanks in part to satellites. In the 1960s, television was the prevailing technology that offered millions of people the opportunity to connect with one another around a collective experience. Neil Armstrong's words from the 1969 Moon landing were beamed around the world and garnered an estimated 530 million viewers worldwide—more than two-and-a-half times the population of the United States at that time (see Figure 2.4.). Today, the Internet is the dominant force connecting more than 2 billion people via an array of Web sites and apps. In 2013, Chris Hadfield, a Canadian astronaut stationed on the ISS, gained more than 18 million views of a single YouTube video and maintains 1 million followers on Twitter. In July 2013, Hadfield visited Twitter's headquarters, an office known for its frequent celebrity visitors, and was said to have brought together a larger audience of Twitter employees than any previous celebrity visitor.[115]

Human spaceflight is seen as *forging* a sense of common destiny. "Shared human destiny" and aspiration constitute a world view that humans are all in this—life, the universe, and everything—together and thus should endeavor to explore new frontiers collectively even if vicariously through the experiences of others. Notably, this rationale is distinguished from survival as a rationale; in this view, collective exploration as part of an intrinsic human experience is separate from and independent of the question of survival.

2.4.7.1 Evaluation of Human Destiny and Aspiration Rationales

Aspirational goals are by nature subjective and therefore unconvincing to those who do not share them. For those who do, however, they typically are strongly held.

---

[111] United Nations, "Treaty on Principles Governing the Activities of States in the Exploration and Use of Outer Space, Including the Moon and Other Celestial Bodies," known as the "Outer Space Treaty of 1967," entered into force October 10, 1967, available at http://www.state.gov/www/global/arms/treaties/space1.html.

[112] President's Science Advisory Committee, "Introduction to Outer Space," March 26, 1958, pp. 1-2, 6, 13-15, http://history.nasa.gov/sputnik/16.html.

[113] The well-known television series in which humans explore on a galactic scale.

[114] D. Day, "Star Trek as a Cultural Phenomenon," Essay, History of Flight, U.S. Centennial of Flight Commission, http://www.aahs-online.org/centennialofflight.net/essay/Social/star_trek/SH7.htm, accessed January 6, 2014.

[115] @mattknox, Twitter, July 26, 2013, https://twitter.com/mattknox/status/360897639409143808.

FIGURE 2.4 Astronaut Neil A. Armstrong, Apollo 11 mission commander, on the first extravehicular activity on the lunar surface on July 20, 1969. Astronaut Edwin E. Aldrin, Jr., took the photograph. SOURCE: Courtesy of NASA. This is a cropped photograph of the original, NASA photograph AS11-40-5886, http://grin.hq.nasa.gov/ABSTRACTS/GPN-2000-001209.html.

Some say that the drive to explore is a fundamental part of what makes us human.[116] In that view, space is a frontier, and the primary rationale for spaceflight is that it is human destiny to explore this frontier as we have explored so many others. Robotic exploration plays a role, but only human spaceflight tackles the enduring questions of how far we can go and what we can discover and do when we get there in that the "we" in these questions specifically means humans rather than their robotic agents. Human spaceflight is an expensive and risky undertaking, and in the judgment of many on this committee only high aspirations such as this can justify taking such risks. Furthermore, this goal and the similarly aspirational goal of human survival are the only goals given as rationales that absolutely require human spaceflight to be reached. For all others, the desired outcome can be supported in multiple ways, human spaceflight among them.

The urge to explore and to reach challenging goals is a common human characteristic. Space is today a major physical frontier for such exploration and aspiration. Some say that it is human destiny to

---

[116] David Dobbs, Restless genes, *National Geographic Magazine,* January 2013, http://ngm.nationalgeographic.com/2013/01/125-restless-genes/dobbs-text.

continue to explore space. While not all share this view, for those who do it is an important reason to engage in human spaceflight.

2.5 ASSESSMENT OF RATIONALES

All of the rationales described above are widely stated by and often represent deeply held convictions of individuals who have professional or avocational connections to space and members of the public. However, the effect of human spaceflight in supporting the domains discussed in previous sections is, in almost every case, difficult to quantify. It should be remembered that lack of validated measurements does not mean that the effects are not real; rather, in many cases the committee can neither provide strong corroborating evidence of effects nor measure their magnitude. For all the pragmatic rationales, the goals could be served by a variety of federal expenditures in addition to (or instead of) expenditures on human spaceflight. However, the lack of quantification applies as well to these alternative investments and thus the committee cannot determine the relative efficacy of investment in the various programs. The aspirational rationales (the last two discussed in the previous section) require human spaceflight if they are to be attained.

Each of the traditional rationales provides some support for continued engagement in human spaceflight. No one of them alone provides an argument that is compelling to all who hear it, and different audiences stress different rationales, as is seen in the inputs from interested audiences discussed at the beginning of this chapter and in the data in Chapter 3 from public surveys and from a stakeholder survey conducted by this committee. Many of the rationales for human spaceflight that are discussed above are not unique to this activity but form much of the justification for other federal programs ranging from the National Institutes of Health to DOD.[117] In other words, many of the rationales for spaceflight discussed above are most appropriately considered in a comparative context in which one tries to assess the effectiveness of an array of federal programs in achieving the goals incorporated in a given rationale. Unfortunately, there are no well-developed analytic methodologies for such assessments.

The fact that no one of these rationales alone provides the key to why the nation should support human spaceflight does not mean that collectively they are not strong. Different individuals may give more weight to one than to another, but few oppose human spaceflight in principle. In these times of tight budgets, many may suggest other expenditures that they favor more strongly. At the same time, few citizens have a clear perspective on the actual budget for human spaceflight, and many take the visibility of NASA programs to mean that the expenditures are much higher than they actually are. As Chapter 3 discusses, such responses depend on the information provided and how the question is asked.

No single rationale seems to justify the value of pursuing human spaceflight.

2.6 VALUE PROPOSITIONS

2.6.1 The Problem with Value Propositions

"However, not everything that can be counted counts, and not everything that counts can be counted."
                                                                            —William Bruce Cameron[118]

---

[117] As John Marburger pointed out in a 2006 speech to the Goddard Memorial Symposium, delivered in his capacity as director of the White House Office of Science and Technology Policy, the 2005 NASA Authorization Act "affirms that 'the fundamental goal of this vision is to advance U.S. scientific, security, and economic interests through a robust space exploration program.' . . . The wording of this policy phrase is significant. It subordinates space exploration to the primary goals of scientific, security, and economic interests. Stated this way, the 'fundamental goal' identifies the benefits against which the costs of exploration can be weighed. This is extremely important for policy making because science, security, and economic dimensions are shared by other federally funded activities. By linking costs to these common benefits it becomes possible, at least in principle, to weigh investments in space exploration against competing opportunities to achieve benefits of the same type" (John Marburger, Keynote Address, 44th Robert H. Goddard Memorial Symposium, Greenbelt, Maryland, March 15, 2006, http://www.nss.org/resources/library/spacepolicy/marburger1.html).

[118] W.B. Cameron, *Informal Sociology: A Casual Introduction to Sociological Thinking,* Random House, New York, 1963 (5th printing, 1967), p. 13.

A value proposition is a statement of the benefits or experiences being delivered by an organization to recipients, together with the price or description of the resources expended for them. From an economic perspective, something is considered to be of value when the worth or benefits of the experience, product, service, or program exceed the costs or resources set forth to obtain it.[119] In a business context, value can be expressed as a ratio of financial benefits to expenditures associated with a given activity or set of activities: a ratio greater than 1.0 denotes positive value. In public administration and policy, however, that ratio is only one of the metrics that address worth.[120]

The value-proposition approach to the assessment of public programs is rooted in large part in the widely remarked differences between private- and public-sector organizations in objectives and in the feasibility of measuring outcomes—there is no obvious "bottom line" for most public programs, which by definition are conducted as not-for-profit activities. The value-proposition framework championed by Moore (1995, 2013)[121] argues that public programs should be evaluated in terms of their ability to achieve a broad set of objectives (in Moore's terms, *values*) in addition to the efficiency with which the objectives are accomplished. The effectiveness of public programs in achieving their broader set of objectives forms the core of value-proposition analysis as applied to public-sector activities. But it is a reflection of the complex environment within which such programs operate that it is difficult to measure objectives, measure progress toward them, and aggregate the various measures of outcomes and progress into any single equivalent to the "bottom-line" measure of profit or loss that figures prominently in private-sector management.

As outlined by Moore (2013), the value-proposition framework lacks any basis for assessing the interdependence among different elements of a given proposition, and this can make it difficult to use as a management or evaluation tool. Management in both the public and private sectors involves establishing priorities and managing tradeoffs among the priorities within a constrained resource environment. Using a value-proposition framework for managing any public program requires an understanding of the extent to which pursuing one objective (or element of the value proposition) affects progress toward others. This need for understanding tradeoffs also applies to program outcomes because the value-proposition framework recommends that managers recognize the priorities assigned by public opinion or by the opinion of key stakeholder groups to different aspects of program outcomes. Here, too, establishing a basis for making tradeoffs is complex and may be even more difficult with a program like NASA human spaceflight, for which mass public opinion reveals broad but lukewarm support (see Chapter 3), with little guidance as to what objectives are valued especially highly.

The emphasis on stakeholder value in governance and management predates Moore and is widely attributed to R.E. Freeman's *Strategic Management: A Stakeholder Approach*,[122] which focused more intensively on private-sector management. Theoretical and empirical work spawned by Freeman's arguments centered on understanding stakeholders, value management, and value delivery and on the ability of an organization to generate, communicate, and transfer value to stakeholders. Like Moore's management approach to value propositions, stakeholder theory is difficult to apply to large government programs in part because different stakeholder groups or individuals may assign different weights to multiple factors that are distributed across the enterprise in assessing value.[123,124] No

---

[119] L. Phillips, Managing customer value when your program's survival depends on it, Paper # AIAA 2007-9928 in *Proceedings of AIAA Space 2007,* September 18-20, Long Beach, Calif., AIAA, Washington, D.C., 2007.

[120] M. Cole and G. Parston, *Unlocking Public Value,* John Wiley & Sons, Hoboken, N.J., 2006, pp. 43-49.

[121] M.H. Moore, *Creating Public Value* (Harvard UP, 1995); M.H. Moore, *Recognizing Public Value* (Harvard UP, 2013). Moore defines a "public value proposition" as follows: "To seek political consensus for the values that they think they ought to be accountable for producing, public managers need to make some kind of public value proposition—a list of the values that would show up on the right-hand side [associated with revenues in a private-sector accounting scheme] of the public value account. . . . A public value proposition might equally fail if it did not connect or resonate with all, or most, or the most important of those in a position to call the agency to account, exposing a public manager to indifference, or angry criticism for neglecting cherished values. Even a public value account conscientiously and meticulously constructed through intensive negotiations with elected overseers and 'market testing' with the wider public can unravel when those who seemed to agree change their mind" (p. 91).

[122] R.E. Freeman, *Strategic Management: A Stakeholder Approach,* Pitman Publishing, Boston, Mass., 2010.

[123] B.G. Cameron, E.F. Crawley, G. Loureiro, and E.S. Rebentisch, Value flow mapping: Using networks to inform stakeholder analysis, *Acta Astronautica* 62:324-333, 2008.

[124] W.K. Hofstetter, "The MIT-Draper CE&R Study: Methodologies and Tools," 2008, http://nia-cms.nianet.org/getattachment/resources/Education/Continuining-Education/Seminars-and-Colloquia/Seminars-2008/CER_NIA_Talk-8-September-2008-final.pdf.aspx, p. 10.

methodological consensus has emerged in stakeholder analysis. However, several recent studies have included value mapping to stakeholder needs or "stakeholder value analysis" with the goal of providing guidance to the designers of NASA programs or components of programs, although it is unclear that the designers incorporated the results of these analyses into the programs or components. The majority of these studies have been led by the Systems Architecture group at the Massachusetts Institute of Technology (MIT).

A joint project between Draper Laboratories and the MIT Department of Aeronautics and Astronautics was conducted in 2004-2005 in support of NASA's Project Constellation Concept Exploration and Refinement contract. The study aimed to develop and refine approaches to human lunar exploration by using stakeholder value analysis to determine program and system objectives. However, the overall inquiry at MIT was aimed at answering a larger question related to the goals of a value-proposition analysis described above: How can we architect a public enterprise that must accommodate numerous (possibly conflicting) views and ideas about how it should achieve its defined mission?[125]

The study began by categorizing NASA's space-exploration constituents into five stakeholder interest groups: exploration, science, economic and commercial efforts, security, and the public. Identification of stakeholder needs, necessary to bound the "value" that might be delivered by exploration programs, was derived from secondary data sources, such as public opinion polls and government reports. These needs were mapped to high-level exploration program objectives provided by NASA. The objectives were then "traded" vis-à-vis candidate exploration mission architectures so that the researchers could obtain relative rankings of architectures that would satisfy stakeholder needs. The rankings coalesced on four dimensions that were believed to contribute to program sustainability: value delivery, policy robustness, risk, and affordability. Those in turn had implications for the development of systems that might be used both to advance Constellation program objectives and to satisfy stakeholder needs.

Relying as it did on indirect measures and researcher rankings, the effort is best described as an early exercise of a developing methodology for stakeholder value analysis rather than as an analysis itself. The authors acknowledged the experimental nature of their work and pointed out several methodological challenges, many of which have continued to plague other models of stakeholder value. First, weighting and setting priorities among program objectives require characterization of the relationship between the organization (NASA) and its stakeholders, including their ability to influence NASA's space-exploration architecture. That in turn requires an empirical method for assessing stakeholder importance or priority, because stakeholders are not "created equal" with regard to their demands on or influence over any organization. Such issues of stakeholder priority were not addressed in the analysis.

A second challenge lies in the assessment of the relationship between stakeholders and the program objectives that are designed to satisfy them. To assess the strength of the relationship, the research team simply rated the relationship on an ordinal scale across several dimensions. An average of all stakeholder scores was then used to provide an overall weight or preference for meeting individual objectives. This approach is indirect, is subjective, requires external validation, and ignores the issue of stakeholder priority described above.

A final challenge pertains to the metrics of stakeholder value, which are at a higher level than those metrics required for engineering design and mission architecture development. The research team attempted to develop "proximate" or intermediate measures that represented steps along a path to delivery of systems that would satisfy both stakeholder needs and program sustainability requirements. However, the measures imposed an additional layer of subjective interpretation of the relationship between stakeholder needs and system design and introduced more noise into the model while suffering from a lack of external or engineering validation and other weaknesses.[126] The authors pointed to the need for additional research.

A later effort to evolve and simplify the entire approach came in 2008 and also focused on NASA programs, specifically the Vision for Space Exploration. Using the answers to a single question—Who are the stakeholders of the space exploration "system of systems" to whom benefit might flow?—the researchers identified nine stakeholder groups: science, security, international partners, economic, executive and Congress, the U.S. public,

---

[125] E.S. Rebentisch, E.F. Crawley, G. Loureiro, J.Q. Dickmann, and S.N. Catanzaro, "Using Stakeholder Value Analysis to Build Exploration Sustainability," paper presented at the AIAA 1st Space Exploration Conference, January 30-February 1, Orlando, Fla., AIAA-2553, 2005, p. 3.

[126] E.S. Rebentisch et al., "Using Stakeholder Value Analysis to Build Exploration Sustainability," 2005.

educators, mass media, and NASA itself. The researchers then identified stakeholder needs by using input and output queries. The input query—Which inputs are required by the stakeholders?—addressed specific stakeholder needs; for example, scientists require science data, and commercial launch providers require customers. The output query was asked about NASA: What are the outputs of the value-creating organization, and who are they provided to? This process generated a total of 48 distinct needs, which were recorded in input-output diagrams centered on each stakeholder and on NASA. These were analyzed further to determine which stakeholders had which effects on others or on NASA. The researchers then connected the inputs and outputs of various stakeholders to each other. Repeated iterations created a "value network" made up of value flows and loops that represented the connection of the output of one stakeholder group to that of another or the provision of value from one stakeholder to another. The ones that explained observable behaviors—for example, "NASA provides launch contracts to the economic community, which provides launch services to the security community, which could provide support for NASA to the executive for NASA funding"—were grouped into six categories representing the following value domains: policy, money, workforce, technology, knowledge, and goods and services. "Inspiration" and "commercial launch" were added after further refinement.

This method generated interactions of stakeholder needs that the researchers described as "common, synergistic, conflicting, or orthogonal." Common needs are ones that are shared among stakeholder groups. Needs are synergistic when the satisfaction of one need results in satisfaction of another need; for example, launching of a spacecraft could satisfy the economic community's need for contracts and the science community's need for data. Conflicting needs often represent external constraints, such as a tension between "gather science data" and "test new technology in space" with fixed funding. Orthogonal needs are independent of other needs.

The researchers described the resulting modeled value-delivery network between stakeholders and NASA as complex, indirect, interactive, and sometimes fragile. The authors offered recommendations about how to architect organizations so that they would be aligned with the creation, communication, and delivery of value to stakeholders. The most important of the recommendations echoes both Freeman and Moore while providing little insight into how to achieve the desired outcome:

> Organizations should be aligned to deliver value—that is to say, the valued outputs created by the organization should be clearly traceable to responsibilities, processes, and incentives within the organization. Recognizing that these outputs constitute the totality of the organization's impact on its environment highlights their importance. Given that these are the products by which an organization will be judged, responsibilities should be clearly delineated and monitored over time.

That conclusion was echoed in 2009 by the Review of U.S. Human Spaceflight Plans Committee in its report *Seeking a Human Spaceflight Program Worthy of a Great Nation*.[127] To fulfill its charge to conduct an independent review of current U.S. human spaceflight plans, that committee developed evaluation criteria for equitable assessment of all the programmatic alternatives under consideration. The criteria clustered around three major dimensions, one of which rested on a value-proposition assessment and sought to characterize benefits of various exploration pathways and program options to stakeholders. Some of the benefits identified by that committee for various exploration destinations are presented in Figure 2.5. Many of the destinations are included in the exploration pathways assessed in Chapter 4.

The Review of U.S. Human Spaceflight Plans Committee developed a list of stakeholder groups on the basis of previous research, a review of relevant policy documents, and public opinion polls. Stakeholders for NASA human spaceflight programs included "the U.S. government, the American public; the scientific and education communities; the industrial base and commercial business interests; and human civilization as a whole."[128] Benefits delivered to stakeholders were "the capability for exploration; the opportunity for technology innovation; the opportunity to increase scientific knowledge; the opportunity to expand U.S. prosperity and economic competitiveness; the opportunity to enhance global partnership; and the potential to increase the engagement of the public in human

---

[127] Review of U.S. Human Spaceflight Plans Committee, *Seeking a Human Spaceflight Program Worthy of a Great Nation,* 2009.
[128] Ibid, p. 77.

Destination	Public Engagement	Science	Human Research	Exploration Preparation
Lunar Flyby/Orbit	Return to Moon, "any time we want"	Demo of human robotic operation	10 days beyond radiation belts	Beyond LEO shakedown
Earth Moon L1	"On-ramp to the inter-planetary highway"	Ability to service Earth Sun L2 spacecraft at Earth Moon L1	21 days beyond the belts	Operations at potential fuel depot
Earth Sun L2	First human in "deep space" or "Earth escape"	Ability to service Earth Sun L2 spacecraft at Earth Sun L2	32 days beyond the belts	Potential servicing, test airlock
Earth Sun L1	First human "in the solar wind"	Potential for Earth/Sun science	90 days beyond the belts	Potential servicing, test in-space habitation
NEO's	"Helping protect the planet"	Geophysics, Astrobiology, Sample return	150-220 days, similar to Mars transit	Encounters with small bodies, sample handling, resource utilization
Mars Flyby	First human "to Mars"	Human robotic operations, sample return?	440 days, similar to Mars out and return	Robotic operations, test of planetary cycler concepts
Mars Orbit	Humans "working at Mars and touching bits of Mars"	Mars surface sample return	780 days, full trip to Mars	Joint robotic/human exploration and surface operations, sample testing
Mars Moons	Humans "landing on another moon"	Mars moons' sample return	780 days, full rehearsal Mars exploration	Joint robotic/human surface and small body exploration

FIGURE 2.5 Benefits of various destinations along the Flexible Path. SOURCE: Review of U.S. Human Spaceflight Plans Committee, 2009, *Seeking a Human Spaceflight Program Worthy of a Great Nation*, p. 41, http://www.nasa.gov/pdf/396093main_HSF_Cmte_FinalReport.pdf.

spaceflight."[129] The stakeholders and benefits accruing to them as a result of each option were considered along with "risk" and "budget realities" as the three major dimensions driving evaluation of program options. No formal value propositions or stakeholder analysis were presented at the conclusion of the study.[130]

In the following year, the MIT team elaborated on its 2008 work: it conducted a quantitative analysis of NASA's space-exploration stakeholder network on the basis of the secondary data sources used in earlier papers. The analysis suggested that the activities of highest stakeholder value in future space-exploration programs would be ones that yielded an opportunity for science returns and opportunities for the public to be virtually present during mission operations by means of the Internet during exploration activities conducted under NASA's Constellation program.[131,132] In 2012, MIT's value network analytic method was applied to data obtained directly from representative stakeholders of the NASA-National Oceanic and Atmospheric Administration (NOAA) Earth Observation Program by means of questionnaires. The value flows and stakeholder priority-setting from that analysis were

---

[129] Ibid.

[130] Additional benefits related to innovation, inspiration, and new means of addressing global challenges are described in a publication of the International Space Exploration Coordination Group, *Benefits Stemming from Space Exploration*, 2013, http://www.globalspaceexploration.org/wordpress/.

[131] The Constellation program was canceled in 2010.

[132] B.G. Cameron, T. Seher, and E.F. Crawley, Goals for space exploration based on stakeholder network value considerations, *Acta Astronautica* 68:2088-2097, 2011.

validated by means of comparison and assessment with other external stakeholders and proxy data sources, such as public opinion polls and literature reviews. The results included recommendations to NASA and NOAA that the Earth Observation program set priorities among its objectives in such a way as to maximize product delivery to scientists, international partners, commercial companies, and the public.[133] No information on whether any of the recommendations have been implemented or to what effect is available.

A final example of a value-proposition analysis of programs related to NASA activities is a 2012 study for the National Geospatial Advisory Committee, The Value Proposition for Ten Landsat Applications.[134] The study calculated the "productivity savings" associated with 10 applications of the Landsat technology by examining 10 "decision processes that would be significantly more expensive without an operational Landsat-like program. Many of these processes are associated with the U.S. government and save significant amounts of money compared to other methods of accomplishing the same objective" (2012, p. 1). Despite the study's title, which implied a value-proposition analysis, it focused exclusively on the financial savings associated with Landsat applications as varied as monitoring of coastal change and fire management. The study's methodology and estimates seem credible although the assessment provided no estimates of the costs associated with achieving the savings; in essence, the study used a cost-benefit framework that focused exclusively on benefits. Nor did the study consider the costs or potential efficiency gains associated with alternative monitoring technologies or related public investment.

Most important, however, the value-proposition analysis did not present an integrated assessment of multiple objectives and other features of the Landsat program, so it seems to have represented a narrow approach to value-proposition analysis in comparison with, for example, Moore's 2013 discussion of Commissioner Bratton and the New York City Police Department. Similarly, the evolving methodology of the MIT stakeholder value-network analysis, the efforts of the Review of U.S. Human Spaceflight Plans Committee, and the efforts of other researchers related to development and integration of large-scale systems in government programs have focused on providing inputs about stakeholder value to systems engineering and program design. These methods have as their goal the development of program objectives for guiding design decisions that can then be translated into value propositions to be delivered to stakeholders by the originating organization.[135,136] Although that represents a multiuser, multiobjective approach that is an improvement on the Landsat effort, it does not provide for definition of the wider range of outcome-related objectives, value delivery, and generation of sustainable stakeholder support for agency missions and leadership.

When applied to NASA human spaceflight, the value-proposition framework that the present committee has developed begins with the set of rationales discussed above, which highlight an array of hypothesized desirable effects of NASA human spaceflight (innovation and economic return, U.S. national security, national stature and international relations, and inspiration of younger citizens to pursue STEM study), all of which might be used to define a set of outcome objectives or "values" for a NASA human spaceflight value proposition. As in most public programs, however, measuring such effects is difficult. Moreover, the effects of NASA human spaceflight on such outcomes as innovation may be even more difficult to measure than those of many other public-sector programs because of the long lags in realizing innovation-related effects.

Beyond measuring those effects, attributing changes in such outcomes as public inspiration, innovative performance, or U.S. national security and prestige to NASA human spaceflight programs is especially difficult. Other rationales discussed above—such as supporting the establishment of human habitation on other planets, the enhanced exploration capabilities associated with space missions that involve astronauts as well as robotic equipment, and the link between space exploration and human destiny—represent motivations for human spaceflight that arguably are unique to NASA. But even defining those rationales (let alone assessing the success of current and planned NASA missions to achieve objectives associated with them) or the tradeoffs among them in terms of

---

[133] T.A. Sutherland, B.G. Cameron, E.F. Crawley, Program goals for the NASA/NOAA Earth Observation Program derived from a stakeholder value network analysis, *Space Policy* 28:259-269, 2012.

[134] National Geospatial Advisory Committee, Landsat Advisory Group, "The Value Proposition for 10 Landsat Applications," 2012, http://www.fgdc.gov/ngac/meetings/september-2012/ngac-landsat-economic-value-paper-FINAL.pdf.

[135] T.A. Sutherland et al., Program goals for the NASA/NOAA Earth Observation Program derived from a stakeholder value network analysis, 2012.

[136] J.M. Brooks, J.S. Carroll, and J.W. Beard, Dueling stakeholders and dual-hatted systems engineers: Engineering challenges, capabilities, and skills in government infrastructure technology projects, *IEEE Transactions on Engineering Management* 58(3):589-601, 2011.

mission priorities or public opinion is extremely difficult, not least because they require decades, or centuries as in the case of permanent off-Earth habitats, to be realized.

Many or most of the challenges of developing and applying value-proposition analysis at the agency level are not unique to NASA. Indeed, the challenges may help to explain the absence of any value-proposition analysis of other federal science and technology programs. None of the various NRC studies of the analysis of federal R&D programs has attempted to develop a value-proposition analysis of these programs, nor does the 2001 National Science Board study of federal R&D programs or the 1991 congressional Office of Technology Assessment study include such an analysis.[137] The present committee sought unsuccessfully to find examples of other publicly available value-proposition analyses of NASA programs, as well as examples of the use of this framework by senior NASA administrators in public statements or by congressional appropriators in decision-making on NASA budgets.

2.6.2 Stakeholder Value and the Impacts of Ending Human Spaceflight

For the reasons described above, a rigorous analysis of the value propositions for NASA human spaceflight at the national level is beyond the capacity of this report—possibly of any report. An alternative way to examine the value proposition of NASA human spaceflight is to consider the effects on various stakeholder groups if the program is terminated. The Review of U.S. Human Spaceflight Plans Committee (2010) characterized benefits of various exploration programs and pathway options to stakeholders in terms of opportunities and potential for value creation, and discovery. A unique perspective on the value proposition for NASA human spaceflight asks, What would happen—and to whom—if those opportunities and potential were no longer available?

NASA human spaceflight stakeholder groups that might benefit from human space exploration have been defined in the statement of task for this committee, which calls for a description of value that takes into account "the needs of government, industry, the economy, and the public good—and in the context of the priorities and programs of current and potential international partners in the spaceflight program." In terms of the rationales discussed above, one can divide government and public-good rationales into distinct interests of different stakeholder groups, namely, national security, international relations, science, and education and inspiration as well as the general public interest. The committee notes that a fully responsive answer to this request would require a value-proposition analysis that goes well beyond any known methodology and applies the concept of value propositions to the human spaceflight enterprise as a whole rather than to individual program designs as described above. What follows is an effort to address the task statement by focusing on losses that would stem from human spaceflight termination and the resulting effect on some stakeholder groups. It should be noted clearly that the committee is not recommending termination. Thinking about that eventuality is merely another way to recognize what is valued about the program from various perspectives. Any potential benefit to any of the stakeholder groups, as previously covered in this chapter's discussion of rationales, will be lost if the program is terminated. To avoid duplication, this section does not restate each of them here; rather, this dicussion highlights where the lens of "what would be lost" adds a perspective that was not captured in the discussion of rationales above.

In a theoretical sense, one should consider that some of the possible losses that would result from termination of the exploration program may be interrelated. The methodology used to evaluate rationales in this report was necessarily unitary; that is, each rationale was considered separately from the others. An alternative view is that

---

[137] See the NRC reports (published by The National Academies Press, Washington, D.C.) *Measuring the Impacts of Federal Investments in Research* (2011), *Allocating Federal Funds for Science and Technology* (1995), *Evaluating Federal Research Programs* (1999), *A Strategy for Assessing Science* (2007); and National Science Board, *Federal Research Resources: A Process for Setting Priorities* (USGPO, Washington, D.C., 2001); U.S. Congress, Office of Technology Assessment, *Federally Funded Research: Decisions for a Decade* (USGPO, Washington, D.C., 1991). It is important to note that these studies do not agree on any alternative analytic framework for evaluating the effects of federal R&D investments, for reasons stated in the National Science Board report: "In many ways, federal research presents greater problems for measurement and benchmarking than does private R&D. A great deal of federally funded research is directed to areas where the market is limited at best. Further, given the types of data available, the returns that result from most calculations must be interpreted as average rather than marginal rates. From a policy perspective, this means that we cannot be certain from this aggregate analysis what the effect of an additional dollar of research expenditure might be. The cost/benefit framework itself may be too restrictive, failing to capture the many benefits that may be derived from publicly-funded basic research. The true effect of such outlays may well be indirect, affecting productivity through changing the returns to private research and development rather than directly as a result of the specific research project" (pp. 78-79).

some of the benefits associated with various rationales are related to each other and that the value flow described by the relationships is irreducible. In the language of the MIT studies, this means that breaking one "value loop" in the network of value delivery may have downstream or corollary effects that are impossible to capture in a discussion of losses to and effects on individual stakeholder groups. Moreover, a loss of opportunity to create value—usually referred to as opportunity cost—is particularly difficult to address because the nature of the value to be created in the future may not be foreseeable.[138]

This observation is bolstered by the presence of a temporal element in the committee's deliberations: We face a future that is unpredictable. In an enterprise like human space exploration, which has a decades-long horizon, loss of value or loss of the opportunity to create value may have a greater effect in the future than current assessments indicate. For human spaceflight to progress, continuing research on risk reduction is necessary—including development of new environmental systems and new launch and transportation technologies, all of which are long-lead items—if technology is to be available at appropriate phases of exploration. Such development is expensive and creates opportunities for pushing the envelope in science, engineering, and operations—with benefits not always identifiable in advance but that would not otherwise be pursued if the program ends.[139]

In examining the possible costs of termination of the exploration program, one must consider the timelines and eventual future of the program for LEO human spaceflight separately from those for beyond-Earth exploration. Termination of one does not imply termination of the other, and each is different from the others; this illustrates the complexity of the issues under consideration. In the discussion that follows, that issue is considered from three perspectives, each of which addresses one or more of the stakeholder groups that the committee was asked to consider:

- Termination of all human spaceflight activities (LEO and beyond LEO)—whenever such termination might occur—is considered with regard to potential effects on the government (national security and international relations) and on the public (national pride and identity).
- Termination of NASA human spaceflight in LEO is considered with regard to science, economics (commerce), and industry.
- Termination of beyond-LEO exploration is discussed with emphasis on international partnerships and with regard to the enduring questions.

2.6.2.1 Ending LEO and Beyond-LEO Human Spaceflight: Effects on the Public Good and the National Interest

As described in the earlier discussion of rationales, human spaceflight has contributed to national pride and stature. As difficult as it is to characterize other benefits of human spaceflight, the cultural "value" of human spaceflight and its role in national pride and identity are even more difficult to assess. National identity has been defined as "the cohesive force that holds nation states together and shapes their relationships with the family of nations" and national pride as "the positive effect that the public feels toward their country as a result of their national identity . . . both the pride or sense of esteem that a person has for one's nation and the pride or self-esteem that a person drives from one's national identity."[140]

Collective experience is represented in and reinforced by national pride and can be reflected in symbols of national experience or achievement with which national pride is strongly correlated.[141] A strong indicator of national pride as reflected in the public's connection to NASA's human spaceflight program surfaced after ter-

---

[138] The discussion in this section of losses in the absence of human spaceflight and the resulting implications for stakeholders has broad similarity to regret theory, which rests on the assumption that under conditions of uncertainty people facing a choice may anticipate regret of their decision. In such cases, a desire to avoid regret may be taken into account in the decision-making process. See G. Loomes and R. Sugden, Regret theory: An alternative theory of rational choice under uncertainty, *The Economic Journal* 92:805-824, 1982.

[139] J. Johnson-Freese, *Space as a Strategic Asset,* 2007, p. 54.

[140] T.W. Smith and S. Kim, National pride in cross-national and temporal perspective, *International Journal of Public Opinion Research,* 18:127-136, 2006.

[141] T.W. Smith, K.A. Rasinski, and M. Toce, *American Rebounds: A National Study of Public Response to the September 11th Terrorist Attacks,* NORC Report, University of Chicago, Ill., 2001.

mination of the Space Shuttle Program in 2011. As described in Chapter 3, public opinion polls done at the time showed that Americans responded to the losses of *Challenger* and *Columbia* and their crews with increased support for NASA's human spaceflight program. That could be attributed to a collective response to tragedy, but as space shuttles were flown to museums beginning in 2012, thousands of people left businesses and homes and got out of their cars, stopping traffic on freeways to watch the space shuttles as they circled Washington, DC, New York, and San Francisco.[142,143] Press reports noted that "hundreds of thousands" lined streets in Los Angeles as *Endeavor* was towed to the California Science Museum preparation site.[144]

National pride is linked to achievements in cultural, nonpolitical activities, including achievements in science and technology.[145] Marketing firms and political campaigns have long made use of symbols that evoke national pride to "brand" products, services, or candidates. Marketing campaigns reflect investment of capital for the purposes of generating revenue. A high recognition value across a number of demographic segments denotes the "branding power" of the icons. The frequent use of icons that have substantial power reflects marketing campaign managers' belief that their products or services will be viewed as attractive as a result of association with the icons.

An indirect indicator of the pervasiveness of NASA human spaceflight in American culture may be found in the use of NASA human spaceflight "brand icons" in advertising campaigns. The extravehicular-activity "spacesuit"— a direct reference to the human in *human spaceflight*—is used regularly in national marketing campaigns.[146] In 2013, spacesuit use by major brands surfaced in advertisements for personal fragrances (Unilever) and automobiles (Kia), as well as for the Make-A-Wish Foundation.[147,148,149]

Those glimpses of persistent connection with and awareness of NASA's human spaceflight program raise a difficult question: What is the actual value of NASA human spaceflight in the national self-image? The committee has heard that the answer may be deeply ingrained in American identity and is a source of national pride, and this is consistent with previous research.[150,151,152] In addition, the stakeholder survey conducted for this study found that "U.S. prestige"—a concept linked to national pride and identity—was expected to suffer the greatest loss if human spaceflight activities in LEO and beyond LEO were terminated (see Table 3.12). Such a loss could be greatest for future generations.

NASA human spaceflight's direct contributions to national security are limited; however, indirect contributions to national security interests have benefit for the government. Indirect evidence of the potential effect of the loss of the NASA human spaceflight workforce on the defense industrial base may be found in a stakeholder survey of 536 companies commissioned by NASA and conducted by the U.S. Department of Commerce's Bureau of Industry and Security Office of Technology Evaluation in 2012. The study examined the effect of space-shuttle and *Constellation* termination on the NASA industrial base and on other U.S. government customers. Eighty-six suppliers reported that their customers would be affected in some way, primarily through loss of experienced

[142] Brian Vastag, "NASA's Discovery Shuttle Wows Washington in 45-Minute Flyover," *Washington Post,* April 17, 2012, http://www.washingtonpost.com/lifestyle/style/space-shuttle-discovery-wows-washington-in-45-minute-flyover/2012/04/17/gIQAKkgFOT_story.html.

[143] CBS San Francisco Bay Area, "Huge Bay Area Crowds Await Shuttle Endeavour," September 21, 2012, http://sanfrancisco.cbslocal.com/2012/09/21/huge-bay-area-crowds-await-shuttle-endeavour/.

[144] CNN Wire Staff, Space shuttle Endeavour rolls into new home as crowds cheer, CNN.com, October 15, 2012, http://www.cnn.com/2012/10/14/us/shuttle-endeavour/.

[145] T.W. Smith and L. Jarkko, *National Pride: A Cross-National Analysis,* NORC Report, University of Chicago, 1998, http://publicdata.norc.org:41000/gss/documents/CNRT/CNR19%20National%20Pride%20-%20A%20cross-national%20analysis.pdf.

[146] Amanda Wills, "Inside the Spacesuit: 10 Rare Views of a NASA Icon," Mashable.com, August 20, 2013, http://mashable.com/2013/08/20/nasa-spacesuit-smithsonian/.

[147] Andrew Adam Newman, Launching a fragrance line (in a manner of speaking), *New York Times,* January 10, 2013, http://www.nytimes.com/2013/01/11/business/media/for-axes-apollo-line-a-campaign-found-in-space.html?_r=0.

[148] Jonathan Brown, "This Super Bowl Kia Commercial Positions Babies as Astronauts," Trendhunter.com, January 30, 2013, http://www.trendhunter.com/trends/super-bowl-kia-commercial.

[149] The Make-A-Wish Foundation public service announcement "Traveler" appeared, for example, in *Central New York Business Journal,* October 26, 2011, p. 12, http://issuu.com/thebusinessjournal/docs/commemorative_issue_flip/13.

[150] Betty Sue Flowers, Panel Discussion, January 18, 2013.

[151] Neil deGrasse Tyson, presentation to the committee, October 23, 2013.

[152] Smith and Jarkko, *National Pride,* 1998.

personnel, increased cost of equipment, potential loss of software and manufactured products, and reduction in R&D expenditures. Twenty-eight companies indicated that their business with the Missile Defense Agency would be affected, primarily in workforce and R&D. Twenty-seven companies indicated that business with the U.S. Air Force and Space and Missile Systems Center would experience cost increases and collateral effects on workforce and innovation.

The small number of suppliers affected were largely small and medium-size businesses and vulnerable to disruption.[153] Previous research by the National Security Space Office, the Air Force, and the Department of Commerce indicated that most space-related R&D spending and innovation is done by companies in those categories.[154] Erosion of the industrial base is a subject of concern for the Department of Defense, which perceives it as a threat to national security.[155] Loss of NASA human spaceflight would probably have a small direct effect given the modest scope of prospective NASA procurement activities related to human spaceflight, but the effect is impossible to assess precisely.

NASA human spaceflight also has been used to promote U.S. geopolitical objectives and as a means of exercising soft power. Both from a geopolitical perspective and from the perspective of possible future commercial exploitation of space resources, that influence is an important element in the value proposition for human spaceflight. Once lost, this influence might not be readily recovered.

2.6.2.2 Ending LEO Human Spaceflight: Commercial and Scientific Effects

In examining the possible costs of termination of the exploration program, one must consider the timelines and eventual future of the program for LEO human spaceflight separately from that for beyond-Earth exploration. Currently, there is an agreement among all the international partners to operate the ISS until 2020. The U.S. administration has announced its intention to extend ISS operations until at least 2024; however, differing national objectives, funding profiles, and space policy make continuation of the ISS to that date with all the original partners less certain.[156] Termination (and deorbit) of the ISS in 2020 will affect NASA's Human Research Program (HRP), which is necessary for continued exploration. The role of the HRP is to "buy down risk" by identifying causal and mitigating factors related to the effects of the space environment on human behavior, performance, and health. Research programs do not generate results on a prescribed timeline, so NASA has identified a set of high-priority risks, including radiation, bone and muscle loss, increased intracranial pressure, and other physiological and psychological responses to the space environment. The studies addressing those issues require sufficient time to unfold to make it possible to characterize risk and determine which of them may create the most important effects on human beings operating in the space environment—and whether and how such risks can be addressed. Although no one knows a date by which the HRP will be able to generate results that build confidence about risks and risk mitigation, termination of the ISS earlier will lead to a higher probability of lower confidence and poorer characterization of effects on human health and behavior than termination at a later date.

As discussed earlier, NASA's LEO human spaceflight program has recently put into place new mechanisms for acquiring space transportation from commercial suppliers with fixed-price contracts. Small and medium-size suppliers and service providers are emerging to support the new transportation systems and prepare for anticipated growth in related economic activities in LEO over the next several decades. The new commercial space companies

---

[153] Department of Commerce (2012). The National Aeronautics and Space Administration's (NASA) Human Space Flight Industrial Base in the Post-Space Shuttle/Constellation Environment (pp. 124-128).

[154] 2007 Defense Industrial Base report.

[155] M. O'Hanlon, "The National Security Industrial Base: A Crucial Asset of the United States, Whose Future May Be in Jeopardy," 21st Century Defense Initiative Policy Paper, Brookings Institute, Washington, D.C., 2012, http://www.brookings.edu/~/media/research/files/papers/2011/2/defense%20ohanlon/ 02_defense_ohanlon.pdf.

[156] A. Krasnov, "Perspectives on the Future of Human Spaceflight," presentation and remarks to the Committee on Human Spaceflight, April 23, 2013.

are offering diverse services, such as spaceflight training, spacesuit design, and vehicle mockups.[157,158,159] In the near future, the companies will depend heavily on activities between Earth and LEO that are centered on utilization of the ISS.[160] Termination of the ISS in 2020 would result in a relatively short period during which NASA would serve as the principal or only customer for the companies, resulting in a very limited opportunity for them to recoup their investment.[161] Government funding outside NASA could be substituted in order to continue development of the systems should they be determined to be in the national interest (for example, to diversify the launch industry), but without the ISS as a driver for U.S. investment in this sector, it is unclear whether the competitive advantages and cost savings justify continuing investment on behalf of the country.[162]

At least one company is currently making use of the ISS as a launch platform for small commercial satellites, and others are exploring commercial R&D, commercial services, and product development pathways under the auspices of the ISS National Laboratory. Other commercial entities have plans in development for LEO, but at present these are still only plans. It should be noted that those possibilities rely on models for commercial market development in LEO that are early in development and speculative.[163,164]

Scientific benefits of the ISS were reviewed above. Terrestrial benefits from the ISS National Laboratory, if any, are probably some years away. Termination of the NASA LEO human spaceflight program by 2020 would cut short the R&D process that the ISS National Laboratory is counting on, and there would be a potential (and unknowable) associated loss of value to science and to commercial entities. Research programs conducted by the international partners would also be severely affected or terminated unless a new partner or partners could be found to make up the operational funding losses created by U.S. withdrawal.

Within a month of the present report's going into NRC review, NASA and the Obama administration announced their intention to continue the U.S. commitment to the ISS until at least 2024. Under those circumstances, the effects of termination in 2020 discussed above would be reduced, and there would be additional time to generate returns from exploration science, the ISS National Laboratory, and commercial activities and operations. However, as discussed in Chapter 4, continuing to operate the ISS beyond 2020 in a flat-budget environment would undercut beyond-LEO activities. The potential value proposition of ISS-based science versus exploration beyond Earth orbit is discussed further in Chapter 4.

2.6.2.3 Ending Beyond-LEO Exploration: Partnerships and the Enduring Questions

The statement of task charges the committee to consider the value of the U.S. human spaceflight program for current and potential international partners. The committee has found that U.S. near-term goals for human exploration beyond LEO are not aligned with our international partners' goals (see Chapter 1), which are focused on Mars (with the Moon as an intermediate goal). Regardless, the *Global Exploration Roadmap* developed by the International Space Exploration Coordination Group (ISECG) is loosely organized around U.S. intentions and

---

[157] The National AeroSpace Training and Research Center (NASTAR), http://www.nastarcenter.com/about-us/etc-and-the-nastar-center, accessed December 21, 2013.

[158] WayPoint2Space, at http://waypoint2space.com, accessed on December 21, 2013.

[159] See "Space Suits," "Space Mockups," and "Space Diving" at Orbital Outfitters, "What We Do," http://orbitaloutfitters.com/what-we-do/. Accessed December 21, 2013.

[160] NASA, *Commercial Market Assessment for Crew and Cargo Systems Pursuant to Section 403 of the 2010 NASA Authorization Act,* issued March 12, 2011, http://www.nasa.gov/sites/default/files/files/Section403(b)CommercialMarketAssessmentReportFinal.pdf.

[161] Recently, NASA and the White House announced their intention to continue the ISS until 2024, in part to enable commercial development in LEO to continue to mature, as well as to maximize science returns.

[162] A recent study commissioned by the Federal Aviation Administration and conducted by the Futron Corporation documented the competitive advantage of the United States in human-spaceflight markets (relative to other nations) as a result of NASA's Commercial Crew Development program. See G. Autry, L. Huang, and J. Foust, *An Analysis of the Competitive Advantage of the United States of America in Commercial Human Orbital Spaceflight Markets,* 2014, https://www.faa.gov/about/office_org/headquarters_offices/ast/media/US_HOM_compet_adv_analysis-Final_1-7.pdf.

[163] See K. Davidian, I. Christensen, D. Kaiser, and J. Foust, "Disruptive Innovation Theory Applied to Cargo and Crew Space Transporation," Paper # IAC-11-E6.3.4 presented at the International Astronautical Congress, Capetown, South Africa, October 2011.

[164] J. Aprea, U. Block, and E. David, "Industrial Innovation Cycle Analysis of the Orbital Launch Vehicle Industry," Paper #IAC-E6.2.6 x19035 presented at the International Astronautical Congress, Beijing, China, September 2013.

contributions, the latter of which far outweigh the total contributions of the other international partners. If the U.S. terminates all government involvement in NASA beyond-LEO exploration—including an asteroid-redirect mission, the Moon, and Mars—human space exploration beyond LEO would probably be delayed by decades, and this would place Mars out of reach until late in the 21st century or early in the 22nd century—and then only if other entities emerge with substantial investment to take the place of U.S. contributions. With regard to the Moon, the situation is less clear because intense commercial development or substantial investment by another country might facilitate lunar exploration, although there is no way to predict that. Termination of U.S. involvement would, in any case, have deleterious effects on the human spaceflight programs and ambitions of many U.S international partners and could complicate future relationships and plans in other areas of joint activity.

The current international partners in the ISS program are not the only nations that are participating in the ISECG. India, the Republic of Korea, Ukraine, and the United Kingdom are also engaged, and China attends the ISECG meetings. All those countries have expressed interest in working with NASA on future opportunities.[165] The largest and most active such programs are in India and China. India has collaborated with the United States for many years in space and has expressed an interest in continuing to do so.[166]

At the end of 2011, China released a white paper detailing its philosophy and programs, updating progress since 2006, and laying out goals for the next 5 years. As described in Chapter 1, China's plans reflect a clear vision, goals, and methodical program development that build on what has come before, with an active robotic program, development of a space station, and study of expeditions to the Moon. Although China wants to engage cooperatively with the United States in human spaceflight, participation in a joint program is unlikely in the near future because of security concerns and political resistance within the United States.[167,168]

Termination of human space exploration beyond Earth orbit could shift the momentum of space exploration to the Asia-Pacific region and specifically to China, which is already creating opportunities for cooperative engagement with other nations, including Russia and such other Western powers as Germany and France.[169] Any potential geopolitical shift among spacefaring nations away from the United States and toward the Asia-Pacific region could have unknown strategic and practical consequences for the United States.

Finally, termination of the NASA human spaceflight program would render the enduring questions unanswerable by the United States—except for provisional information that can be developed via robotic exploration. Two questions—How far from Earth can humans go? What can humans discover and achieve when we get there?—are intended to create a framework for guiding human space-exploration program development and execution leading toward Mars within the pathways and decision rules outlined in Chapter 4. Termination of the human spaceflight program before that goal would render pursuit of answers to the enduring questions beyond the reach of this nation until such time as the country might decide to return humans to space, leaving it to other countries to engage in and manage the activities required to answer them. The committee believes that such an eventuality is not in the best interests of the United States.

2.7 CONCLUSIONS ON THE BENEFITS OF HUMAN SPACEFLIGHT

The current practical benefits of human spaceflight, although they have meaning for specific stakeholder groups, do not rise to the level of compelling justification for human spaceflight. Aspirational and inspirational rationales and value propositions, however, are most closely aligned with the enduring questions and when "added" to the practical benefits do, in the committee's judgment, argue for continuation of NASA human spaceflight programs, provided that the pathways approach and decision rules described in Chapter 1 are applied.

---

[165] Logsdon, 2010.

[166] G.S. Sachdeva, Space policy and strategy of India, pp. 303-321 in *Space Strategy in the 21st Century* (E. Sadeh, ed.), Routledge (Taylor & Francis), London, 2013.

[167] S. Chen, U.S. and China partner on small-scale space projects, *South China Morning Post,* September 30, 2013, http://www.scmp.com/news/china/article/1321102/us-and-china-partner-small-scale-space-projects.

[168] L. David, Security fears impede U.S. space cooperation with a rising China, *Space News,* December 2, 2013.

[169] L. David, China invites foreign astronauts to fly on future space station, *Space.com,* September 28, 2013, http://www.space.com/22984-china-space-station-foreign-astronauts.html.

The NRC report *American's Future in Space: Aligning the Civil Space Program with National Needs* (2009), noted that "the U.S. civil space program has long demonstrated a capacity to effectively serve U.S. national interests" but recommended future alignment of space-program capabilities and plans with "high-priority national imperatives, including those where space is not traditionally considered,"[170] including climate monitoring and change, development of advanced technologies, and international relations. Of these, that committee noted that the imperative of international relations is most closely aligned with human spaceflight and can serve the interests of the United States by "inviting emerging economic powers to join with us in human spaceflight adventures."[171] The report also noted that human spaceflight activities "should be prioritized by their potential for and likelihood of producing a transformative cultural, scientific, commercial, or technical outcome" although it cautioned that such activities require many years and a "long-term commitment to come to fruition."[172] Such outcomes, although they may be possible, represent future objectives of human spaceflight, requiring not only commitment but investment, planning, and management designed to realize them in alignment with the pathways and decision rules developed in the present report.

The analytic and other challenges (e.g., defining and measuring objectives, attributing effects to NASA programs, and evaluating tradeoffs among the various elements of the NASA value proposition and rationales associated with value-proposition analysis) mean that a definitive assessment of the "NASA human spaceflight value proposition" is beyond the capability of the present committee and, in the committee's judgment, that of most other objective observers. These programs include an array of activities whose outcomes are difficult to monitor, that are affected by multiple other federal programs, and whose monitoring requires detailed data on outcomes that may span decades. Finally, the question What would be lost?—although it offers a different perspective from which to view the benefits of human spaceflight for various stakeholders—does not lead to any change in the conclusions that the committee has developed on the basis of its analysis of the rationales.

---

[170] NRC, *America's Future in Space: Aligning The Civil Space Program With National Needs,* The National Academies Press, Washington, D.C., 2009, p. 59.

[171] Ibid., p. 63.

[172] Ibid.

3

Public and Stakeholder Attitudes

Controversies over the nature and function of public opinion have always been a part of discussions about the democratic process. Some theorists have argued that politically engaged citizens are essential for a true democracy, and others have emphasized the strengths of representative democracy, which relies on delegation of decision-making to elected representatives on many issues. It is beyond the scope of the present report to discuss the vast literature on the normative, philosophical, or empirical aspects of the role of public opinion in policy-making. However, in matters that are of particularly low salience to the public, policy decisions often result from interactions between decision-makers at the federal level and groups of relevant policy leaders, especially when there is a high level of agreement among the decision-makers and policy leaders.[1] Because of that pattern of decision-making, in addition to a review of existing public opinion data on the topic of space exploration, the Public and Stakeholder Opinions Panel (referred to hereafter as the panel) also sought the input of those close to the policy process—those with a stake in, but not necessarily advocates of, human spaceflight.

This chapter discusses public opinion and stakeholder views on space exploration and human spaceflight, including rationales of and support for various programs. It first reviews public opinion data collected over the years by the nation's major polling organizations and then discusses findings of a survey conducted as part of the present study to assess the views of key stakeholder groups.

3.1 PUBLIC OPINION

The discussion of public opinion in this chapter is based on a large collection of studies, datasets, and papers related to public opinion on space exploration and human spaceflight. Considerable data on the topic have been collected over the course of several decades. This discussion of public attitudes relies on surveys that used probability sampling. The number of surveys based on nonprobability methods has grown considerably over the past few years (due, in part, to concerns about declining response rates and rising data collection costs), and they include some of the recent surveys conducted on the topic of space exploration. The accuracy of surveys based on nonprobability samples varies, and many of the newer techniques have not been adequately evaluated.[2] Although

---

[1] G.A. Almond, *The American People and Foreign Policy,* Harcourt, Brace and Company, New York, 1950; J.D. Miller, *Space Policy Leaders and Science Policy Leaders in the United States: A Report Submitted to the National Aeronautical and Space Administration,* Northwestern University, Chicago, Ill., 2004.

[2] American Association for Public Opinion Research, *Report on the AAPOR Task Force on Non-Probability Sampling,* Deerfield, Ill., http://www.aapor.org/Reports1/6877.htm#.U3u7S9fMcfU, June 2013.

BOX 3.1
Survey Sampling and Concerns Related to Non-Probability Sampling

Most well-respected public opinion surveys are based on probability sampling, which is a sampling technique that assures that everyone in the population of interest has a known probability of selection into the sample. The statistical theories underlying probability sampling enable researchers to quantify the accuracy of the estimates made about the population of interest.

Non-probability surveys include participants without a known probability of selection. The participants could be selected by the researchers, such as in a medical research study, or self-selected, as in opt-in online panels. The lack of a clear relationship between the sample and the target population in the case of non-probability samples makes it more difficult to measure the accuracy of the estimates and makes the calculation of "sampling error" inappropriate. Even if the data are weighted to reflect the demographic composition of the target population, calculations of a sampling error can be misleading.

SOURCE: American Association of Public Opinion Research, "Opt-in Surveys and Margin of Error," http://www.aapor.org/Opt_In_Surveys_and_Margin_of_Error1.htm#.UtmdwvMo6os, accessed January 2014.

---

the use of nonprobability surveys, such as opt-in on-line panels, can be appropriate in some circumstances, the review in this section is focused on public opinion data that were collected with probability techniques that are most broadly accepted by the survey community. (For additional information about probability sampling, see Box 3.1.)

It is important to note that attitudes on topics that are of relatively low interest may be more difficult to measure than attitudes on other topics. Some have argued that survey respondents are often reluctant to acknowledge that they have no opinion on an issue and that many, under perceived pressure to respond to a question, will select an answer option that does not necessarily reflect an existing attitude.[3] In some cases, respondents form a preference only when asked and tend to express views that are based on considerations that happen to be most salient to them at the moment, often because they are mentioned in the question or preceding questions in the survey.[4]

Appendix B includes further information about the methods of the surveys that were included in this review, including the wording of the questions asked. Most of the data cited in this report are available on line from the Roper Center for Public Opinion Research at the University of Connecticut, the Inter-university Consortium for Political and Social Research at the University of Michigan, or the Web sites of the organizations that collected the data.

This section describes data on public interest in and awareness of space exploration and human spaceflight. It then discusses views of government funding for space exploration and support for specific human spaceflight missions, such as the Moon, the space shuttle, the space station, and Mars. It also discusses perceptions of the U.S. role as a leader in human spaceflight, how the public feels about international competition and cooperation, and the role of the government and the private sector in the future of space exploration and funding. Various rationales historically given for space exploration are examined, and the section concludes with a discussion of group differences (related to education, race, sex, and age) in attitudes toward space exploration.

3.1.1 Interest in Space Exploration and the Attentive Public

National survey data collected as part of the National Science Board's *Science and Engineering Indicators* (a compilation by the National Science Foundation of quantitative data available on science and engineering) show that public interest in space exploration increased gradually through the 1980s and that about one-third of American

---

[3] P. Converse, The nature of belief systems in mass publics, in *Ideology and Discontent* (D. Apter, ed.), Free Press, New York, 1964.

[4] J.R. Zaller, *The Nature and Origins of Mass Opinion,* Cambridge University Press, Cambridge, U.K., 1992.

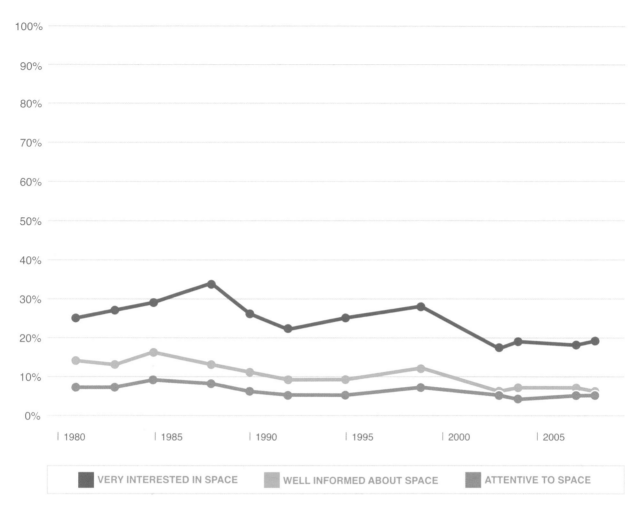

FIGURE 3.1 Public interest in, sense of being informed about, and attentiveness to space exploration, 1981-2008. SOURCE: 1981-2000: NSF Surveys of Public Attitudes; Science News Study, 2003-2007; American National Election Study, 2008

adults said that they were "very interested" in space exploration in 1988 (Figure 3.1). The loss of the space shuttle *Challenger* in January 1986 may have resulted in a brief boost in public interest in space exploration. Interest declined in the years after the return to flight that followed the space-shuttle accident but began to increase again in the late 1990s as the first parts of the International Space Station (ISS) were being assembled.

On the average, over the past 3 decades, about one-fourth of Americans had a high level of interest in space exploration although most Americans described themselves as interested to at least some degree. The most recent General Social Survey (GSS 2012) estimated that 21 percent of the American public was "very interested" in space exploration (an additional 44 percent was "moderately interested").

In a landmark study, Gabriel Almond argued that citizen engagement in policy issues depends on a combination of interest in a topic and a sense of being adequately informed about it.[5] That framework is useful for understanding public engagement in low-salience issues, such as space exploration. There is an "attentive public" for most policy issues, and those who fall into this category tend to follow an issue in the news, have more developed cognitive schemas about it, and retain more information on it than on issues to which they are less attentive.

---

[5] Almond, *The American People and Foreign Policy*, 1950.

Because the "attentive public" is defined as those who are both very interested and well informed about a topic, the attentive public in connection with space exploration is much smaller than the interested public. During the past 3 decades, far fewer Americans felt well informed about space exploration than highly interested in it (see Figure 3.1). The proportion of American adults who were attentive to space exploration (both very interested and well informed) has been in single digits, rising to 9 percent in 1985 just before the *Challenger* disaster and dropping to 5 percent in recent years.

Public interest in space exploration is modest relative to that in other public policy issues (Figure 3.2). The 2012 GSS found the proportion "very interested" in space exploration tied with the proportions "very interested" in international and foreign-policy issues; these topics were at the bottom of 10 issues asked about and trailed new inventions and technologies (42 percent) and new scientific discoveries (41 percent).

3.1.2 Support for Spending on Space Exploration

Public opinion of NASA has been relatively positive and stable over the years. An October 2013 study by the Pew Research Center found that NASA is one of the government agencies with the most favorable views among the public: 73 percent of respondents had a "very favorable" or a "mostly favorable" view of NASA (Pew 10/13).

Despite favorable attitudes toward NASA, there is relatively little public support for increased spending for space exploration (Figure 3.3). According to GSS data, over the past 40 years, about 10-20 percent of the general public thought that we were spending too little on space exploration, and higher percentages (about 30-60 percent) regarded spending as too high, although the size of this group has declined. The gap between the too-little and too-much groups was more than 50 percentage points in 1973 and had shrunk to about 10 percentage points by 2012. In the most recent survey, 22 percent of respondents said that we were spending too little on space exploration and 33 percent too much.

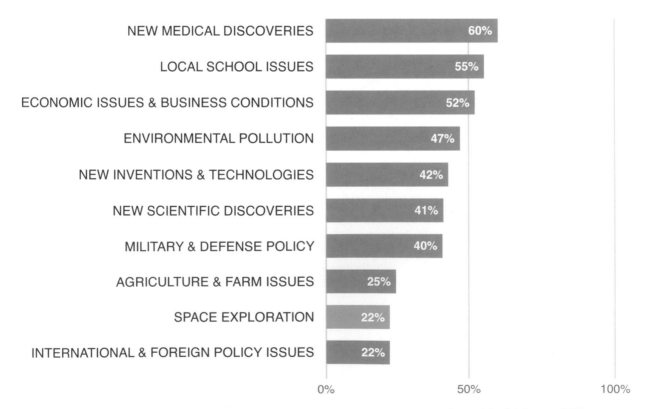

FIGURE 3.2 Percentage of respondents "very interested" in various issues. SOURCE: General Social Survey, 2012.

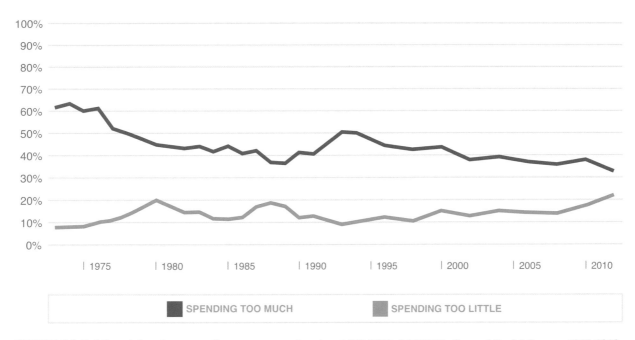

FIGURE 3.3 Public opinion about spending on space exploration, 1972-2012. SOURCE: General Social Survey, 1972-2012.

Although only a minority of the public expresses a desire to increase spending on space exploration, support for increased spending is higher among those who are interested in space exploration. In the 2012 GSS, 45 percent of those who said that they were very interested in space exploration said that we were spending too little on space exploration (compared with 11 percent of the moderately interested and 6 percent of those who were not at all interested).

In some cases, support for spending on space exploration appears to depend on how the question is asked. For example, a Gallup survey conducted in 2006 for the Space Foundation (Gallup 8/06) said to respondents that "NASA's budget request this year is under one percent of the federal budget which would amount to approximately $58 per year for the average citizen. Do you think the nation should continue to fund space exploration?" and then asked whether the nation should fund space exploration at increased or decreased levels. About 31 percent said that funding should be increased, including 9 percent who said "significantly" increased. The questions come from different surveys and the difference in wording is not the only difference in the survey method, but support for increased spending in the Gallup survey is double the 15 percent who said that we were spending too little on space in the GSS survey of the same year. Those types of differences, which depend on whether cost or other relevant considerations are mentioned in a survey question and how the cost information is framed, are common in attitude measurement.

Space exploration generally fares poorly in comparison with other possible spending priorities. The 2012 GSS asked about 18 national problems, and space exploration ranked 16th in the proportion of respondents who thought that the government was spending too little. Only foreign aid and welfare spending were less popular than space exploration.

That is not a recent development. In a 2004 Pew Research Center study about priorities for the president and Congress, just 10 percent said "expanding America's space program" should be "top priority," putting it last among 22 choices (Pew 1/04). The next-lowest items and percentages were "reforming the campaign finance system" with 24 percent and "dealing with global trade issues" with 32 percent. Other polls that asked respondents over the years to attach spending priorities to federal programs have typically found funding for space exploration near the bottom.[6]

---

[6] R.D. Launius, Public opinion polls and perceptions of U.S. human spaceflight, *Space Policy* 19:163-175, 2003; S.A. Roy, E.C. Gresham, and C.B. Christensen, The complex fabric of public opinion on space, *Acta Astronautica* 47:665-675, 2000.

3.1.3 Trends in Support for Specific Human Spaceflight Missions

Relatively few people say that they are very interested in the topic of space exploration and even fewer feel well informed about it, but a higher proportion of the public has expressed support for specific human spaceflight programs over the years. This section discusses support for the Apollo program, the space shuttle, and a Mars mission.

3.1.3.1 The Apollo Program

Launius[7] reviewed poll results of the question "Should the government fund human trips to the Moon?" at the height of the space race in the 1960s and found that the American public was hesitant about pursuing the Apollo program. Surveys that mentioned a cost for the Apollo program found relatively little support for it. A Gallup poll in May 1961, shortly before the program began, found just one-third willing to spend "40 billion dollars—or an average of $225 a person" to send a man to the Moon (Gallup 5/61). A Harris Poll conducted in July 1967 found just one-third saying that they felt that it was worth spending $4 billion a year for the next 10 years to do so (HI 7/67). A Gallup poll in 1967 found that just one-third thought that it was important to send a man to the Moon before Russia did (Gallup 2/67).

Although spending on a Moon mission was not very popular during the 1960s and 1970s, in hindsight the views of the general public about the Apollo program have become more favorable (Figure 3.4). When asked in 2009 to look back, 71 percent of the respondents to a CBS News poll said that the program had been worth it. About 15 years earlier, in 1994, the poll had put the number at 66 percent; 15 years before that, in 1979, just 47 percent said that it had been worth it (CBS 7/09; CBS/NYT 6/94, 7/97).

3.1.3.2 The International Space Station and the Space Shuttle

The first component of the ISS was put into place in late 1998. Attitudes of Americans toward the planned construction of a large space station were favorable during the decade preceding the construction (Table 3.1). In 1988, slightly more than 70 percent of American adults agreed or strongly agreed that the space station should be built. In 1992, support had dipped to 58 percent; then it grew again slightly closer to the launch of the initial component of the current space station (NSF Surveys of Public Attitudes 1988, 1992, 1997, 1999).

Figure 3.4 shows two time series of support for the Space Shuttle Program: one based on a question asking whether the space shuttle was a good investment for the country (CBS/NYT 1/87, 1/88, 10/88; CBS 12/93, 8/99, 7/05) and the other based on a question asking whether the space shuttle was worth continuing. The final two data points in each series still show a majority favoring the Space Shuttle Program. In the earlier decades, both sets of surveys show 60-70 percent of the public supporting the program.

3.1.3.3 A Mars Mission

A CBS News poll in 2009 found more respondents favoring than opposing "the U.S. sending astronauts to explore Mars" by a margin of 51-43 percent (CBS 7/09). Other readings on the same question in earlier polls showed somewhat stronger support in the 1990s (Favor-Oppose): 1994, 55-40 (CBS/NYT 6/94); 1997, 54-41 (CBS 7/97); 1999, 58-35 (CBS 8/99) (see Figure 3.4).

The distribution of the responses shifts when cost is factored in. A Gallup survey in 2005 asked whether respondents would favor or oppose "setting aside money for such a project" and found 40 percent in favor and 58 percent opposed (Gallup/CNN/USA Today 6/05). That distribution is roughly the same as when Gallup asked the same question in 1969 and 1999 (Gallup 7/69, 7/99). A 2004 Associated Press (AP)-Ipsos survey asked "As you may have heard, the United States is considering expanding the space program by building a permanent space station on the Moon with a plan to eventually send astronauts to Mars. Considering all the potential costs and benefits, do you favor expanding the space program this way or do you oppose it?" It found 48 percent on each

---

[7] R.D. Launius, Why go to the Moon? The many faces of lunar policy, *Acta Astronautica* 70:165-175, 2012.

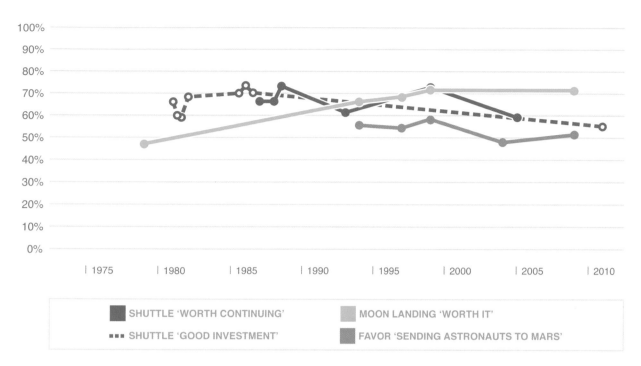

FIGURE 3.4 Public support for the space shuttle, Moon landing, and Mars mission, 1979-2011. SOURCE: Shuttle continuation: CBS/NYT (1987, 1988), CBS (1993, 1999, 2005); Shuttle investment: NBC/AP (1981,1982), NBC/WSJ (1985,1986), Pew (2011); Moon landing: CBS/NYT (1979, 1994), CBS (1997, 1999, 2009); Mars: CBS/NYT (1994), CBS (1997, 1999, 2004, 2009).

TABLE 3.1 Support for Building "a space station large enough to house scientific and manufacturing experiments," 1988-1999

| | Year | | | |
	1988	1992	1997	1999
Strongly agree	11%	10%	12%	9%
Agree	61	48	51	55
Not sure	5	5	6	6
Disagree	22	31	26	27
Strongly disagree	1	6	5	3

SOURCE: National Science Foundation Surveys of Public Attitudes Toward and Understanding of Science and Technology.

side of the issue (AP/Ipsos 1/04). (Half the sample received a slightly different version of the question, which replaced "the United States" with "the Bush Administration." The latter wording found 43 percent in support and 53 percent opposed.)

The 2007 Science News Study included a question on support for a crewed Mars mission. Respondents were asked to agree or disagree with the statement "The United States should begin planning for a manned mission to Mars in the next 25 years." Fifty years after the launch of Sputnik, 40 percent agreed with the statement, and 58 percent disagreed (Science News Study 2007). An NSF survey asked the question with similar wording in 1988, and attitudes were more favorable then, with 51 percent agreeing and 41 percent disagreeing (NSF Surveys of Public Attitudes 1988).

3.1.4 Human Versus Robotic Missions

Apparent support for human space exploration drops greatly when cost savings associated with robotic missions are mentioned. For example, the Gallup Organization in 2003 asked this: "Some people feel the U.S. space program should concentrate on unmanned missions like Voyager 2, which will send back information from space. Others say the U.S. should concentrate on maintaining a manned space program like the space shuttle. Which comes closer to your view?" Human space exploration was preferred over robotic missions by a margin of 52-37 percent (Gallup/CNN/USA Today 2/03). But in an AP-Ipsos poll the next year, which prefaced the question by stating that "some have suggested that space exploration on the Moon and Mars would be more affordable using robots than sending humans," answers tilted heavily in the other direction—a preference for robots by a margin of 57-38 percent (AP/Ipsos 1/04).

Risk does not appear to play a central role as a reason not to send humans into space. Most of the public seems to accept that there are inherent dangers in exploring space. Public support for NASA and space exploration *increased* after the *Challenger* disaster. Shortly after the *Columbia* accident, a 2004 AP-Ipsos survey asked whether human spaceflight should be continued "in light of the space shuttle accident last February [2003] in which seven astronauts were killed," and 73 percent said that the United States should continue to send humans into space (AP/Ipsos 1/04).

3.1.5 NASA's Role, International Collaboration, and Commercial Firms

3.1.5.1 American Leadership and Cooperative Space Exploration

When asked in a 2011 Pew Research Center survey whether they thought that it was essential for the United States to "continue to be a world leader in space exploration," 58 percent of the respondents said that it was essential (Pew 6/11). The percentage has fluctuated. A Time/Yankelovich poll in 1988 found that 49 percent of Americans thought it was "very important" for "this country to be the leading nation in space exploration" (Time/Yankelovich 1/88). Sixteen years later, in 2004, an AP-Ipsos poll found that 38 percent thought that it was important for the United States to be the leading nation in space (AP/Ipsos 1/04).

Although a majority of respondents in 2011 said that it was "essential for the United States to continue to be a world leader in space," a July 2011 survey conducted by CNN/ORC found just 38 percent saying that it was "very important" for the United States "to be ahead of Russia in space exploration" (CNN/ORC 7/11). In a different era, a Gallup poll in June 1961 put that number at 51 percent (Gallup 6/61). In March 2006, Gallup asked this: "A number of Asian and European countries now have space programs of their own or have announced plans for space activities and exploration. As more countries embark on space programs, how concerned are you that the U.S. will lose its leadership in space?" (Gallup 3/06). In response, just 13 percent told Gallup that they were very concerned about the possibility, and only another 22 percent were somewhat concerned. In other measures on the topic, just 11 percent were "very concerned" that China would become the new leader in space exploration in a 2008 Gallup survey, and only another 21 percent were "somewhat concerned" (Gallup, 4/08). Two-thirds of the American public said that they would not be concerned to see this happen.

3.1.5.2 International Competition and Collaboration

Few recent surveys have explored international collaboration in depth, but the available data suggest that the public is generally in favor of it. Even in the late Cold War period of 1988, a survey by Time/Yankelovich found 71 percent of Americans saying that it would be a good idea for the United States and the Soviet Union to undertake cooperative space efforts, such as going to Mars (Time/Yankelovich 1/88). A Harris poll in July 1997 found 77 percent of the public saying that they favor "joint space missions involving Americans, Russians and people from other countries," and 66 percent favored putting a joint U.S. and international space station in orbit (HI 7/97). A CBS survey later in 1997 found two-thirds of respondents saying that the United States should work with Russia on space missions (CBS 7/97). Finally, even when told in 2008 that the United States would have a

5-year gap between the space shuttle's last mission and new programs and that it would depend on Russia to get to and from the space station, only 13 percent that said they were "very concerned" (Gallup 4/08).

3.1.5.3 The Role of the Private Sector

There is little in the survey literature about the public's views about the roles of government and the private sector in the exploration of space or human spaceflight; this reflects both the low salience of space exploration and the relatively recent emergence of private space activities. A 1997 survey by Yankelovich found that 53 percent thought that the space program should be funded and managed by the government, whereas 30 percent favored private business (Yankelovich 12/97). A 2011 CNN question found that 54 percent of American adults thought that the United States should rely more on private companies to run human space missions in the future, compared with 38 percent who wanted to keep human spaceflight primarily a government function (CNN/ORC 7/11). Those two results suggest that the public is becoming more receptive to private commercial activity in space, but it is difficult to draw firm conclusions from so few survey results.

3.1.6 Rationales for Support of Space Exploration

A few surveys have probed for rationales underlying public support of space exploration. Most of them have offered specific rationales in closed-ended questions. Not surprisingly, the apparent level of support for many rationales is less when an open-ended format is used than when the rationales are explicitly mentioned in the question. Therefore, when analyzing responses to closed-ended questions, it is useful (where possible) to examine the proportion who *strongly* agree with a rationale as a potentially more valid indicator of the depth of support for the rationale. However, surveys have used different forms of closed-ended questions, and there is no comparable series with which to evaluate changes in public support of one or another rationale. Taking the available surveys together, the conclusion seems to be that no rationale garners overwhelming support from the public.

In response to an open-ended CBS/NYT poll in 1994, 56 percent of the public said that the best reason for exploring space was to increase knowledge or to search for other life forms (CBS/NYT 6/94). No other reason was offered by more than 7 percent of the respondents; economic benefits and national security were both cited by 3 percent, and national pride and leadership in space by just 1 percent. About one-fourth of the respondents said that they did not know or provided no reason.

A June 2004 Gallup survey used a closed-ended question to ask respondents to choose among various rationales to indicate what they considered to be the main reason for continuing to explore space (Gallup 6/04). Some 29 percent chose the rationale that it is human nature to explore, 21 percent the need to maintain our status as an international leader in space, 18 percent the benefits on Earth, 12 percent keeping our nation safe, and 10 percent the idea that space exploration inspires us and motivates our children. (This survey did not include the rationale of increasing knowledge.)

As Table 3.2 shows, a 2011 Pew survey that used three closed-ended questions found that 34-39 percent of the public agreed that the space program "contributed a lot" to each of the following: encouraging people's interest in science and technology, scientific advances, and national pride and patriotism (Pew 2011).

TABLE 3.2 Support for Rationales for the Space Program, 2011

How much does the U.S. space program contribute to . . .	A lot	Some	Not much/ Nothing	No Opinion
Encouraging people's interest in science and technology	39%	35%	22%	4%
Scientific advances	38	36	22	5
This country's national pride and patriotism	34	34	28	5

SOURCE: Pew Research Center.

TABLE 3.4 Proportion Saying That "We spend too little/too much on space exploration," by Demographic Group, 2012

	Spending Too Little	Spending Too Much
Total U.S.	**22%**	**32%**
Sex		
Male	28	28
Female	18	35
Race/Ethnicity		
White	26	29
Black	11	49
Other	13	33
Age, years		
18-29	19	24
30-49	24	34
50-64	20	35
65 and over	24	35
Education		
H.S. graduate or less	15	38
Some college	21	30
College graduate or more	32	26
Ideology		
Conservative	26	29
Moderate	19	36
Liberal	22	30
Working		
Full-time	23	31
Part-time	19	31
Not working	22	35

SOURCE: General Social Survey.

TABLE 3.5 Development of Interest in Space Exploration, 1987-2011

	Grade/Year				
	Grade 10	Grade 11	Grade 12	2008	2011
Interest in space exploration					
Very interested	16%	20%	17%	18%	24%
Moderately interested	48	51	53	52	48
Not interested	36	29	30	30	28
Informed about space exploration					
Very well informed	11	13	11	6	6
Moderately well informed	50	53	51	42	43
Not well informed	39	34	38	52	51
Attentiveness to space exploration					
Attentive	6	7	6	4	4
Interested	10	13	11	13	17
Residual	84	80	83	83	79

SOURCE: Longitudinal Study of American Youth.

much appears to have been declining. Support for increased spending is higher among those who are interested in space exploration. Among the general public overall, support for space-exploration funding tends to be low, especially in comparison with support for other possible spending priorities. Those trends—generally positive views of space exploration and human spaceflight programs but low support for funding increases and low levels of public engagement—have held true for a few decades. Although the level of interest spiked around the time of some major space events, it never fluctuated by more than about 10 percentage points.

No particular rationale for space exploration consistently attracts support from a clear majority of the American people. One survey discussed above found that a little over half the respondents gave reasons for space exploration that could be described as "increasing knowledge and searching for other life forms," but most surveys found responses divided among several rationales, none of them garnering strong agreement from a majority of the respondents.

3.2 STAKEHOLDER SURVEY

Although space exploration has had a substantial effect on American society, space policy can be described as an issue that has low visibility in public discourse.[8] Because relatively few members of the public pay close attention to space exploration, policy leaders and stakeholder groups are likely to have more effect than the general public on policy decisions in this realm.

In 2003, Miller conducted a survey of policy leaders on the topic of space exploration.[9] To construct a sampling frame for the survey, Miller identified the positions likely to be influential in the formulation of space policy and then identified the individuals who occupied these positions. Miller argued that because space policy is closely intertwined with science policy, in addition to space policy leaders, policy leaders in non-space-related fields also have substantial influence on decisions related to space policy. On the basis of that conceptualization, his study included leading scientists and engineers in research universities and selected corporations; the leadership of major universities, corporations, and organizations active in scientific or space-related work; scientific, engineering, and other professional societies relevant to space science and engineering; and the leadership of relevant voluntary associations.

To learn more about the views of stakeholders, the panel conducted a survey that relied on Miller's general approach for building a sampling frame. For the purposes of this study, stakeholders were defined as those who may reasonably be expected to have an interest in NASA's programs and to be able to exert some influence over its direction. In addition to the groups typically viewed as NASA's stakeholders,[10] scientists from non-space-related fields were added as a group of interest because of their influence on the formulation of space policy, as described above. A brief description of each of the eight stakeholder groups is provided in Table 3.6.

The sampling frame for the stakeholder survey was built by identifying leadership positions within each of the stakeholder groups of interest and then identifying the individuals who were occupying the positions. Within each group, a systematic random sample was drawn for the survey. Table 3.6 shows the initial sampling frames and sample sizes and the response rates for each group. Because "NASA's stakeholders" are not a clearly defined population and because the stakeholders were selected by using sampling frames that were reasonable and convenient rather than comprehensive, the results from this survey cannot be generalized to all stakeholders. In addition, no attempt was made to weight the data to compensate for differences in the sizes of the various groups or overlaps in their composition. As a result, combining the responses from the groups does not represent the universe of all NASA stakeholders. Nevertheless, the sample for each group was a probability sample, and this method provides a broader and more diverse perspective on stakeholder views than a nonprobability sample would. A detailed description of the methods used to conduct the survey, including more information about the sampling frame, can be found in Appendix C.

---

[8] Miller, *Space Policy Leaders and Science Policy Leaders in the United States,* 2004.

[9] Ibid.

[10] B. Cameron and E.F. Crawley, "Architecting Value: The Implications of Benefit Network Models for NASA Exploration," paper presented at the AIAA SPACE 2007 Conference & Exposition, Long Beach, Calif., 2007.

TABLE 3.6 Stakeholders Included in the Survey

Stakeholders	Description	Size of Sampling Frame	Size of Sample	Number of Complete Cases	Response Rate (AAPOR 3),[a] %
Economic/ industry	For-profit companies that interact directly or indirectly with NASA (e.g., contractors and aerospace firms); states that have an economic interest in the issue	573	384	104	28.6
Space scientists/ engineers	Top scientists and engineers in relevant fields	919	395	261	67.1
Young space scientists and engineers	Up-and-coming scientists and engineers in relevant fields	549	195	90	49.7
Other scientists and engineers	Top scientists and engineers in non-space-related fields	6,106	396	201	51.3
Higher education	Deans and heads of academic departments that could reasonably be expected to have some students with an interest in space	766	399	294	74.1
Security/ defense/ foreign policy	Top experts and researchers working in fields related to national security/defense who can reasonably be expected to have an interest in space issues	115	110	71	66.4
Space writers and science popularizers	Space writers, science journalists, bloggers, planetarium and public observatory directors	1,096	99	53	56.4
Space advocates	Officers and board members of space-advocacy groups	267	96	46	51.7
Total (not including duplicates across strata)		10,391	2,054	1,104	55.4

[a] The AAPOR 3 response rate estimates what proportion of cases of unknown eligibility is actually eligible and is based on a formula recommended by the American Association for Public Opinion Research, http://www.aapor.org/AM/Template.cfm?Section=Standard_Definitions2&Template=/ CM/ContentDisplay.cfm&ContentID=3156.

This section discusses the findings of the stakeholder survey, including stakeholder views about the rationales for space exploration in general and human spaceflight specifically, and possible directions for NASA's human spaceflight program. Appendix D contains the paper questionnaire used for the survey (further information about the data-collection instruments, including the two versions of the questionnaires, is included in Appendix C with the detailed description of the survey methods). We do not include here a detailed discussion of each question by stakeholder group but instead summarize the data to highlight comparisons guided by the content of the questions. Appendix E includes the frequency distributions of the responses to each survey question by stakeholder group.

3.2.1 Characteristics of the Respondents

Because the stakeholder survey focused on the leadership of relevant organizations and leading scientists in the field, the respondents tended to be older than the general population, to be more likely to have advanced degrees, and to be more likely to be male (Table 3.7). About half the respondents were academics, and about one-fourth held managerial or professional positions. About 11 percent described themselves as nonteaching scientists and about 9 percent as engineers.

Respondents were asked whether they were involved in space-related work, and about half said that they were very involved or somewhat involved. About 20 percent of the sample members were scientists who were selected specifically from non-space-related fields, but the sample also included others who were part of the leadership of organizations that do space-related work, regardless of whether the respondents personally characterized themselves as involved in space-related work. The sample also included deans, chairs, and department heads of academic programs in fields related to space but not necessarily only space. Many of these respondents were not conducting space-related work themselves. About 40 percent of those who said that they were involved in space-related work said that they were very or somewhat involved in work related to human spaceflight.

TABLE 3.7 Characteristics of the Respondents

Characteristics	Count	%
Age, years		
Under 40	127	12
40-49	109	10
50-59	292	26
60-69	273	25
70 or over	277	25
No answer	26	2
Sex		
Male	936	85
Female	152	14
No answer	16	2
Education attainment		
High school or some college	12	1
Bachelor's degree	95	9
Master's degree	147	13
Professional degree	46	4
Doctorate	791	72
No answer	13	1
Current position		
Postsecondary educator	460	50
Scientist in a nonteaching position	105	11
Engineer	82	9
Managerial or professional	224	24
Other	53	6
Not employed/retired	158	14
No answer	20	2
Involved in space-related work		
Very involved	233	21
Somewhat involved	325	29
Not involved	532	48
No answer	14	1
Involved in work related to human spaceflight		
Very involved	75	13
Somewhat involved	150	27
Not involved	333	60
No answer	0	0

Unless otherwise noted, for the purposes of the analyses presented in this report, the term *space scientist/ engineer* will be used to refer to space scientists and engineers who reported in the survey that they were involved in space-related work (n = 198), regardless of whether they were initially included in the sample as part of the "space scientists and engineers" group or the "other scientists and engineers" group. The term *nonspace scientist/ engineer* will be used to refer to scientists and engineers in the "space scientists and engineers" and "other scientists and engineers" groups who said that they were not involved in space-related work (n = 264). Although the decision to reassign some cases from one stakeholder group to another on the basis of self-reported responses might seem somewhat unconventional, it helps to compensate for the sparseness of the information that was available from the initial sampling frame.

3.2.2 Rationales for Space Exploration and Human Space Exploration

One of the primary goals of the survey was to understand stakeholder views on the rationales for space exploration and human space exploration. Before being presented with a list of rationales that commonly surfaces in the literature and discussions in the space community, respondents were asked what they considered to be the main reasons for (and against) space exploration and human spaceflight in an open-ended format. The results are presented in Table 3.8.

In the case of space exploration, reasons related to pursuing knowledge and scientific understanding were the most frequently mentioned. Technological advances and arguments related to a basic human drive to explore new frontiers were also mentioned spontaneously by about one-third of the respondents.

In terms of *human* space exploration, the rationales were more divided. The three most frequently mentioned arguments, each offered by about one-third of the respondents, were that humans are able to accomplish more than robots in space, that humans have a basic drive to explore new frontiers, and that human space exploration

TABLE 3.8 Reasons for Space Exploration and Human Spaceflight (Open-Ended, All Mentions)

Reason for Space Exploration	% Mentioned	Reason for Human Spaceflight	% Mentioned
Knowledge and scientific understanding	78	Humans can accomplish more than robots	32
Technological advances	35	Basic human drive to explore new frontiers	30
Basic human drive to explore new frontiers	32	Knowledge and scientific understanding	28
Human economic activity beyond Earth	11	Future settlements in space	18
Future settlements in space	8	Technological advances	14
U.S. prestige	7	Public support	9
Careers in science, technology, mathematics, and engineering	5	U.S. prestige	8
Search for signs of life	5	Careers in science, technology, mathematics, and engineering	6
National security	5	Human economic activity beyond Earth	5
Prevent threats from space	3	International cooperation	2
International cooperation	3	National security	2
Commercial space travel	0	Search for signs of life	2
Other	8	Commercial space travel	0
None/No compelling reason for space exploration	0	Prevent threats from space	0
		Other	7
		None/No compelling reason for human space exploration	6

can improve knowledge and scientific understanding. Working toward the establishment of future settlements in space was mentioned by about one-fifth of the respondents.

For both space exploration and human spaceflight, the same pattern of rationales emerged from the responses of those who said that they were involved in space-related work and from those who said that they were not.

To understand the pattern and strength of the rationales, for each response to the open-ended questions about the main reasons for space exploration and human spaceflight the panel recorded the rationale mentioned first (Table 3.9). For space exploration, in 60 percent of cases, the first rationale provided contained a reference to knowledge and scientific understanding. In the case of the main reasons provided for human spaceflight, the same pattern emerged for first mentions as for responses to the question overall: first mentions were divided among the idea that humans can accomplish more than robots (23 percent), a basic human drive to explore (21 percent), and knowledge and scientific understanding (19 percent).

After being asked to describe the rationales in an open-ended format, respondents were presented with a list of rationales historically given for space exploration and were asked to indicate for each whether they consider it "very important," "somewhat important," "not too important," or "not at all important" as a reason for space exploration in general and for human spaceflight in particular.

"Expanding knowledge and scientific understanding" emerged as the rationale that the overwhelming majority of the respondents felt was a "very important" reason for space exploration (Table 3.10). A majority of the respondents also considered driving technological advances, inspiring young people to pursue careers in STEM fields, and satisfying a basic human drive to explore new frontiers to be very important. Except for the STEM-careers rationale, that pattern is comparable with the one that emerged when respondents were asked to describe the rationales in an open-ended format. Well over half the respondents (62 percent) considered space exploration's potential to inspire STEM careers "very important" when presented with the list, but only 5 percent mentioned it spontaneously.

TABLE 3.9 Reasons for Space Exploration and Human Spaceflight (Open-Ended, First Mentions)

Reason for Space Exploration	% Mentioned First	Reason for Human Spaceflight	% Mentioned First
Knowledge and scientific understanding	60	Humans can accomplish more than robots	23
Basic human drive to explore new frontiers	21	Basic human drive to explore new frontiers	21
Technological advances	9	Knowledge and scientific understanding	19
Future settlements in space	2	Future settlements in space	10
U.S. prestige	2	None/No compelling reason for human space exploration	6
Other	2	Public support	6
Human economic activity beyond Earth	2	Other	4
Careers in science, technology, mathematics, and engineering	1	Technological advances	4
Search for signs of life	1	U.S. prestige	4
National security	1	Careers in science, technology, mathematics, and engineering	2
Prevent threats from space	0	Human economic activity beyond Earth	1
International cooperation	0	National security	0
None/No compelling reason for human space exploration	0	International cooperation	0
Commercial space travel	0	Search for signs of life	0
		Commercial space travel	0
		Prevent threats from space	0

TABLE 3.10 Reasons for Space Exploration and Human Spaceflight (% Very Important)

Reason for Space Exploration	% Very Important	Reason for Human Spaceflight	% Very Important
Expanding knowledge and scientific understanding	84	Inspiring young people to pursue careers in science, technology, engineering, and mathematics	47
Driving technological advances	66	Satisfying a basic human drive to explore new frontiers	45
Inspiring young people to pursue careers in science, technology, engineering, and mathematics	62	Driving technological advances	40
Satisfying a basic human drive to explore new frontiers	60	Expanding knowledge and scientific understanding	38
Maintaining our national security	41	Paving the way for future settlements in space	31
Creating opportunities for international cooperation	33	Enhancing U.S. prestige	27
Enhancing U.S. prestige	29	Creating opportunities for international cooperation	26
Paving the way for future settlements in space	22	Paving the way for commercial space travel	21
Extending human economic activity beyond Earth	21	Extending human economic activity beyond Earth	18
Paving the way for commercial space travel	17	Maintaining our national security	17

The stakeholder groups differed in their overall enthusiasm for the rationales, but the same four rationales received the most support in all the groups, and "expanding knowledge and scientific understanding" received more support in all the groups than any of the other rationales.

In the case of *human* spaceflight, no rationale drew a majority of responses as very important even when respondents were given the list in a closed-ended format. Inspiring young people to pursue STEM careers and satisfying a basic human drive to explore new frontiers were the rationales that were most frequently cited as very important, followed by driving technological advances and expanding knowledge and scientific understanding, but no rationale was viewed as a very important reason for human spaceflight by a majority of the respondents. Again, the top rationales tended to be the same among the different stakeholder groups, although space scientists and engineers were more likely to consider enhancing U.S. prestige a very important reason (33 percent said this), and space advocates were more likely to consider paving the way for future settlements in space a very important reason (61 percent) than expanding knowledge and scientific understanding.

Those who selected more than one rationale as "very important" were asked to indicate which was "most important." For space exploration in general, 58 percent of these respondents chose "expanding knowledge and scientific understanding" and mentioned all the other rationales far less frequently (Table 3.11). In the case of human spaceflight, views of the "most important" rationale were again divided: 22 percent of the respondents said "satisfying a basic human drive to explore new frontiers," 18 percent said "inspiring young people to pursue careers in science, technology, math and engineering," and 16 percent said "expanding knowledge and scientific understanding."

To probe the rationales from an additional perspective, the panel asked respondents to describe in an open-ended format what they thought *would be lost* if NASA's human spaceflight program were terminated (Table 3.12). The most frequently cited argument (by one-fourth of respondents) was that U.S. prestige would suffer. About 15 percent of the respondents provided answers that could be summed up as "nothing would be lost." Nonspace scientists/engineers were most likely to say that nothing would be lost (29 percent), and young space scientists were least likely to say that nothing would be lost (1 percent).

Respondents were given an opportunity to voice arguments *against* both space exploration and human spaceflight. About one-fourth either did not provide a response to the question about reasons against space exploration or said that there are no reasons (or no compelling reasons) against it. Almost all the remaining responses focused on the costs involved, either in an absolute sense or compared with other potential uses for the money. For human space exploration, about 23 percent of the respondents either declined to provide an answer or argued that there are

TABLE 3.11 Reasons for Space Exploration and Human Spaceflight (% Most Important)

Reason for Space Exploration	% Most Important	Reason for Human Spaceflight	% Most Important
Expanding knowledge and scientific understanding	58	Satisfying a basic human drive to explore new frontiers	22
Satisfying a basic human drive to explore new frontiers	11	Inspiring young people to pursue careers in science, technology, engineering, and mathematics	18
Driving technological advances	11	Expanding knowledge and scientific understanding	16
Inspiring young people to pursue careers in science, technology, engineering, and mathematics	8	Paving the way for future settlements in space	14
Maintaining our national security	5	Driving technologic advances	11
Paving the way for future settlements in space	3	Creating opportunities for international cooperation	4
Extending human economic activity beyond Earth	3	Enhancing U.S. prestige	4
Creating opportunities for international cooperation	1	Paving the way for commercial space travel	4
Enhancing U.S. prestige	0	Extending human economic activity beyond Earth	3
Paving the way for commercial space travel	0	Maintaining our national security	3

TABLE 3.12 What Would Be Lost If NASA's Human Spaceflight Program Were Terminated (Open-Ended, All Mentions)

What Would Be Lost	% Mentioned
U.S. prestige	26
Knowledge and scientific understanding	20
Basic human drive to explore new frontiers	17
Technological advances	16
Public support	11
Investment we have made so far	10
Science, technology, engineering, and mathematics careers	10
Future settlements in space	6
National security	4
Ability to accomplish what robots cannot	3
Human economic activity beyond Earth	3
International cooperation	3
Search for signs of life	0
Commercial space travel	0
Prevent threats from space	0
Other	14
Nothing would be lost	15

no reasons against human space exploration. The majority of the respondents (60 percent) focused on the costs of human space exploration, 39 percent mentioned the risks involved, and 28 percent argued that it would be better to focus on robotic space exploration.

3.2.3 Views on a Course for the Future

Respondents were asked to consider what goals a worthwhile and feasible U.S. human space-exploration program might work toward over the next 20 years. They were given a list of possible projects that NASA could pursue and asked to indicate how strongly they favored or opposed each of them. Although describing all the nuances associated with the options was not feasible in the survey format, the options were presented with approximate overall costs to provide some context in terms of the scale of the projects.

Overall, the option that received the most "strongly favor" responses was continuing with low-Earth orbit (LEO) flights to the ISS until 2020, followed by extending the ISS to 2028 and conducting orbital missions to Mars to teleoperate robots on the surface (Table 3.13). When the "strongly favor" and "somewhat favor" responses are combined, the same three options emerge as the top three preferences, with no difference between the degree of support for extending the ISS to 2028 and conducting orbital teleoperated missions to the Mars surface.

Tables 3.14 through 3.16 compare the preferences of the main stakeholder groups. In Tables 3.15 and 3.16 scientists in space-related fields include space scientists/engineers, young space scientists, and academics who said that they were involved in space-related work. Scientists in non-space-related fields are respondents in the same three groups who said that they were not involved in space-related work.

Those who said that they were involved in space-related work, and in particular those involved in human spaceflight-related work, were generally more likely to favor most programs strongly, but continuing with LEO flights to the ISS until 2020 was the option that received the most "strongly favor" responses in all three groups (Table 3.14). Overall, extending the ISS to 2028 was the second option with the most "strongly favor" responses, although those involved in space-related work that did not include human spaceflight were equally likely to favor conducting orbital missions to Mars to teleoperate robots on the surface strongly. Conducting orbital missions to Mars to teleoperate robots on the surface was the third-most popular option among those who were not involved in space-related work as well. Those who were involved in human-spaceflight-related work were more likely to favor landing humans on Mars strongly than conducting orbital missions to Mars to teleoperate robots on the surface.

Among scientists in space-related fields and non-space-related fields, the option that received the most "strongly favor" responses was continuing with LEO flights to the ISS until 2020, followed by extending the ISS to 2028 and conducting orbital missions to Mars to teleoperate robots on the surface (Table 3.15). Among the industry respondents, the options that received the most support were continuing with LEO flights to the ISS until 2020 and extending the ISS until 2028, followed by establishing outposts on the Moon, conducting orbital missions to Mars to teleoperate robots on the surface, and landing humans on Mars. The priorities in the defense community generally reflected the ones indicated by scientists/engineers. Space advocates and science popularizers were generally more likely to favor most options strongly, but respondents in this group were also more likely to support continuing with LEO flights to the ISS until 2020. Space advocates and science popularizers were about as likely to support extending the ISS to 2028 as landing humans on Mars; with more than half the respondents in this group strongly favored these options.

When the "strongly favor" and "somewhat favor" responses are combined (Table 3.16), similar patterns emerge, especially with respect to the program options that received the most support overall (continuing with LEO flights to the ISS until 2020, followed by extending the ISS to 2028 and conducting orbital missions to Mars to teleoperate robots on the surface).

Those who were under 40 years old were generally more likely to favor most projects strongly than those who were 40 years old and older (Table 3.17). Continuing with LEO flights to the ISS until 2020 and extending the

TABLE 3.13 Goals for NASA's Human Spaceflight Program over the Next 20 Years

	Strongly Favor, %	Strongly Favor or Somewhat Favor, %
Continue with LEO flights to the ISS until 2020	45	79
Extend the ISS to 2028	37	67
Conduct orbital missions to Mars to teleoperate robots on the surface	31	66
Land humans on Mars	25	48
Establish outposts on the Moon	22	49
Return to the Moon and explore more of it with short visits	22	56
Send humans to a near-Earth asteroid in its native orbit	19	50
Establish a human presence (base) on Mars	12	30

TABLE 3.14 Goals for NASA's Human Spaceflight Program over the Next 20 Years by Involvement in Space-Related Work

	Strongly Favor, %			Strongly Favor or Somewhat Favor, %		
	Involved in human-spaceflight-related work (n = 225)	Involved in space-related work but not human spaceflight (n = 333)	Not involved in space-related work (n = 532)	Involved in human-spaceflight-related work (n = 225)	Involved in space-related work but not human spaceflight (n = 333)	Not involved in space-related work (n = 532)
Continue with LEO flights to the ISS until 2020	55	45	42	82	78	80
Extend the ISS to 2028	51	33	35	75	65	68
Send humans to a near-Earth asteroid in its native orbit	20	20	18	55	51	49
Return to the Moon and explore more of it with short visits	28	20	21	64	56	54
Establish outposts on the Moon	40	20	17	66	50	42
Conduct orbital missions to Mars to teleoperate robots on the surface	34	33	30	74	65	66
Land humans on Mars	48	21	18	70	47	41
Establish a human presence (base) on Mars	28	12	6	53	29	22

TABLE 3.15 Goals for NASA's Human Spaceflight Program over the Next 20 Years: Options Strongly Favored among Main Stakeholder Groups

	Strongly Favor, %				
	Scientists/engineers in space-related fields (n = 373)	Scientists/engineers in non-space-related fields (n = 464)	Industry (n = 104)	Defense (n = 71)	Space advocates and science popularizers (n = 99)
Continue with LEO flights to the ISS until 2020	43	40	56	47	65
Extend the ISS to 2028	33	35	51	38	57
Send humans to a near-Earth asteroid in its native orbit	21	18	21	11	19
Return to the Moon and explore more of it with short visits	18	19	25	21	41
Establish outposts on the Moon	22	16	35	18	44
Conduct orbital missions to Mars to teleoperate robots on the surface	30	30	34	30	41
Land humans on Mars	25	16	35	27	57
Establish a human presence (base) on Mars	13	6	20	7	38

TABLE 3.16 Goals for NASA's Human Spaceflight Program over the Next 20 Years: Options Favored (Strongly Favored and Somewhat Favored) among Main Stakeholder Groups

	Strongly Favor or Somewhat Favor, %				
	Scientists/ engineers in space-related fields (n = 373)	Scientists/ engineers in non-space-related fields (n = 464)	Industry (n = 104)	Defense (n = 71)	Space advocates and science popularizers (n = 99)
Continue with LEO flights to the ISS until 2020	76	78	83	82	86
Extend the ISS to 2028	62	66	80	73	81
Send humans to a near-Earth asteroid in its native orbit	52	48	50	42	58
Return to the Moon and explore more of it with short visits	56	52	64	54	72
Establish outposts on the Moon	51	40	60	45	73
Conduct orbital missions to Mars to teleoperate robots on the surface	65	64	80	63	76
Land humans on Mars	50	38	63	52	74
Establish a human presence (base) on Mars	32	20	43	31	60

TABLE 3.17 Goals for NASA's Human Spaceflight Program over the Next 20 Years: Those Under 40 Years Old vs Those 40 Years Old and Older

	Strongly Favor, %	
	Under 40 years old (n = 127)	40 years old and older (n = 951)
Continue with LEO flights to the ISS until 2020	59	44
Extend the ISS to 2028	56	35
Send humans to a near-Earth asteroid in its native orbit	34	17
Return to the Moon and explore more of it with short visits	24	22
Establish outposts on the Moon	40	20
Conduct orbital missions to Mars to teleoperate robots on the surface	35	31
Land humans on Mars	42	23
Establish a human presence (base) on Mars	25	11

ISS to 2028 were the two options with the most "strongly favor" responses among respondents who were under 40 years old, followed by establishing outposts on the Moon and landing humans on Mars.

To probe priorities from an additional perspective, respondents were asked to rate the importance of several possible projects or activities for NASA over the next 20 years. Overall, those who were involved in space-related work, especially in human-spaceflight-related work, rated most items higher than those who were not (Table 3.18).

Those who were involved in human-spaceflight-related work were more likely to say that it was very important to make the investment necessary to sustain a vigorous program of human space exploration than those who were not. Among those who were involved in human-spaceflight-related work, 56 percent felt that making the investment

TABLE 3.18 Important for NASA to Do over the Next 20 Years: Those Who Are Involved in Space-Related Work vs Those Who Are Not

	Very Important, %		
	Involved in human-spaceflight-related work (n = 225)	Involved in space-related work but not human spaceflight (n = 333)	Not involved in space-related work (n = 532)
Make the investments necessary to sustain a vigorous program of human space exploration	56	29	26
Make the investments necessary to sustain a vigorous program of robotic space exploration	57	76	62
Maintain the ISS as a laboratory for scientific research	46	33	36
Limit human space exploration to Earth-orbit missions while maintaining robotic missions for exploring in and beyond the solar system	16	37	30
Plan for a manned mission to Mars	37	20	15
Improve orbital technologies, such as weather and communication satellites	47	71	70
Maintain world leadership in human space exploration	53	35	31
Expand space exploration collaboration with other countries	44	47	41

necessary to sustain a vigorous program of human space exploration was very important, and about as many felt that way about the investment necessary to sustain a vigorous program of robotic space exploration. Among those who were involved in space-related work but not human spaceflight, 29 percent said that the investment in human space exploration was very important and 76 percent said that the investment in robotic space exploration was very important. The responses of those who were not involved in space-related work were closer to the responses of those who were involved in space-related work, but not with human spaceflight.

Those who were involved in human-spaceflight-related work were more likely to say that maintaining the ISS as a laboratory for scientific research was very important and that planning a crewed mission to Mars was very important. They also considered maintaining world leadership in human space exploration to be more important than did those who were involved in space-related work but not human spaceflight or those who were not involved in space-related work. However, those involved in space-related work but not human spaceflight and those who were not involved in space-related work were more likely to rate improving orbital technologies, such as weather and communication satellites, very important. Expanding space collaboration with other countries was the item that received comparable levels of support in all three of those groups.

3.2.4 Other Findings

Some other noteworthy findings of the survey include the following:

- *International collaboration.* A majority (59 percent) said that NASA should conduct human space-exploration missions beyond LEO mainly or exclusively as part of an international collaboration that includes both current partners and new and emerging space powers, very few (8 percent) said that NASA should conduct human space-exploration missions beyond LEO mainly or exclusively as U.S.-only missions, 13 percent said that NASA should do it mainly or exclusively in collaboration with current international partners (such as ISS partners), and 17 percent said that NASA should not conduct human space-exploration missions beyond LEO at all.

- *Robotic versus human missions.* The majority (64 percent) felt that NASA should focus on a combination of both robotic and human missions, and 34 percent said that NASA should focus mainly or exclusively on robotic space exploration.
- *Role of the private sector.* The overwhelming majority of respondents thought that NASA should take the lead in space exploration for scientific research but that the private sector should take the lead in space travel by private citizens (Table 3.19). The majority of the respondents said that the private sector should also take the lead in extending human economic activity beyond Earth. About half the respondents said that NASA should take the lead in working toward establishing an off-planet human presence, about one-third felt that neither should do this, and 20 percent said that the private sector should take the lead.

3.2.5 Correlates of Support for Human Spaceflight

The panel created a scale to measure overall support for human spaceflight on the basis of responses to five of the questionnaire items: about support for establishing outposts on the Moon (question 10e), landing humans on Mars (10g), establishing a human presence on Mars (10h), sustaining a vigorous program of human space exploration (11b), and maintaining world leadership in human space exploration (11e). Responses to those five items were highly intercorrelated. Table 3.20 shows the averages on this scale for various subgroups of those who completed the survey. The averages are on a 4-point scale, with 4 representing the highest possible level of support for human spaceflight and 1 representing the lowest level.

Support for human spaceflight goes up with involvement in human space exploration and declines steadily with age. A multivariate model shows that those two variables are significantly related to overall support for human spaceflight. In addition, the stakeholder groups differ in their support for human spaceflight: advocates and popularizers are highest in support of human spaceflight, and non-space scientists/engineers are least supportive.

3.2.6 Summary of Findings of the Stakeholder Survey

One of the primary goals of the survey was to understand stakeholder views of the rationales for space exploration and human spaceflight. There was substantial agreement among the respondents on the rationales for space exploration, but views of the rationales for human spaceflight were more divided.

For space exploration, "expanding knowledge and scientific understanding" emerged as the rationale shared by the overwhelming majority of the respondents. Although there were a few other rationales that more than half the respondents agreed were "very important" reasons for space exploration, when asked what they considered to be the *most important* rationale they mentioned "expanding knowledge and scientific understanding" by far the most frequently. That was also the case when respondents were asked to describe the reasons for space exploration in their own words, and the rationale was equally dominant among those who were involved in space-related work and those who were not.

None of the rationales traditionally given for *human* spaceflight garnered agreement from a majority of the respondents. When asked what the most important rationale was from among the rationales traditionally given for human spaceflight, "satisfying a basic human drive to explore new frontiers" was the reason selected by the

TABLE 3.19 Who Should Take the Lead on Each of These Activities over the Next 20 Years

	NASA, %	Private Sector, %	Neither, %
Space exploration for scientific research	95	2	1
Extending human economic activity beyond Earth	16	68	14
Space travel by private citizens	1	85	12
Establishing an off-planet human presence	48	20	30

TABLE 3.20 Mean Levels of Support of Human Spaceflight, by Selected Subgroups

	Mean[a] (n)
Group	
Scientists/engineers in space-related fields	2.55 (370)
Scientists/engineers in non-space-related fields	2.32 (461)
Industry	2.91 (98)
Defense	2.55 (64)
Space advocates and popularizers	3.22 (95)
Involvement with human space flight	
Not involved in space exploration	2.36 (530)
Not involved in human space exploration	2.50 (333)
Somewhat involved in human space exploration	2.95 (150)
Very involved in human space exploration	3.22 (75)
Age, years	
39 or under	3.10 (127)
40-49	2.89 (109)
50-59	2.63 (292)
60-69	2.56 (273)
70 or over	2.05 (273)
Education	
Bachelor's degree or less	3.09 (107)
Master's or professional	2.79 (192)
Doctorate	2.41 (788)
Sex	
Men	2.53 (932)
Women	2.65 (152)

[a] Higher numbers indicate greater support (1 represents the lowest possible value, 4 the highest).

highest number of respondents, but fewer than one-fourth agreed that it was the most important reason; the rest of the responses were split among a number of other rationales.

When asked to describe the reasons for human spaceflight in the form of an open-ended question, about one-third of the respondents provided reasons that can be summarized as "humans can accomplish more than robots in space." Somewhat fewer than one-third argued that the reason was to satisfy a basic human drive to explore new frontiers. "Expanding knowledge and scientific understanding" was mentioned by about one-fourth as a reason for human spaceflight.

Those who were involved in space-related work were more likely to provide additional rationales for human spaceflight in their responses to the open-ended question, but the rationales were endorsed by a relatively small percentage of respondents overall (even among those who were involved in space-related work). Rationales that were mentioned by less than 10 percent of the respondents spontaneously included the following: reduced public support for space-exploration programs in general without the visibility of the human-spaceflight program, enhancing U.S. prestige, national security, international cooperation, preventing threats from space, inspiring STEM careers, expanding human economic activity beyond Earth, searching for signs of life, and paving the way for commercial space travel. Some of those were mentioned spontaneously by only a handful of respondents.

To provide an additional perspective on the reasons for human spaceflight, respondents were asked in an open-ended format what they thought would be lost if NASA's human spaceflight program were terminated. There was no majority agreement among the respondents when the question was asked that way either. The most commonly mentioned loss was national prestige, which was mentioned by only one-fourth of respondents; 15 percent said that nothing would be lost.

When asked about possible goals for the next 2 decades of NASA's human space-exploration program in the context of costs, the projects that were favored by a majority of the respondents included LEO flights to the ISS until 2020, extending the ISS to 2028, conducting orbital missions to Mars to teleoperate robots on the surface, and returning to the Moon to explore more of it with short visits.

Overall, large majorities thought that it was very important for NASA to be involved in nonhuman spaceflight, such as robotic space exploration and improving orbital technologies, but responses to another question indicate that the majority favored a program that includes both robotic and human space exploration. A majority said that NASA should conduct human space-exploration missions beyond LEO mainly or exclusively as part of an international collaboration that includes both current partners and new and emerging space powers.

Key predictors of support for human spaceflight are age and whether a person is involved in work related to human spaceflight. In addition, the stakeholder groups differed in their support of human spaceflight: advocates and popularizers were highest in their support of human spaceflight, and non-space scientists/engineers were lowest.

4

Technical Analysis and Affordability Assessment of Human Exploration Pathways

4.1 INTRODUCTION AND OVERVIEW

Sending humans to destinations beyond low Earth orbit (LEO) is a technologically, programmatically, and politically complicated endeavor. As discussed elsewhere in this report, there are frequently cited but difficult-to-substantiate arguments that, despite the cost and risk to human life, such endeavors are justified by various national and more general societal benefits. In any resource-constrained environment, it is responsible to attempt not just to identify the potential benefits but also to understand the level of proposed or required investments and the difficulty of the proposed tasks.

Given the complexity of human space exploration—that is, human spaceflight beyond LEO—and the fact that U.S. goals in human spaceflight have changed on timescales much shorter than the time it would take to accomplish the goals, it makes sense to decompose a human spaceflight program into smaller building blocks. This would allow the building blocks to be assembled in various configurations that allow the changed goals to be addressed without analysis ab initio.

A "capability-based" approach to space exploration focuses research and technology development resources on systems and capabilities that are expected to be of value in the future with no particular mission or set of missions in mind. The process of selecting future missions would then tend to favor those missions that could make use of the systems and capabilities that have been developed. The Asteroid Redirect Mission (ARM) is an example of this process in action. A "flexible-path" approach is a more sophisticated version of capability-based planning that considers what destinations might be desirable. In contrast, a "pathway-based" approach would commit the U.S. human spaceflight program to a pathway with a specific sequence of missions, usually of increasing difficulty and complexity, that target specific exploration goals that are typically tied to various destinations that humans may explore. A pathways approach would facilitate continuity of development of required systems for increased capability and efficiency.

NASA is developing modular and general-purpose systems. Although NASA has characterized this approach as capability-based, the systems may or may not be supportive of a pathway-based approach. Instead of pursuing a "capability-based" or "flexible path" approach where no specific sequence of destinations is specified. The committee's prioritized recommendations (see Chapter 1), if adopted, would lead to a NASA commitment to design, pursue, and maintain the execution of an exploration pathway beyond LEO that leads toward a clear "horizon goal" (that is, the most distant destination that is considered feasible in the foreseeable future). The committee is not recommending one pathway over another, but it does recommend (see Chapter 1) that NASA "maintain long-term focus on Mars as the 'horizon goal' for human space exploration."

The rest of this chapter contains three major sections:

4.2 Technical Requirements
4.3 Technology Programs
4.4 Key Results

The Technical Requirements section begins by defining possible destinations for human spaceflight. For the foreseeable future, limitations of human physiology and space exploration technology limit potential destinations to the Earth-Moon system, Earth-Sun Lagrange points, near-Earth asteroids, and Mars. Design reference missions (DRMs) to these destinations have been prepared by NASA, the International Space Exploration Coordination Group (ISECG), and others.

This chapter assesses three specific, notional pathways as examples to illustrate the various tradeoffs among schedule, development risk,[1] affordability, and decommissioning date of the International Space Station (ISS). (The analysis presented in this chapter was completed before the administration announced its intention to extend the operation of the ISS to 2024; the bounding cases of 2020 and 2028 presented in this report nonetheless well illustrate the impact of the ISS decommissioning date on human spaceflight beyond LEO.) Although other pathways are possible, the ones chosen for exposition here span the likely programmatic space well enough to provide insight into affordability and technical difficulty.

All three of the pathways terminate with a human mission to the most challenging destination that is still technically feasible: the Mars surface. Depending on practical factors, an actual human spaceflight program might have to take an off-ramp to an intermediate destination before the final destination is reached.

Each pathway to Mars includes three to six different DRMs, as follows:

- *ARM-to-Mars pathway*
  — ARM
  — Martian moons
  — Mars surface
- *Moon-to-Mars pathway*
  — Lunar surface sortie
  — Lunar surface outpost
  — Mars surface
- *Enhanced Exploration pathway*
  — Earth-Moon L2[2]
  — Asteroid in native orbit
  — Lunar surface sortie
  — Lunar surface outpost
  — Martian moons
  — Mars surface

Completing any of the above pathways would require a variety of mission elements. For example, a human mission to the Mars surface would require development of 11 primary mission elements, such as heavy-lift launch vehicles, deep-space habitats, and pressurized surface mobility systems. Eight additional mission elements would be necessary to complete all six of the other DRMs that appear in one or more of the roadmaps. Three of the additional mission elements are transitional in that they contribute directly to the development of one of the 11 primary

---

[1] Development risk is the risk that a program will encounter large increases in cost and/or schedule during the program's development phase.

[2] As described later in the chapter, the Earth-Moon L2 point is a particular point in space defined by its position relative to the Earth and Moon.

mission elements. For example, the lunar orbital outpost is not needed for a Mars surface mission, but it would contribute directly to the development of the deep-space habitats that would be needed. There are also five dead-end mission elements. Although necessary for completing one or more of the DRMs, the advanced capabilities of these mission elements have little or no applicability to the Mars surface mission.

Requirements for completing the pathways can also be framed as capabilities. A wide range of capabilities was assessed in terms of the technical challenges, capability gap, regulatory challenges, and cost and schedule challenges that would need to be overcome for their development to be completed. This assessment determined that current research and development programs would need to address the 10 capabilities below as a high priority, with particular emphasis on the first three:

- Mars entry, descent, and landing (EDL)
- Radiation safety
- In-space propulsion and power
- Heavy-lift launch vehicles
- Planetary ascent propulsion
- Environmental control and life support system
- Habitats
- Extravehicular activity (EVA) suits
- Crew health
- In-situ resource utilization (ISRU) (with the Mars atmosphere as a raw material)

For a pathway to be affordable, its cost profile must fit within the projected human spaceflight budget profile. Given the uncertain nature of federal budget projections and to show the impact of budget profile on other pathway characteristics, the affordability of each pathway is assessed in terms of three scenarios:

- A schedule-driven affordability scenario, in which the pace of progress is determined by the rate at which necessary development programs can be completed. This scenario would require rapid growth in NASA's human spaceflight budget.
- A budget-driven affordability scenario, in which the pace of progress is limited by a human spaceflight budget that grows with inflation. This scenario would result in an operational tempo (in terms of the overall launch rate of crewed and cargo missions and/or the frequency of crewed missions in particular) that is far below historical norms.
- An operationally viable affordability scenario, in which the pace of progress reflects a compromise between the human spaceflight budget and operational tempo. This scenario would require a human spaceflight budget that increases faster than inflation but not as fast as for the schedule-driven scenario. It would also result in an operational tempo that is below historical norms but not nearly as much as the budget-driven scenario.

Having examined mission elements, capabilities, and the affordability of the pathways, the Technical Requirements section concludes with an assessment of each pathway in terms of six desirable properties:

- The horizon and intermediate destinations have profound scientific, cultural, economic, inspirational, or geopolitical benefits that justify public investment.
- The sequence of missions and destinations permits stakeholders, including taxpayers, to see progress and develop confidence in NASA's ability to execute the pathway.
- The pathway is characterized by logical feed-forward of technical capabilities from one mission to subsequent missions.
- The pathway minimizes the use of dead-end mission elements that do not contribute to later destinations on the pathway.
- The pathway is affordable without incurring unacceptable development risk.

- The pathway supports, in the context of the available budget, an operational tempo that ensures retention of critical technical capability, proficiency of operators, and effective use of infrastructure.

The Technology Programs section briefly summarizes noteworthy programs and plans related to human spaceflight that are under way by NASA, industry, the Department of Defense, and the international community. A later discussion of robotic systems describes the importance of evolutionary improvements in robotic capabilities to the future of human spaceflight and the possibility that rapid improvements in robotics in the coming decades will open new pathways to Mars and beyond.

The Key Results section contains seven statements that flow from the panel's assessment of the following topics:

- Feasible destinations for human exploration
- Pace and cost of human exploration
- Human spaceflight budget projections
- Potential cost reductions
- Highest-priority capabilities
- Continuity of goals
- Maintaining forward progress

The chapter finishes by emphasizing that without a strong national (and international) consensus about which exploration pathway to pursue and without the discipline needed to maintain course over many administrations and Congresses, the horizons of human existence will not be expanded beyond LEO—at least not by the United States.

4.2 TECHNICAL REQUIREMENTS

4.2.1 Possible Destinations in the Context of Foreseeable Technology

NASA's vision includes "expanding human presence" in the solar system. This concept was first made part of U.S. national space policy by President Reagan in the classified *Presidential Directive on National Space Policy*.[3] All but one of the succeeding administrations have articulated similar goals, and various committees and commissions outlined proposed pathways and projected budgets for achieving the goals.[4,5,6,7,8] President Obama commissioned the Review of United States Human Spaceflight Plans Committee[9] (also known as the Augustine Committee), which concluded in 2009 that "the ultimate goal of human exploration is to chart a path for human expansion into the solar system" but noted the need for both physical and economic sustainability. The report of the Augustine Committee detailed various options, such as "Mars First," "Moon First," and "Flexible Path," potentially involving missions to lunar orbit, Lagrange points, near-Earth objects, the moons of Mars, the surface of the Moon, and the surface of Mars. Ultimately, President Obama declared that the near-term goal for U.S. human spaceflight beyond the Earth-Moon system would be exploring a near-Earth asteroid, which would lead to human

[3] *Presidential Directive on National Space Policy*, February 11, 1988, available in NASA Historical Reference Collection, History Office, NASA, Washington, D.C., http://www.hq.nasa.gov/office/pao/History/policy88.html.

[4] NASA, *Report of the 90-Day Study on Human Exploration of the Moon and Mars,* November 1989, available in NASA Historical Reference Collection, History Office, NASA, Washington, D.C., http://history.nasa.gov/90_day_study.pdf.

[5] Advisory Committee on the Future of the U.S. Space Program, *Report of the Advisory Committee on the Future of the U.S. Space Program,* U.S. Government Printing Office, Washington, D.C., December 1990.

[6] Executive Office of the President, "President Bush Announces New Vision for Space Exploration Program: Remarks by the President on U.S. Space Policy, NASA Headquarters," Washington, D.C., January 14, 2004, http://history.nasa.gov/Bush%20SEP.htm.

[7] NASA, *The Vision for Space Exploration*, NASA, Washington, D.C., February 2004, http://history.nasa.gov/Vision_For_Space_Exploration.pdf.

[8] Executive Office of the President, "U.S. National Space Policy," August 31, 2006, available at http://history.nasa.gov/ostp_space_policy06.pdf.

[9] Review of U.S. Human Space Flight Plans Committee, *Seeking a Human Spaceflight Program Worthy of a Great Nation*, October 2009, http://www.nasa.gov/pdf/396093main_HSF_Cmte_FinalReport.pdf.

orbital missions to Mars and a Mars landing thereafter.[10] In contrast, the Clinton administration's space policy committed the United States only to establish a permanent human presence in Earth orbit, where "the International Space Station will support future decisions on the feasibility and desirability of conducting further human exploration activities."[11] Thus, since 1988, except for the Clinton administration, presidential space policy has *advocated* expanding human presence in some form throughout the solar system; however, since President Reagan, the United States for practical purposes has *executed* the Clinton administration space policy.

Congressional guidance for NASA has, in general, been less expansive than the executive branch's human spaceflight vision. The appetite for ambitious, Apollo-style goals beyond LEO and the attendant budgets has been notably lacking. Even so, the National Aeronautics and Space Administration Authorization Act of 2010,[12] which mandated this report, called explicitly for the development of a heavy-lift launch vehicle capable of supporting human spaceflight beyond the ISS and LEO, with a focus on cislunar space in the near term. Indeed, this congressional direction constitutes an important boundary condition for the report's analysis, which is based on the assumption that the Space Launch System (SLS) would be the primary launcher that enables exploration beyond LEO.

It is critically important for stakeholders of U.S. human spaceflight to understand, however, that currently understood physiological limitations of human beings to endure the space radiation and zero gravity of space for long periods, the inadequacy of foreseeable technological and medical countermeasures in addressing many of those limitations, and the performance of future in-space propulsion systems, severely limit the destinations in the solar system to which humans may travel. For example, using traditional standards for lifetime cancer risk, the NASA Human Research Program has established that the risk of cancer induced by galactic cosmic radiation (GCR) exceeds current guidelines for missions longer than 615 days.[13,14] This is for the most optimistic case with 55-year-old men who have no previous radiation exposure and have never smoked, assuming complete engineered protection (such as shielding crew quarters with water) from solar particle events (SPEs). For female astronauts, for astronauts younger than 55 years old, for astronauts who have had previous radiation exposure, and for missions that do not take place during solar maximum (when the Sun's magnetic field provides substantial protection from GCR), the permissible durations are much shorter, approaching 6 months in many cases. It is also probable that other factors—such as noncarcinogenic effects of GCR, musculoskeletal degeneration in zero g, ocular impairment, and psychosocial effects—could further limit permissible mission durations. Although NASA's Human Research Program is working efficiently to quantify the health risks of deep-space exploration, remaining uncertainties are considerable. In addition, the number of potential health problems associated with human spaceflight continues to increase. Ocular impairment and increased intracranial pressure were identified as a potentially serious problem only in 2011,[15] and the potential for microgravity to have adverse effects on the development of endothelial cells, which line the interior of blood vessels, was identified only in 2013.[16] Apart from risks that might limit permissible cruise durations, astronauts will face, for example, such environmental factors as dust or perchlorates from lunar or martian soil, respectively. The impact of psychosocial factors on long-duration spaceflight has also been an issue of concern for some time.[17] Finally, it is worth noting that the number of evidence books that the

---

[10] *Remarks by the President on Space Exploration in the 21st Century*, John F. Kennedy Space Center, Florida, April 15, 2010, http://www.whitehouse.gov/the-press-office/remarks-president-space-exploration-21st-century.

[11] National Science and Technology Council, *Presidential Decision Directive/NSC-49/NSTC-8*, September 14, 1996, available at http://www.fas.org/spp/military/docops/national/nstc-8.htm.

[12] Public Law 111-267, October 11, 2010.

[13] National Research Council, *Managing Space Radiation Risk in the New Era of Space Exploration*, The National Academies Press, Washington, D.C., 2008.

[14] Steve Davison, NASA Headquarters, "Crew Health, Medical, and Safety: Human Research Program," briefing to NRC Committee on Human Spaceflight Technical Panel, March 27, 2013.

[15] J. Fogarty et al., *The Visual Impairment Intracranial Pressure Summit Report*, NASA/TP–2011-216160, NASA Johnson Space Center, October 2011.

[16] S. Versari et al., The challenging environment on board the International Space Station affects endothelial cell function by triggering oxidative stress through thioredoxin interacting protein overexpression: The ESA-SPHINX experiment, *Journal of the Federation of American Societies for Experimental Biology* 27:4466-4475, 2013.

[17] Institute of Medicine, Chapter 5 in *Safe Passage: Astronaut Care for Exploration Missions*, The National Academies Press, Washington, D.C., 2001.

Human Research Program has accumulated (more than 30) highlights the complexity of assessing the safety of long-duration human spaceflight.[18]

For distant destinations, limitations on the duration of human exploration missions impose minimum spacecraft velocities that will be hard to achieve. Example mission designs for Mars suggest that chemical propulsion (the only technology that has been used for human spaceflight) *might* enable getting humans to Mars orbit and back if substantial progress is made in storing cryogenic propellants for long periods with minimal loss. Other types of propulsion systems—such as solar electric propulsion (SEP), nuclear electric propulsion (NEP), and nuclear thermal propulsion (NTP)—could be used to support a human mission to the Mars surface. However, developing high-power operational systems that use any of these concepts would be a major undertaking.

The development challenges associated with any solar system destinations beyond the Earth-Moon system, Earth-Sun Lagrange points, near-Earth asteroids, and Mars are profoundly daunting, involve huge masses of propellant, and have budgets measured in trillions of dollars.[19]

4.2.2 Design Reference Missions

With an initial reconnaissance of the Moon completed, the maturation of the ISS, and robotic exploration of Mars supporting NASA's search for signs of past or present life elsewhere in the solar system, human exploration of Mars seems to be a logical goal for U.S. human spaceflight given its prominence in the national space policies of multiple administrations, including the current one. To achieve a human mission to the Mars surface, advanced capabilities and new technologies will be needed. NASA has outlined the challenge of a Mars surface mission with the Mars Design Reference Architecture (DRA) 5.0, which lays out the major mission requirements for a Mars surface landing and exploration.[20] The DRA 5.0 report summarizes the tradeoffs among energy expenditure, surface time, and cruise duration for a human mission to the Mars surface. However, Mars is not the only potential destination beyond LEO. A number of scientifically and technically interesting destinations exist between Earth and the surface of Mars. Missions to such intermediate destinations may also demonstrate capabilities and improve technologies needed for a Mars surface mission. NASA and others have examined the potential destinations,[21,22] and reasonable mission concepts have been formulated.

DRMs are point designs that provide an overview of how a mission goal could be achieved. DRMs serve two purposes during initial conceptual design. First, they provide an overview of the mission and mission needs, so that mission requirements can then be used to generate system requirements. Second, they provide a benchmark that can be used for comparison with alternative mission concepts for achieving mission goals. This study used a set of representative DRMs to define the three pathways and to assist in evaluating the challenges of expanding human spaceflight beyond LEO as far as the "horizon goal" of a human mission to the Mars surface. DRMs to other feasible destinations, such as the Earth-Moon L1 point and the Earth-Sun L1 and L2 points,[23] could also have been used, but pathways that use the selected DRMs are sufficient for assessing the full scope of the technical and affordability challenges faced by human space exploration. The definitions of the DRMs are based on recent

---

[18] Evidence books are collections of risk reports and journal articles that address a particular human health issues. A list of Human Research Program evidence books appears at http://humanresearchroadmap.nasa.gov/Evidence/, accessed April 20, 2014.

[19] R. McNutt et al., Human missions throughout the outer solar system: Requirements and implementations, *Johns Hopkins APL Technical Digest* 28(4):373-388, 2010.

[20] B. Drake, "Design Reference Architecture 5.0," http://www.nasa.gov/pdf/373665main_NASA-SP-2009-566.pdf, 2009.

[21] NASA, Human Spaceflight Exploration Framework Study, Washington, D.C., January 11, 2012, http://www.nasa.gov/pdf/ 509813main_Human_Space_Exploration_Framework_Summary-2010-01-11.pdf.

[22] Chris Culbert and Scott Vangen, NASA Human Space Flight Architecture Team, "Human Space Flight Architecture Team (HAT) Technology Planning," NASA Advisory Council briefing, March 6, 2012, http://www.nasa.gov/pdf/629951main_CCulbert_HAT_3_6_12=TAGGED.pdf.

[23] Lagrangian points, also referred to as L points or libration points, are five relative positions in the co-orbital configuration of two bodies, of which one has a much smaller mass than the other. At each of the points, a third body with a mass that is much smaller than either of those two will tend to maintain a fixed position relative to the two. In the case of the Earth-Moon system, L1 is a position between the Moon and Earth where a spacecraft could be placed, and L2 is a position beyond the Moon that would be similarly fixed in orientation to the Earth and Moon. See "The Lagrange Points," http://wmap.gsfc.nasa.gov/mission/observatory_l2.html.

publications from NASA (for the ARM,[24] Lunar Surface Outpost,[25] Asteroid in Native Orbit,[26] Mars's Moons,[27] and Mars Surface[28] DRMs) and the ISECG *Global Exploration Roadmap*[29] (for the Earth-Moon L2, Lunar Surface Sortie, and Lunar Surface Outpost DRMs).[30]

The DRMs presented below are notional and representative of possible missions, but they are neither comprehensive nor final designs. The DRMs for each major destination beyond LEO demonstrate the challenges associated with various pathways to human exploration of Mars.

The DRMs below can be divided into two groups: cislunar and deep-space missions. Cislunar missions include missions to the vicinity of the Moon as well as lunar surface missions. Cislunar missions have the advantage of remaining close to Earth. This reduces mission risk by allowing abort contingencies with a relatively quick return to Earth.[31] Cislunar missions would also cost substantially less and be more affordable than deep-space missions. Autonomy requirements are also reduced because of short time delays in communications between spacecraft and Earth. Cislunar missions that do not include a lunar landing also minimize propulsion requirements by avoiding interactions with strong gravitational bodies. Lunar surface missions are a good analogy for Mars surface operations and some of the associated constraints on hardware and human physiology that will need to be overcome. Deep-space missions include the horizon goal of a human mission to the Mars surface, missions to asteroids in their native orbits, and missions to the moons of Mars. The asteroid and Mars Moons missions would allow demonstration of spacecraft vehicles and systems and validate the ability to sustain human health during long-duration missions that are similar in scale to a Mars surface mission.

4.2.2.1 The Space Launch System and the Design Reference Missions

The SLS is a heavy-lift launch vehicle that is being developed by NASA to support human space exploration beyond LEO. The need for a heavy-lift launch vehicle has been noted in many blue ribbon studies of human spaceflight. The planned SLS payload capacity of 70-130 metric tons (MT) depending on the version of the SLS, and the large shroud would reduce the number of launches required for human exploration missions beyond LEO. The Falcon Heavy launch vehicle, which is being developed by SpaceX, will have a payload capacity of up to 53 MT. The Augustine study suggested that human exploration could be accomplished with a 50-MT launch system if the architecture relied on advanced capabilities, such as in-space refueling and expandable habitats.[32] Smaller alternatives to the SLS would require more launches, more time in orbit, and more docking events, which might reduce mission reliability. The effect on costs is hard to predict. On the one hand, the increased launch rate and the potential for commonality with other launch systems that use smaller vehicles might reduce development and operational costs for the launch vehicle. On the other hand, the development cost for other mission elements would increase because of the need for additional technologies and interface hardware. In addition, increasing the amount of in-orbit assembly would tend to increase operational costs. For simplicity and consistency of presentation, the analysis of all the DRMs in this report assumes the use of the SLS as the launch vehicle.

---

[24] NASA, "Asteroid Initiative Related Documents," http://www.nasa.gov/content/asteroid-initiative-related-documents/.

[25] L. Toups and K. Kennedy, "Constellation Architecture Team-Lunar Habitation Concepts," in *AIAA Space 2008 Conference and Exposition,* American Institute of Aeronautics and Astronautics, 2008, http://arc.aiaa.org/doi/abs/10.2514/6.2008-7633.

[26] B.G. Drake, "Strategic Considerations of Human Exploration of near-Earth Asteroids," paper presented at the Aerospace Conference, 2012 IEEE, March 3-10, 2012.

[27] D. Mazanek et al., "Considerations for Designing a Human Mission to the Martian Moons," paper presented at the *2013 Space Challenge,* California Institute of Technology, March 25-29, 2013.

[28] B. Drake, "Design Reference Architecture 5.0," 2009, http://www.nasa.gov/pdf/373665main_NASA-SP-2009-566.pdf.

[29] International Space Exploration Coordination Group (ISECG), *The Global Exploration Roadmap,* NASA, Washington, D.C., August 2013, https://www.globalspaceexploration.org/.

[30] Detailed analyses of the DRMs that in some cases added to or modified the information from these sources were performed to ensure that each DRM was based on a consistent set of assumptions concerning, for example, the performance of major system elements. As indicated, the Lunar Outpost DRM is based on information from NASA and the ISECG.

[31] For some emergencies, such as a subsystem failure that would doom a Mars-bound crew, a safe return from cislunar space might be possible. For other emergencies, such as an acute medical emergency, the quicker return from cislunar space would still be problematic.

[32] Review of U.S. Human Space Flight Plans Committee, *Seeking a Human Spaceflight Program Worthy of a Great Nation,* 2009.

The business model and schedule for the SLS are almost totally driven by the projected costs and the flat budget profile established for the SLS program. A system that, like the SLS, is derived from mature systems could reach full operational capability in less time and at less total cost than currently planned if the funding profile resembled that for a normal development (that is, if funding ramped up as the program progressed from program initiation into advanced development and production). Much of the necessary design work was done under the Constellation program. Before that, vehicles derived from the space shuttle had been proposed several times, and considerable preliminary design work had been done. Nonetheless, the designs for the SLS were announced in September 2011 with a projected development cost (including ground-based infrastructure) of $12 billion through first flight in late 2017 and an additional $6 billion for development of the Orion Multi-Purpose Crew Vehicle.[33,34] Orion is the crew capsule being developed in concert with the SLS to support human space exploration beyond LEO. Since then, NASA's budget uncertainty has increased, and this has put both the launch date and the cost at high risk.[35]

4.2.2.2 Asteroid Redirect Mission

The ARM envisions a crew of two briefly interacting with a 7- to 10-m asteroid while avoiding the longer travel times in deep space required to reach an asteroid in its native orbit. A precursor robotic mission using advanced SEP would be sent to an as-yet-unspecified near-Earth asteroid with appropriate mass and orbital properties. Depending on the mission concept ultimately selected, the robotic spacecraft would either capture a small asteroid in entirety or retrieve a boulder from the surface of a larger asteroid[36] and then transport its payload to an orbit in the lunar vicinity. After the multiyear asteroid redirect phase, the crew would be launched on the SLS with the Orion capsule, rendezvous with the asteroid in its new orbit, and dock to the attached robotic spacecraft. The mission is ultimately constrained by the current capability planned for the Orion capsule, but a two-person crew could spend up to 6 days at the asteroid, collecting samples during EVA before returning to Earth.

4.2.2.3 Earth-Moon L2

The Earth-Moon L2 reference mission would demonstrate long-term human habitation and operations in deep space. To achieve this goal, a habitation module and supporting systems would be developed, built, and launched by an SLS. The spacecraft would transit to either an Earth-Moon Lagrange point or a stable lunar orbit. This minimal space station would be capable of supporting a crew of four or more for at least 6 months. The crew would reach the habitat using the SLS and Orion systems. The primary goal would be to develop technologies and techniques that could enable crews to survive and function on long-duration deep-space missions while maintaining a mission-abort capability with a quick return to Earth. Other activities, such as observations of the far side of the Moon or support of lunar surface activities, could occur. This mission would require a substantial improvement in the reliability and sustainability (operation without resupply) of environmental control and life support systems (ECLSS) compared to the ECLSS on the ISS. In addition, radiation exposure from SPEs during extended stays outside Earth's magnetosphere would have to be mitigated without exceeding mass and volume constraints. (No mitigation of GCR currently appears to be practical.) The ISECG has proposed a mission of this type.[37]

4.2.2.4 Lunar Surface Sortie

The Lunar Surface Sortie reference mission would leverage substantial prepositioned assets in lunar orbit, and it is similar to the proposal for lunar exploration in *The Global Exploration Roadmap* recently updated by the

---

[33] NASA, "NASA Announces Design For New Deep Space Exploration System," Press Release 11-301, September 14, 2011.

[34] M.S. Smith, "New NASA Crew Transportation System to Cost $18 Billion Through 2017," posted September 14, 2011, updated December 5, 2011, http://www.spacepolicyonline.com/news/.

[35] M.K. Matthews, "New NASA rocket faces delays," *Orlando Sentinel,* September 6, 2013 (quoting NASA Deputy Administrator Lori Garver).

[36] NASA is studying two mission concepts for robotic the ARM mission. The reference robotic mission concept would capture a small asteroid. The alternative mission would capture a boulder.

[37] ISECG, *The Global Exploration Roadmap,* 2013.

ISECG.[38] This mission would sustain a crew of four on the lunar surface for 28 days. Predeployed pressurized lunar surface mobility units would be positioned using SLS launch vehicles and reusable lunar ascent-descent vehicles. The crew would be launched using SLS and Orion and then rendezvous with a permanent lunar orbital facility. This facility would serve as the staging point for the crew and the ascent-descent stage, which would be augmented by a low-cost disposable deceleration stage. Scientific exploration would be conducted using surface mobility units that also operate as habitats. The exploration range would be limited to a reasonable return distance to the ascent-descent stage. Beyond the attendant lunar surface science, this DRM would demonstrate surface operations, surface habitation, and surface mobility required for partial-gravity environments where dust and other potential contaminants are present.

4.2.2.5 Lunar Surface Outpost

The Lunar Surface Outpost DRM is an extension of the Lunar Surface Sortie mission and requires the deployment of long-duration surface assets. These assets would be delivered using a similar architecture of SLS launch vehicles, reusable lunar ascent-descent vehicles, a staging orbital facility, and disposable propulsion stages. The additional assets would extend the surface mission duration from 28 days to as much as 6 months. The mobile assets would allow extended sortie missions from the outpost to scientifically diverse sites, while the outpost itself would be used for scientific experimentation and testing of Mars-focused technologies, such as long-life high-capacity power generation systems and operations planning for long-duration surface stays.

4.2.2.6 Asteroid in Native Orbit

The Asteroid DRM is a deep-space mission beyond cislunar space for a crew of four to a near-Earth asteroid. The asteroid for the mission would be selected on the basis of scientific interest (or relevance to planetary defense) and Earth-asteroid alignment to allow the crew to transfer to and from the asteroid within a 270-day total mission duration. An Orion vehicle, a deep-space propulsion unit, a deep-space habitation module, and a space exploration vehicle would be launched on SLS vehicles and rendezvous in LEO before the transfer to the asteroid. On arrival at the asteroid after a 60- to 130-day transit, the crew would transfer to the space exploration vehicle and perform close-proximity operations and EVAs to collect samples and perform experiments on the asteroid. After a 14-day mission at the asteroid, the crew would return to Earth in the space habitat on a 70- to 160-day journey before performing a direct entry using the Orion vehicle. This DRM features deep-space habitation capability for more than the 180 days needed for a transit to or from Mars, deep-space navigation, low-gravity foreign-body exploration, and potentially important scientific returns.

4.2.2.7 Mars Moons

The Mars Moons DRM is similar to the Asteroid DRM in that it is an exploration of a low-gravity body in deep space using space exploration vehicles and EVAs for crewed exploration. The major distinguishing factor is the location of the low-gravity body. A crewed mission to Phobos and Deimos in Mars orbit would include many elements of a crewed mission to Mars but without the challenge of EDL and ascent from Mars. After departure from Earth, the mission would attain Mars orbit insertion and then use orbital maneuvering units to spend up to 60 days at Phobos and Deimos. The mission length would increase from less than 1 year for the Asteroid in Native Orbit mission to almost 2 years for the Mars Moons mission. The increases in mission duration and propulsion requirements result would require an advanced in-space propulsion system. The current design baseline is NTP, although NASA was still examining the propulsion trade space as of the end of 2013. Two space exploration vehicles and small propulsion stages for martian orbital maneuvering would be predeployed at Mars using an advanced propulsion stage. The crew would rendezvous with their long-duration, deep-space habitat and advanced propulsion stage in LEO before transferring to Mars. The habitat would need to protect the crew from the deep-space environment

---

[38] ISECG, *The Global Exploration Roadmap*, 2013.

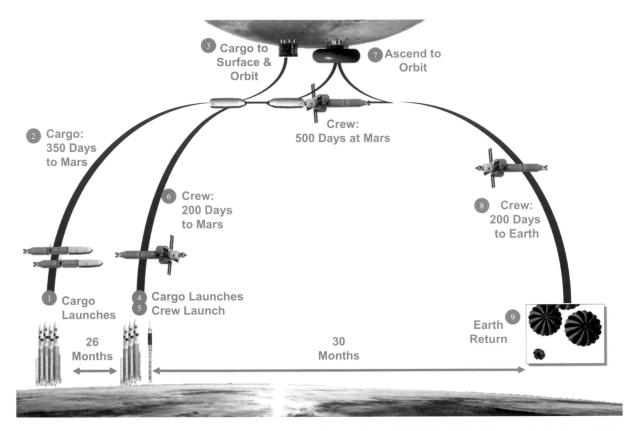

FIGURE 4.1 Design Reference Architecture 5.0 Human Landing on Mars. SOURCE: NASA, Human Exploration of Mars Design Reference Architecture 5.0, 2009.

throughout the 700-day mission (except for the 2 months spent in Mars orbit). The in-space duration for this mission is more than 3 times that expected for the Mars surface mission, and deep-space habitation and logistics for a 700-day in-space mission may not be feasible, depending on the challenges of GCR.

4.2.2.8 Mars Surface

The horizon goal for human spaceflight is the human exploration of the Mars surface. Numerous concepts for surface exploration missions have been described in various documents; the analysis in this report is based on NASA's Mars DRA 5.0. The mission is based on sending three different vehicles to Mars, as shown in Figure 4.1. Multiple SLS launches would be required to place both the cargo and crewed portions of the mission in LEO.

The cargo portion of the mission would use two vehicles to carry support equipment and travel to Mars during a planetary alignment before the crew transit.[39] This would allow verification that the support equipment had arrived successfully and that the crew's ascent vehicle has landed on Mars with sufficient time to use in-situ resources to produce propellant. The mission concept relies on the crew landing on Mars close to the predeployed ascent vehicle and its support equipment. The cargo missions would use a minimal-energy 350-day trajectory from Earth, and they would enter Mars orbit using aerocapture technology. One of the predeployed vehicles would then perform aero-assisted EDL and initiate preliminary robotic efforts to prepare the landing site for the crewed mission. The second predeployed vehicle would wait in Mars orbit for the arrival of the crewed vehicle.

---

[39] Transit times and propulsion requirements for missions to Mars are minimized when Earth and Mars are favorably aligned in their orbits. Such alignment occurs every 26 months or so.

The crewed portion of the mission would depart in a long-duration habitat from a LEO staging point using NTP on a 6-month transfer to Mars orbit. The crewed system would be propulsively captured into Mars orbit and rendezvous with the predeployed vehicle. The crew of six would then transfer to the predeployed vehicle that contains the surface habitat, which would transfer them to the planet's surface. The EDL system would land the habitat close to the predeployed assets, and the crew would then be able to conduct mobile scientific exploration of Mars. The Mars surface mission would last for about 500 days, and the crew would then board the ascent stage and return to the deep-space habitat and propulsion system that remained in Mars orbit. The surface assets would continue autonomous missions and data collection for possible use by future Mars exploration crews. On crew transfer to the deep-space habitat, which would have been in standby mode, the crew would jettison the ascent vehicle and return to Earth on another 6-month transfer and a direct Earth entry using the Orion vehicle.[40]

The crew and cargo portions of the mission would be sufficiently massive to require advanced propulsion stages, currently modeled using NTP, for the transfer to Mars orbit from a LEO staging point. Because NTP may not be feasible because of technical, financial, and/or political factors, NASA is still evaluating other advanced in-space propulsion options. The Mars DRA 5.0 study suggests that an NTP-based system would result in the lowest life cycle cost and mission risk, assuming a sustained campaign of many Mars missions. Other propulsion options, such as SEP and advanced cryogenic chemical propulsion, may have lower cost if only a few missions are planned. The Mars surface elements would require a highly reliable power source that is capable of generating a total of 30 kW or more. Mars DRA 5.0 concluded that a fission reactor would have about one-third the mass of a solar-based surface power system and that deploying large solar arrays robotically in the high wind and dust environment of Mars would be very challenging.

4.2.3 Potential Pathways

Now that the various stepping-stone destinations and associated mission concepts have been highlighted, potential pathways, all ending with a Mars surface mission, can be defined. As described above, each of the three pathways is a series of human spaceflight missions to various destinations. The first pathway (ARM-to-Mars) is essentially the current administration's proposed U.S. human spaceflight program. The Moon-to-Mars pathway makes use of the Moon as a testing and development destination to mature Mars-oriented technology, while revisiting the lunar environment for more in-depth scientific study than was possible during Apollo. This pathway is also consistent with the goals of the United States' traditional international space partners and the ISECG, of which NASA is a member.[41] Finally, the Enhanced Exploration pathway essentially exhausts the potentially feasible classes of destination through a Mars surface landing, and it allows exploration of essentially all destinations that humans can explore, given the current understanding of physiology and technology.

The three pathways are used to compare and illustrate the challenges of sending humans as far as the surface of Mars. Although the specific destinations on the journey to Mars are few and each requires development and demonstration of hardware elements of various categories (such as propulsion systems, power systems, and habitation systems) as well as research related to human health issues, NASA, in concert with other international and domestic organizations, could further define, mature, and analyze a broad range of detailed conceptual pathways to Mars.

Table 4.1 and Figure 4.2 define and illustrate the three representative pathways to the horizon goal of a human mission to the Mars surface. Table 4.1 defines the specific DRMs of each pathway, while Figure 4.2 illustrates each of the stepping-stone destinations and the three pathways to the Mars surface. The first pathway, ARM-to-Mars, leverages the initial demonstration of the SLS and Orion systems in cislunar space via the ARM and then proceeds directly to activity in the Mars vicinity by focusing on exploring the moons of Mars, followed by a Mars landing.

The second pathway, Moon-to-Mars, first focuses on missions in the lunar vicinity and surface to demonstrate longer-duration in-space habitats and complex propellant staging in lunar orbit. These missions would also develop

---

[40] The DRA 5.0 architecture outlined here is for the long-surface-duration, so-called conjunction mission. Given the orbital mechanics of Earth and Mars, a second, short-stay "opposition" class mission is also possible. This alternate mission has a transit time to Mars of about 200 days, a surface stay of about 60 days, and a return transit of 400 days.

[41] ISECG, *The Global Exploration Roadmap*, 2013.

TABLE 4.1 Pathway Definition and Associated DRMs

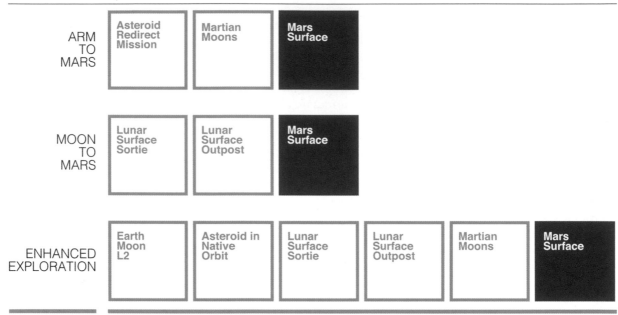

NOTE: The sources used to characterize the DRMs are listed above in the section "Design Reference Missions."

critical partial-gravity surface habitats, both fixed and mobile, and long-term, reliable power generation systems to maintain these assets. The focus in this pathway is to develop the required assets and techniques for martian surface exploration using the nearest and most easily accessible celestial body, relatively short return times, and open launch windows. To proceed to Mars as soon as is practical, after an appropriate time for examining hardware, operational, and human health issues, the lunar assets would be retired from government service and optionally maintained and leveraged by future commercial endeavors. Alternatively, if the actual crewed exploration of Mars became infeasible for financial, technical, or crew health reasons, the Moon-to-Mars pathway would constitute a natural off-ramp, leaving the United States to lead global exploration and exploitation of the Moon.

The third pathway, Enhanced Exploration, presents a potentially lower risk than the other pathways, but it is also a longer-duration pathway, exploring several destinations while slowly increasing the capability of key mission elements needed for a Mars surface mission. It begins by focusing on the challenges of a long-term in-space habitat with a mission to Earth-Moon L2 and a native asteroid. That would be followed by a focus on the Moon to develop partial-gravity surface operations capabilities. Finally, there is the development and use of the advanced in-space propulsion systems with missions to the Mars moons followed by Mars surface operations. Human health issues would also be investigated during this pathway as more challenging missions were completed.

Each of the three pathways is described in more detail and assessed below.

4.2.4 Drivers and Requirements of Key Mission Element Groups

Each DRM that appears in one or more of the three pathways discussed above requires the development of key mission element groups to support the specific mission. These element groups are launch, in-space transportation, habitation, EDL and ascent, and destination systems. A brief description of the performance requirements that would drive the development of these groups, with a focus on the Mars surface mission as the horizon destination or goal, is presented below. Although the groups encompass most of the basic needs for successful missions

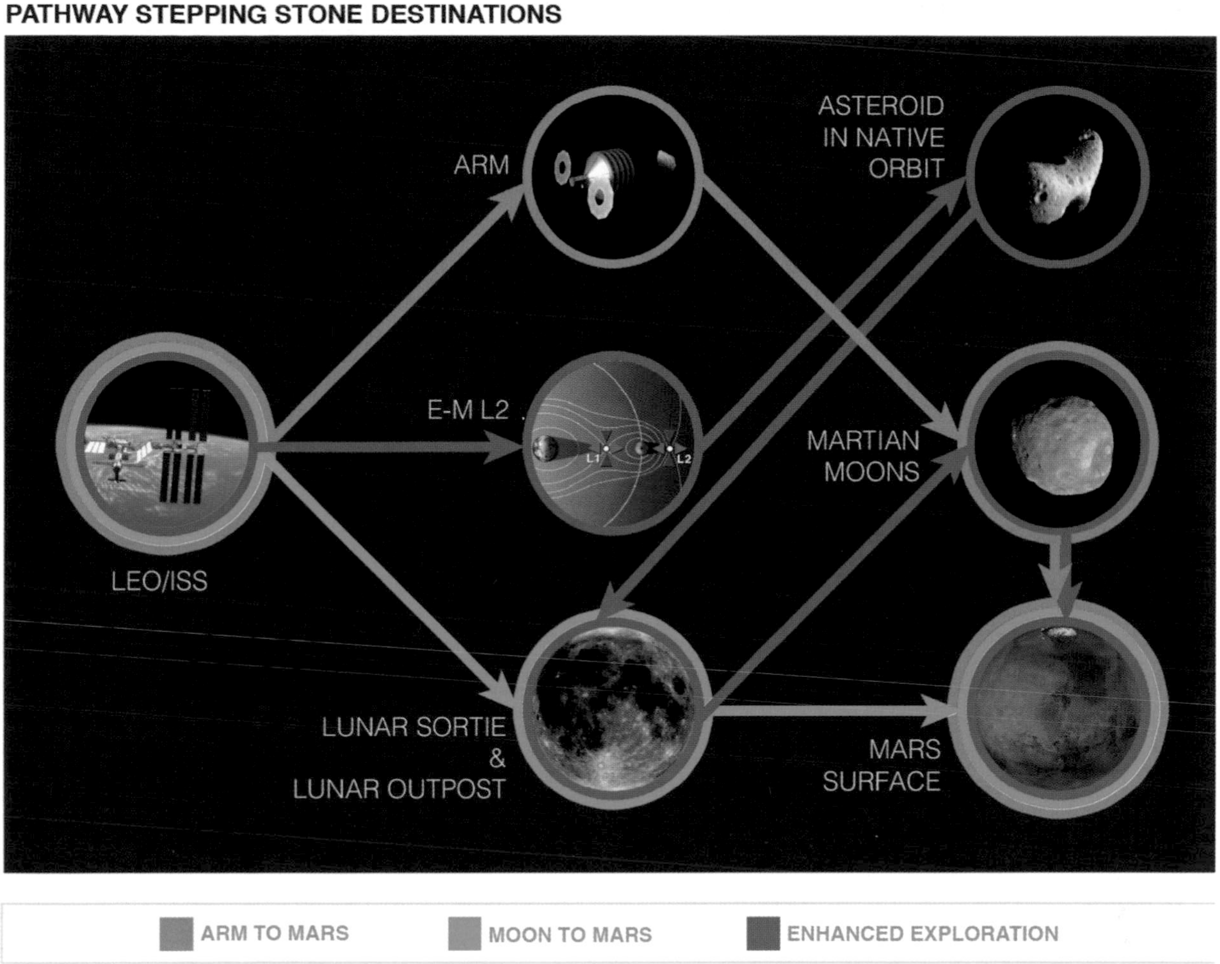

PATHWAY STEPPING STONE DESTINATIONS

FIGURE 4.2 Pathway stepping-stone destinations.

beyond LEO, other assets, such as improved communication networks and advanced mission operations, will probably need to be developed.

4.2.4.1 Launch

Launch system requirements for human space exploration are driven by the total payload mass required in a specific orbit (typically LEO), payload diameter and volume, and reliability requirements. State-of-the-art U.S. launch systems can place 23 MT of payload in LEO with a 5-m-diameter fairing, but no U.S. launch systems are human-rated. In any case, a much larger launch system would be needed for the human surface mission in Mars DRA 5.0. This mission would require payload diameters of about 10 m with total lift capability to LEO of 105-130 MT. NASA is developing the SLS, which eventually will be human-rated, to meet the Mars surface mission requirements. The initial version of the SLS will have only a 70-MT capability, but planned development includes 105-MT and 130-MT variants. Although the total mass needed for a Mars surface mission could be launched with more launches using lower-capability vehicles, the associated increase in launches, system complexity due to assembly, and operational complexity may reduce total mission reliability to below an acceptable level.

TABLE 4.2 Approximate Initial Mass in Low Earth Orbit (IMLEO) Required for the DRMs That Appear in One or More of the Pathways

DRM	Approximate IMLEO (metric tons)
ARM	100
Earth-Moon L2	200
Asteroid in native orbit	200
Lunar sortie	400
Lunar outpost	900
Mars moons	400-800
Mars surface	900-1,300

NOTE: For each of the destinations shown, the values include all launch mass needed up to and including the first human launch in accordance with the relevant DRMs. For the lunar missions, this includes a habitat in lunar orbit that would serve as a waypoint and docking station for a reusable lunar lander. An IMLEO for a Mars mission is shown as a range of values; the actual value would depend on the propulsion systems chosen for different mission elements (such as nuclear thermal, solar electric, and chemical). For comparison, the mass of the ISS, which is in LEO, is 420 MT.

An analysis of the DRMs that appear in one or more of the pathways allows the determination of the total initial mass in LEO required to accomplish each mission (see Table 4.2). As discussed above, the payload capacity requirements of the launch vehicles used for these missions could be traded off with factors such as risks of in-orbit assembly or fueling. In principle, none of the missions beyond ARM can be accomplished using a single SLS launch. Either multiple SLS launches would be required or one or more SLS launches could be replaced by multiple launches of smaller rockets that could carry the human crew, smaller subsystems, and/or bulk consumables.

4.2.4.2 In-Space Transportation

Human spaceflight missions would require in-space transportation systems to perform major propulsion burns that are not provided by the launch vehicle, often at multiple points during a mission. These transportation systems would consist of high-performance, multiple-restart engines with large fuel tanks that are capable of long-term propellant storage and management. In-space transportation systems with these characteristics do not exist on the scale necessary for human spaceflight missions beyond cislunar space.

For a space propulsion system, thrust is directly proportional to specific impulse (I_{sp})[42] and mass flow rate of engine exhaust gases. Current chemical storable propellants have low propulsion efficiency (that is, low I_{sp}), resulting in high mass. Cryogenic fuels for deep-space missions would require advanced thermal control and boil-off management.

SEP offers much higher I_{sp}, but it is limited by relatively low propellant flow rates, which results in lower thrust, lower velocities, and longer trip times. For crewed missions, this would increase the exposure of the crew to space radiation en route. As a result, SEP is not considered as a viable option for crewed missions, although it may be suitable for cargo vessels that would preposition food, water, propellant, and/or landing and return vehicles in Mars orbit. However, the SEP systems needed for an appropriately sized cargo vessel would require massive solar arrays with a generating capacity at least 3 times the power of the ISS arrays to achieve sufficient thrust.

NTP offers a relatively modest increase in I_{sp} while maintaining higher thrust capability and shorter flight durations. Nuclear thermal rockets were developed as part of Project Rover from 1955-1972. After NASA was formed in 1958, Project Rover was administered by NASA, with the nuclear reactor portion falling under the Atomic Energy Commission. Ultimately, several NTP systems (collectively known as NERVA—Nuclear Engine for Rocket Vehicle Application) were successfully ground tested. The NERVA program was canceled by the Nixon

---

[42] I_{sp} is the force produced by a propulsion system divided by the mass flow rate of the fuel that is consumed to produce that force. The higher the specific impulse, the lower the propellant mass flow rate required to achieve a given level of thrust. Thus, in some sense, I_{sp} is a measure of propulsion-system efficiency.

administration in 1973, in part because of environmental and hazard concerns but primarily because the intended application, a crewed Mars mission, was considered too expensive.[43]

NEP systems would avoid problems associated with the large solar arrays of SEP systems, but they face the complications associated with development, production, and operation of nuclear systems. In addition, NEP systems appropriate for space travel have never been developed.

The tradeoff among in-space transportation options—namely chemical propulsion, SEP, NTP, and NEP, alone or in combination—is complex and ongoing. Total life cycle cost considerations are heavily influenced by the total number of flights to the various destinations and by the time and cost needed for development and testing. Substantial advances in technology are required for the use of any of these technologies for the aforementioned pathways, and the final selection of one technology over the others will probably be driven by many factors over the next several decades.

4.2.4.3 Habitation

Human presence aboard the ISS and its predecessors has demonstrated in-space habitation capabilities since the 1970s. The ISS has been consistently occupied for more than a decade, and individual stays have reached 180 days or longer. Some Russian cosmonauts were on the Russian space station Mir for more than a year. One astronaut and one cosmonaut will undertake a year-long stay on the ISS beginning in 2015. Unlike space stations in LEO, habitats for deep-space missions will not be able to take advantage of periodic resupply, the removal of waste products, and the radiation protection provided by Earth's magnetosphere. The Mars DRA 5.0 surface mission includes two periods of travel in deep space of 6 months or more with a crew of six. The deep-space habitat must meet stringent volume and mass constraints for launch and must provide a highly reliable ECLSS with closed or near-closed loops for air, water, and food. The habitat will need to accommodate all crew needs so that they arrive at their destinations physically and mentally fit for their exploration tasks and are subsequently returned to Earth in a similar state. The habitat will need to protect the crew from SPEs. As discussed below in section 4.2.6.1.2 ("Radiation Safety"), protection from GCR by the habitat is impractical and most likely will be accomplished primarily by limiting mission duration. The in-space habitat will also need to operate in a dormant state with no crew during the 500-day surface mission. A mission to Mars that does not include a stay on the surface of Mars, such as the Mars Moons DRM, will need an in-space habitation system that can maintain crew health continuously for 2 years or longer. Such extremely long missions may require habitats to have artificial gravity generated through centripetal acceleration to maintain crew health.

New habitation elements will also be needed when the crew is on the surface of the Moon. In addition to the needs of in-space habitats, surface habitats for the Moon or Mars will need to operate in partial gravity environments and have mitigation strategies for additional hazards, such as potentially toxic dust. Mars crews will need to live on the surface of Mars for more than 1.5 years, spending most of their time in stationary habitats with excursions of weeks or longer in mobile habitats.

4.2.4.4 Entry, Descent, Landing, and Ascent

To conduct a human mission to the Mars surface, the landing system must be capable of placing individual payloads of about 40 MT with accuracy to within hundreds of meters of the targeted landing point. This is well beyond current capabilities at every stage of EDL. The recent landing of the Mars rover Curiosity demonstrated the ability to land a payload of about 1 MT with a landing error ellipse of tens of kilometers. In addition, the EDL approach used for Curiosity applied *g*-loads to the payload that are inconsistent with human passengers. Thus, the Curiosity EDL system cannot simply be scaled up for human missions. In addition, NASA DRA 5.0 calls for the crew to land very close to the ISRU plant predeployed to the Mars surface, so the error ellipse achieved by Curiosity is far large for a human mission. The size and mass of the payloads for a Mars surface mission demand

---

[43] B. Fishbine et al., "Nuclear rockets: To Mars and beyond," in *National Security Science*, Issue 1, LALP-11-015, 2011, http://www.lanl.gov/science/NSS/issue1_2011/story4full.shtml.

a more advanced thermal protection system, more advanced hypersonic and supersonic deceleration systems, and more advanced terminal landing systems to survive the passage through the thin martian atmosphere. Each of the EDL phases also represents potential single point failure opportunities with no abort options. Such a system would most likely rely on a combination of aerodynamic and propulsive deceleration techniques and may require rapid changes in configuration during the flight sequence. An advanced aerodynamic shield would also be required to perform an aerocapture maneuver to place spacecraft into Mars orbit.

The first human landing on Mars will be a monumental occasion for expanding human presence to another planet, but equally important is returning the crew to Earth at the end of the mission. Propellant for the ascent from the planetary surface and return of the crew to Earth requires either the production of resources on the surface or the deployment of a fully fueled system from Earth. The crew ascent vehicle would be predeployed and pre-fueled and probably launched from Earth 26 months in advance of the crew. This would allow in situ production of the oxygen portion of the propellant by separation from the martian carbon-dioxide atmosphere. Alternatively, the complete ascent vehicle and all its associated resources could be launched from Earth, although this would require additional launch assets and more capable EDL systems. The ascent vehicle carrying a crew of six would need to launch from the Mars surface to orbit and rendezvous with the waiting deep-space habitat.

The entry capsule, not used since the crew launched from Earth approximately 30 months earlier and thus requiring long-duration standby abilities, would return the crew to Earth. The entry from a Mars mission would have to survive the highest-velocity crewed entry ever attempted, at more than 13 km/s. The entry shield would have to withstand temperatures of at least 3,000°C, whereas a spacecraft on a lunar return trajectory would experience peak temperatures of about 2,750°C.[44]

4.2.4.5 Destination Systems

Not since the Apollo lunar landings has a spaceflight crew performed surface exploration. Most current space-based tools and human interface systems are focused on maintenance and repair of the ISS and crew and cargo transportation systems to support the ISS. Surface exploration missions will require specialized suits, tools, and vehicles that need to be developed for the destinations of interest. Deep-space space suits and surface spacesuits would be necessary for the EVAs during space and surface operations. Surface EVA suits would require much greater dexterity than any suits built to date. Large rovers capable of carrying multiweek habitation units would be essential for expanding the range of exploration outside the near vicinity of the stationary habitat. Robotic assistants may be required to support the crew by performing tasks and going into environments for which the crew is not suited. Destination systems need to be developed for the environments of specific DRMs while keeping in mind the potential for reuse of the design or function during subsequent missions in a particular pathway.

Mars surface systems would require a continuous supply of 30 kW or more of power. Power would need to be supplied throughout the year, day and night, and during dust storms. The power generation system would need to be predeployed with the ascent stage to support ISRU operations. A nuclear fission reactor is used as the baseline surface power system in DRA 5.0 because of its expected lower mass, reduced volume, and greater reliability compared to the alternative of a surface solar array. On the Moon, due to the 14 days of darkness during surface eclipse, nuclear fission reactors are also the most practical method of providing power to a lunar outpost unless it is in specific locations near the lunar poles.

Additional excursion vehicles would be used for missions to bodies that have near-zero gravity (e.g., an asteroid in its native orbit or the moons of Mars). The excursion vehicles would have the short-term habitation systems of the surface pressurized rovers mated to a zero-gravity mobility system. The vehicles would offer the crew greater protection than an EVA suit when working in close proximity to an asteroid or the moons of Mars. They would also allow the in-space transportation and deep-space habitats to be kept a safe distance from any debris generated by the crew during interactions with celestial bodies.

---

[44] NASA, "To the Extreme: NASA Tests Heat Shield Materials," February 3, 2009, http://www.nasa.gov/mission_pages/constellation/orion/orion-tps.html.

PRIMARY MISSION ELEMENTS

HEAVY LIFT LAUNCH VEHICLE (SLS 100+ MT)

ADVANCED PROPULSION

TELE-ROBOTIC ROVERS

SURFACE NUCLEAR POWER

ADVANCED EVA

PRESSURIZED SURFACE MOBILITY

AEROASSIST SYSTEM

CREW COMMAND & SERVICE MODULE (ORION)

DEEP SPACE HABITATION

LONG DURATION SURFACE HABITAT

MARS ASCENT VEHICLE

TRANSITIONAL MISSION ELEMENTS   **DEAD-END MISSION ELEMENTS**

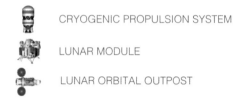 

CRYOGENIC PROPULSION SYSTEM

LUNAR MODULE

LUNAR ORBITAL OUTPOST

ASTEROID RETRIEVAL VEHICLE

MULTI-YEAR DEEP SPACE HABITAT

LARGE STORABLE STAGE

MARS ORBIT TRANSFER VEHICLE

SPACE EXPLORATION VEHICLE

FIGURE 4.3 Primary mission elements for a DRA 5.0 human mission to the Mars surface along with transitional mission elements and dead-end mission elements and their associated icons.

4.2.5 Contribution of Key Mission Elements to the Pathways

To achieve the horizon goal of a human mission to the Mars surface, all the aforementioned element groups are necessary. More specifically, 11 individual mission elements required for the DRA 5.0 Mars surface mission have been identified. These 11 are the primary mission elements shown in Figure 4.3. Only two of the 11 are currently funded for development of operational flight hardware (SLS and Orion). DRMs to destinations other than the surface of Mars might not directly use the primary mission elements but instead use elements that are in the technological development path of a primary element. These are the three transitional mission elements identified in Figure 4.3. Also shown in Figure 4.3 are five dead-end mission elements. Beneficial technology may arise from the development of the dead-end mission elements, but for the most part they have requirements other than what is needed for a Mars surface mission and are extraneous to the direct goal of landing humans on Mars. To understand how each mission builds toward the horizon goal, the progression of element use in each pathway is presented below. It is important to note that the element development for each pathway highlights the location of major jumps in capability between various DRM stages and shows the need for nonessential mission elements that may take additional funding away from the future Mars surface horizon goal.

The buildup of mission critical elements for the ARM-to-Mars pathway is shown in Figure 4.4. The figure shows the mission elements used for each of the three DRMs in this pathway, starting with the ARM, then the Mars

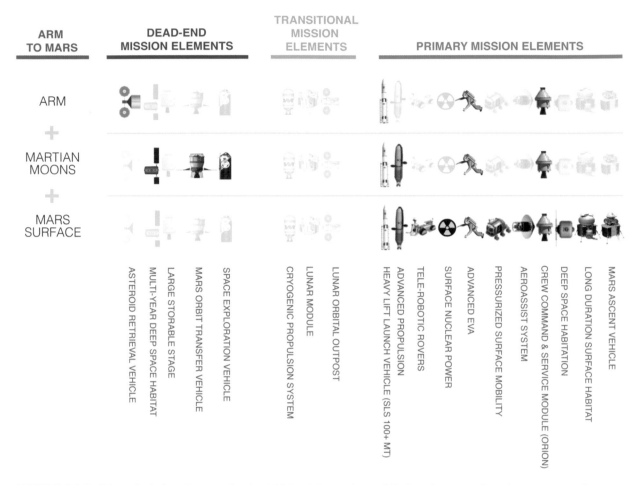

FIGURE 4.4 Buildup of mission elements for the ARM-to-Mars pathway. Mission elements whose icons are grayed out are not required for the relevant missions.

Moons mission, and finally the Mars Surface mission. The ARM mission uses only three of the key Mars surface elements, and each element must be further improved before the Mars Moons mission. Advanced EVA capabilities may be needed for the ARM mission, and the Mars Moons mission may leverage some of the capabilities. Because of the assumption of an NTP-based advanced propulsion system for the Mars Surface mission, the ARM robotic asteroid-redirect vehicle is considered a dead-end mission element, inasmuch as its advanced SEP capabilities are not leveraged in future missions as currently envisioned. In the event that NTP in-space propulsion is not used to transfer cargo for a Mars Surface or Mars Moons mission, SEP could be used for this function. However, the 50-kW SEP system required for ARM is about an order of magnitude below what is required to carry cargo to Mars for the Mars Moons and Mars Surface missions.

The Mars Moons mission would demonstrate all the in-space elements required for a Mars Surface mission and would require major advances in in-space habitation and propulsion. In fact, the in-space habitation requirements for the Mars Moons mission vastly exceed those for the Mars Surface mission. The additional habitat requirements and the need for the space exploration system and orbital maneuvering elements, needed to reach the Mars moons from Mars orbit, lead to additional dead-end mission elements and development. This pathway leaves the development of the six critical surface elements until its final step in the pathway without previous transitional development.

MOON TO MARS	DEAD-END MISSION ELEMENTS	TRANSITIONAL MISSION ELEMENTS	PRIMARY MISSION ELEMENTS

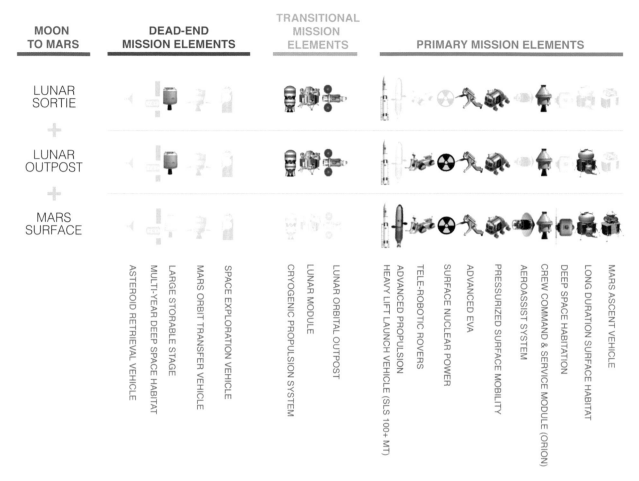

FIGURE 4.5 Buildup of mission elements for the Moon-to-Mars pathway.

The element buildup for the Moon-to-Mars pathway is shown in Figure 4.5. The Lunar Sortie mission requires the development of a number of elements, most of which provide some advance in capabilities that will be needed for the Mars Surface mission. The cryogenic propulsion system will advance propellant management and storage for an NTP-based in-space propulsion system. The reusable lunar lander will advance terminal landing and ascent operations. The disposable descent stage does not provide any significant advance in technology that supports the Mars Surface mission, so it is shown as a dead-end mission element.

The Lunar Outpost mission requires the development of many more of the mission elements needed by a human mission to the Mars surface than any of the other missions included in the pathways.[45] It entails long-term surface operations in a dust-prone partial-gravity environment and would demonstrate habitation, robotic augmentation, and nuclear power generation technologies and systems that would be required for the Mars Surface mission.[46]

---

[45] Three core primary mission elements are part of every mission: the heavy-lift launch vehicle (SLS), the crew command and service module (Orion), and advanced EVA. The lunar-outpost mission requires the development of and provides the opportunity for operational demonstrations of four additional primary mission elements. In contrast, the DRMs for the other five intermediate destinations would demonstrate no or one primary mission element in addition to the core three.

[46] Certainly, there are important differences between the surface environments of the Moon and Mars with regard to, for example, atmosphere (or the lack thereof), gravity, day-night cycle, and dust properties. Nevertheless, the Moon provides the best opportunity to test surface systems and human physiology in a partial-gravity environment before committing astronauts to a Mars surface mission.

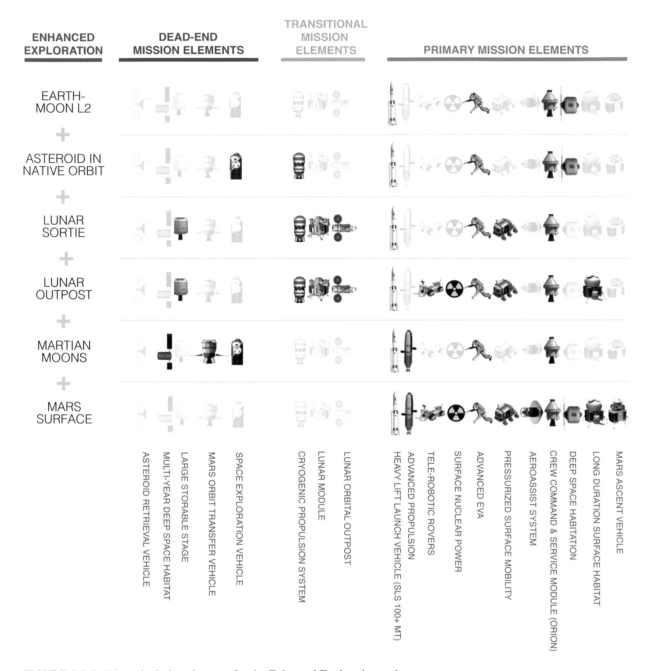

FIGURE 4.6 Buildup of mission elements for the Enhanced Exploration pathway.

Important advances can be made in short-term habitation in lunar orbit, but the longer-term habitat would have to be demonstrated as part of the Mars Surface mission. The Moon-to-Mars pathway would also leave the major efforts of in-space propulsion and EDL until the Mars Surface mission.

The Enhanced Exploration pathway element buildup shown in Figure 4.6 illustrates the more gradual pace and lower risk of this pathway, with each of the six DRMs incrementally developing required hardware. The in-space habitation capabilities begin with a short-duration capability that uses a more efficient and more reliable ECLSS

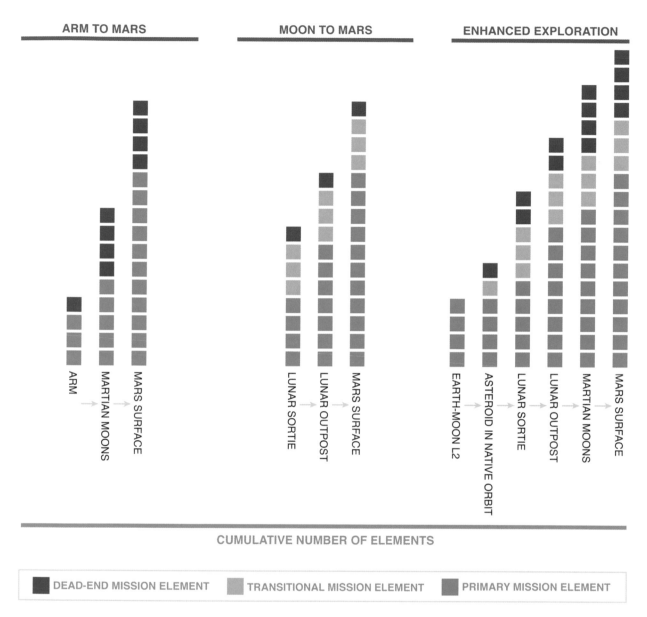

FIGURE 4.7 Comparison of the mission-element buildup for each pathway.

starting with the Earth-Moon L2 mission, with increasing duration and volume for the Native Asteroid mission, and culminating in the Mars Moons mission. The surface exploration capabilities are matured on the lunar missions, leaving the advanced in-space propulsion system as the only significant development for the Mars Moons mission. The only completely new development for the Mars Surface mission is the one capability that cannot be demonstrated anywhere else—Mars EDL.

A simplified summary of the critical element progression for each of the three pathways is shown in Figure 4.7. Although the full interaction of the technology advances and precursor missions required to achieve the capability to land humans on the Mars surface is complex, a simple count of the critical elements for each DRM in each of the three pathways shows how they bound the problem.

All pathways have the goal of landing humans on Mars, which is consistent with the DRA 5.0 architecture, so each pathway in Figure 4.7 indicates a cumulative total of 11 primary mission elements (which are depicted by green squares). The ideal progression would be a smooth transition, with minimal jumps, from a few to the 11 final green elements. This progression would minimize both technical risk and the need for major, temporary increases in human spaceflight funding, and it would spread major technical challenges out in time. For efficiency, there should be few, if any, red or dead-end mission elements.

4.2.6 Challenges in Developing Key Capabilities

Developing the capabilities needed for a human mission to the Mars surface will require considerable resources and technological innovation in many disciplines to accommodate the environments to be encountered in space and during surface operations.[47,48,49] Technology has made huge leaps since the early days of human spaceflight, as has understanding of the risks and challenges posed by the space environment. The ISS has proved to be an essential platform for investigating and enhancing the ability of humans to survive most of the hardships of space exploration. However, some unknowns remain, particularly with regard to the long-term effects of space radiation and the partial gravity present on the Moon and Mars. Enabling humans to land on the surface of Mars and return safely requires advanced capabilities in many areas, including the following:

- Launch vehicles capable of placing large masses in LEO at minimum cost.
- Reliable power generation for deep-space and surface operations.
- Efficient in-space propulsion to increase transit time and payload capacity, while reducing human exposure to the deep-space environment.
- Habitats, systems, and procedures to ensure the health of human explorers and preserve their physical and mental capabilities during long stays in space and on the surface of Mars.
- EDL systems to land very large payloads on Mars and subsequently return them to Earth.
- Vehicles and systems for landing large masses on planetary bodies, lifting an ascent vehicle back into Mars orbit, and returning the crew to Earth.
- Systems for surface operations, including science instruments, robotic vehicles, crewed rovers, spacesuits, and ISRU systems to produce oxygen and, if possible, other consumables and materials.

Making the necessary advances in some of the above areas will be more challenging than in others. To determine which capabilities should have the highest priority for current research and development programs, the Technical Panel assessed a wide variety of capabilities in terms of four parameters:

- Technical challenges
- Capability gap
- Regulatory challenges
- Cost and schedule challenges

Mission need was not used as an evaluation parameter because it did not help in differentiating capabilities. A great many technical capabilities—not just those ranked as having high priority—are essential for the success of a human mission to the Mars surface.

Based on the expertise of the Technical Panel members and additional information reviewed by the panel, the difficulty of making needed advances for each capability was ranked as high, medium, or low for each of the four parameters. The criteria for assigning these rankings are listed in Figure 4.8.

---

[47] NRC, *NASA Space Technology Roadmaps and Priorities: Restoring NASA's Technological Edge and Paving the Way for a New Era in Space*, The National Academies Press, Washington, D.C., 2012.

[48] NRC, *Microgravity Research in Support of Technologies for the Human Exploration and Development of Space and Planetary Bodies*, National Academy Press, Washington, D.C., 2000.

[49] NRC, *Recapturing a Future for Space Exploration: Life and Physical Sciences Research for a New Era*, The National Academies Press, Washington, D.C., 2011.

	HIGH	MEDIUM	LOW
TECHNICAL CHALLENGES	Technical solution currently unknown or unattainable with current technology	Solution is known but not well understood	Solution is well understood with current or previous relevant research
CAPABILITY GAP	No relevant systems exist or have existed at the appropriate scale	Systems exist or have existed that are scalable to mission needs	Systems exist that are translatable or are easily scalable to mission needs
REGULATORY CHALLENGES	Current regulations impose significant restrictions and will be difficult to change	Current regulations impose a challenge	No regulatory issues
COST & SCHEDULE CHALLENGES	Development to operational capability is on the order of previous large, national programs (Shuttle Orbiter)	On the order of Apollo Heat Shield or Orion ECLSS	< 5 years development with < 50 person team

FIGURE 4.8 Capability assessment criteria.

The capability assessment ranked the following capabilities as a high priority:

- Mars EDL
- Radiation safety
- In-space propulsion and power[50]
  — Fission power
  — In-space cryogenic propulsion
  — NEP
  — NTP
  — SEP
- Heavy-lift launch vehicles
- Planetary ascent propulsion
- ECLSS
- Habitats
- EVA suits
- Crew health
- ISRU (Mars atmosphere)

Advances in many other capabilities will be essential for human exploration beyond LEO. These are addressed below in section 4.2.6.2 ("Additional Capabilities").

In 2010, the National Research Council issued a report that assessed and prioritized the space technologies that were included in a comprehensive set of draft roadmaps prepared by the NASA Office of the Chief Technologist.[51] All but one of the high-priority capabilities listed above are closely linked to one or more of the high-priority technologies identified in the roadmaps report; the lone exception is heavy-lift launch vehicles.

Figures 4.9 and 4.10 show the relationship of the above capabilities to the Mars mission elements (see Figure 4.3) that are needed to complete the missions in one or more of the pathways.

---

[50] As development of NEP, NTP, SEP, and cryogenic propulsion technologies proceeds, a down-selection will be required.
[51] NRC, *NASA Space Technology Roadmaps and Priorities*, 2012.

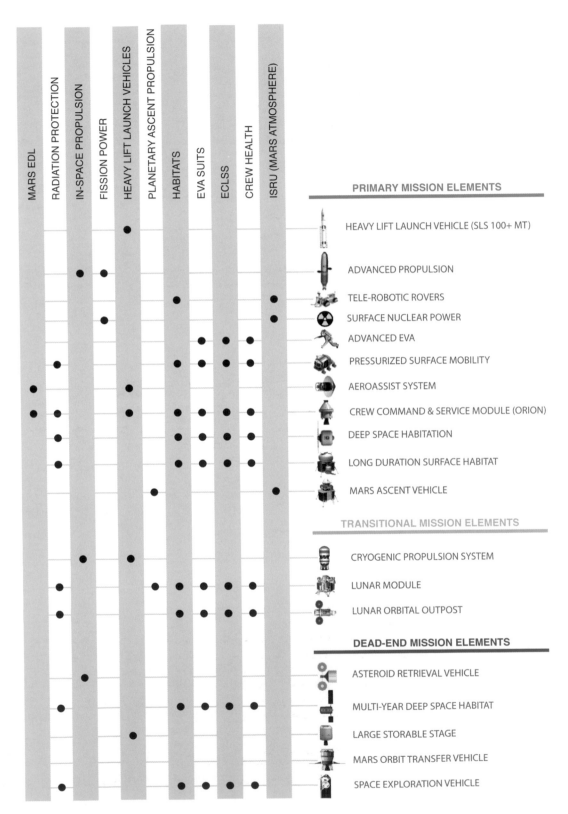

FIGURE 4.9 Relationship of mission elements to high-priority capabilities.

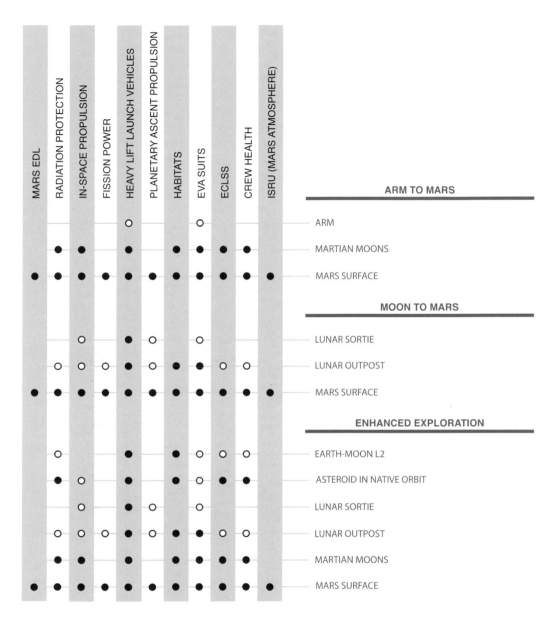

FIGURE 4.10 Relationship of missions to high-priority capabilities.
O For a given capability, a substantial development effort would be required to execute a particular mission, at which point the development effort would need to continue essentially unabated to prepare for a Mars surface mission.
● For a given capability, a substantial development effort would be required to execute a particular mission, at which point minimal additional development would be needed to prepare for a Mars surface mission.

4.2.6.1 High-Priority Capabilities

This section summarizes the assessments of the high-priority capabilities. Although all these capabilities were ranked as a high priority, the first three—Mars EDL, radiation safety, and in-space propulsion and power—are in a class by themselves. In particular, the cost and schedule of developing a Mars EDL capability will be extraordinary because of the physical size of the operational system, the need to test at least one operational system in the atmosphere of Mars, the narrow windows to launch test systems, and the extreme cost and schedule delays that would result if the operational test failed. In addition, an approach for overcoming the technical challenges of Mars EDL for a crewed mission remains to be determined. Radiation safety is unique in that conventional approaches to reducing radiation exposure (i.e., shielding) would *increase* radiation exposure to astronauts while they are in transit and on the surface of Mars, and the efficacy of unconventional technical and biological solutions remains speculative. As with Mars EDL, the large size of in-space propulsion systems greatly increases the cost of developing, manufacturing, and testing operational systems. That cost will be further increased for systems that include nuclear power. The cost of developing chemical in-space propulsion systems will be modest by comparison, but the cost of using chemical propulsion systems on a per mission basis will be so high that it would threaten the sustainability of the Mars program. The only other options for in-space transportation are NEP and SEP, but because of the inherent limitations of electric propulsion NEP and SEP may not be feasible for crewed vehicles. The cost of in-space propulsion and power capability will be further increased by the need to develop a surface power system. All 10 of the high-priority capabilities are essential, and all will be challenging to develop, but none is comparable either with Mars EDL and in-space propulsion and power in terms of cost or with radiation safety and Mars EDL in terms of technical challenges and capability gap.

4.2.6.1.1 Mars EDL

On August 6, 2012, the world watched as the Mars Science Laboratory (MSL) completed an autonomous landing on Mars. The fate of the most complex machine that humans had ever sent to another planet rested on an innovative 7-minute landing sequence that had been years in the making. EDL encompasses mission design, software, systems development, operations, and integration processes. For MSL, hundreds of people worked for about 8 years to get the job done. During EDL, the MSL spacecraft had to autonomously perform six configuration changes, complete 76 pyrotechnic events, and execute approximately 500,000 lines of computer code with almost no margin for error.

Prior to MSL, the United States had successfully landed six robotic systems on Mars. Those systems all had landed masses of less than 600 kg and landing errors of hundreds of kilometers in diameter.[52] MSL raised the bar considerably, safely delivering a 900-kg rover to the Mars surface within a landing error of just 20 km. Even so, compared with the capability required to deliver humans to the Mars surface, MSL was a baby step.

A human mission to the Mars surface could require landing payloads of 40-80 MT in close proximity (tens of meters) to pre-positioned assets.[53] Because existing EDL technologies do not scale up to payloads of this size, the EDL systems required for a Mars surface mission would have little resemblance to those in use today. Multiple EDL concepts for a human surface mission have been proposed, including supersonic retropropulsion, slender-body aeroshells, inflatable aerodynamic decelerators, and advanced thermal protection systems. However, an EDL system that incorporates such advanced concepts would be extremely complex, and the feasibility of these concepts remains to be proved.

EDL technologies are highly interdependent and are generally validated in the context of the sequence of events associated with a particular mission. Before a human mission to the Mars surface, Mars EDL systems would probably require precursor flight tests in the atmospheres of both Earth and Mars. Developing new EDL technologies and systems and flight tests on Earth and at Mars would require substantial resources and time. NASA's *Entry, Descent, and Landing Roadmap*, published in 2012, concluded that for a human landing on Mars in the 2040s the United States would need to begin EDL technology development within the next few years.[54]

Advances in EDL technology that are developed to support a human mission to the Mars surface would likely facilitate robotic missions to Venus, Titan, or the gas giants (Jupiter, Saturn, Uranus, and Neptune). The development of Mars EDL systems would also facilitate the development of EDL systems for spacecraft returning to Earth from remote destinations in the solar system on high-velocity trajectories.

The assessment of EDL systems needed for a human mission to the Mars surface is summarized in Figure 4.11. Technical challenges are ranked high because the technologies needed for a Mars EDL system that would be

---

[52] R.D. Braun and R.M. Manning, "Mars Exploration Entry, Descent and Landing Challenges," paper presented at the Aerospace Conference, 2006 IEEE, March 4-11, 2006, doi:10.1109/AERO.2006.1655790.

[53] Braun and Manning, "Mars exploration entry, descent and landing challenges," 2006.

[54] NASA, Entry, Descent, and Landing Roadmap, Technology Area 09, NASA, Washington, D.C., 2012, http://www.nasa.gov/offices/oct/home/roadmaps/, p. TA09-1.

MARS EDL (ENTRY, DESCENT, LANDING)

FIGURE 4.11 Assessment of EDL systems for a human mission to the Mars surface.

capable of handling large payloads have yet to be identified. The capability gap is ranked high because the necessary payload capacity of the EDL systems is far beyond the capability of existing EDL systems. Regulatory challenges are ranked low because no regulatory changes are needed. Cost and schedule challenges are ranked high because extraordinary resources and time would be needed to identify suitable technologies, scale them up to the requisite size, and conduct flight testing in the atmosphere of Earth and/or Mars to build confidence that they are safe enough for use on a crewed mission.

4.2.6.1.2 Radiation Safety

Space radiation in the form of ionizing radiation, SPEs, and extremely high-energy GCR would be a serious threat to crew health on long-duration missions beyond LEO.[55] Overcoming this threat would require advances in all aspects of radiation safety: prediction, risk assessment modeling, total exposure monitoring, and protection. Radiation safety systems reduce or counteract the effects of radiation, and associated standards limit the total dose of a given type of radiation that people are authorized to accumulate over a given period. Shielding incorporated into the design of vehicles and habitats could effectively reduce the exposure of astronauts to SPEs. However, because of secondary radiation produced when primary penetrating particles interact with spacecraft structures, shielding, or other materials, conventional shielding has not been shown to be an effective countermeasure for GCR. In addition, spacesuits do not effectively shield astronauts from SPEs or GCR during EVAs.[56] Shielding astronauts using spacecraft-generated electromagnetic fields has been proposed, but such systems would carry severe power and mass penalties, new health issues could arise from the exposure of the crew to powerful electromagnetic fields, and the systems are well beyond the technology horizon considered in this report.[57]

The overall goal of radiation safety research is to reduce radiation exposure to acceptable levels with as little impact as possible on spacecraft and habitat mass, cost, complexity, and so on. Adequate technical, biological, and/or pharmacological solutions have yet to be identified, and there is a large gap between current capabilities and what is needed to provide adequate safety.

The ISS is substantially protected from space radiation (especially SPEs) by Earth's magnetic field, so the space radiation environment in which the ISS operates is more benign than it is above LEO, although passage through the South Atlantic Anomaly increases the crew's radiation exposure. With longer ISS tours planned and because radiation effects are cumulative, crew radiation exposure on the ISS is becoming a matter of greater concern. Because the ISS does not provide an environment typical of deep space, NASA's Human Research Program

---

[55] NASA, *Human Exploration Destination Systems Roadmap*, Technology Area 07, NASA, Washington, D.C., 2012, http://www.nasa.gov/offices/oct/home/roadmaps/, p. TA07-8.

[56] NRC, *NASA Space Technology Roadmaps and Priorities*, 2012.

[57] NRC, *Managing Space Radiation Risk in the New Era of Space Exploration*, 2008.

TABLE 4.3 LEO Exposure Limits in Sieverts

LEO 10-Year Career Whole-Body Effective Dose Limits (Sv)		
Age (years)	Male	Female
25	0.7	0.4
35	1.0	0.6
45	1.5	0.9
55	3.0	1.7

SOURCE: National Council on Radiation Protection and Measurement (NCRP), *NCRP-132: Recommendations of Dose Limits for Low Earth Orbit*, Bethesda, Md., 2000.

conducts radiation studies using animal models in ground particle accelerators, but these do not achieve the energies of some of the GCR particles that are of concern for astronauts.

The National Council on Radiation Protection and Measurements has recommended career exposure limits for astronauts in LEO (see Table 4.3). The limits are based on the probability of a 3 percent excess cancer mortality for the type of radiation experienced in LEO. For longer-duration missions outside LEO, it may be necessary to re-evaluate the limits on the basis of tradeoffs between the difficulty of meeting existing limits, the applicability of the limits to the deep-space radiation environment, the potential health risks and regulatory challenges associated with modifying the limits, and a better understanding of the noncarcinogenic effects of GCR, such as cumulative neural degeneration, which may prove to be more limiting than carcinogenic effects.

Based on current estimates of the space environment, existing radiation limits would likely be exceeded after about 600 days in space, even with the most permissive crew composition (never-smoking men more than 55 years old with no previous radiation exposure) and with the assumption that carcinogenesis is the only radiation risk that needs to be controlled. More conservative but realistic assumptions might lead to considerably shorter permissible durations. However, it may be possible to increase the above limits safely by reducing current uncertainties, such as the risk of adverse biological effects and the efficacy of possible radiation countermeasures. More accurate predictions of SPEs and solar storms associated with intense periods of ionizing radiation would provide more time to prepare for these events and reduce false alarms and thereby improve mission effectiveness.[58] Finally, astronaut selection and mission assignment may ultimately involve consideration of individual susceptibility to radiation as inferred from genome analysis.[59] Note, however, that use of personal susceptibility information to inform mission assignment would appear to violate the Genetic Information Nondiscrimination Act of 2008, from which only the military is exempt.

The assessment of the ability to ensure radiation safety for a human mission to the Mars surface is summarized in Figure 4.12. Technical challenges are ranked high because a suitable approach for providing adequate radiation safety has yet to be identified.[60] The capability gap is ranked high because the ability to provide the level of radiation safety required for a human mission to the Mars surface is so far beyond the state of the art. Regulatory challenges are ranked medium because part of the solution may be to relax current radiation exposure limits (based on greater knowledge of the human health effects of the radiation environment in space and on the Mars surface and/or a reconsideration of the level of acceptable risk). Cost and schedule challenges are ranked medium because the time and resources necessary to develop adequate radiation safety systems are substantial—although not of the same order as, for example, those required to develop Mars EDL systems.

4.2.6.1.3 In-Space Propulsion and Power

Once a spacecraft is launched into Earth orbit, in-space propulsion systems are used to move it to its intended destination and to return it to Earth (for crewed missions and robotic sample return missions). Most of the bio-

---

[58] NRC, *NASA Space Technology Roadmaps and Priorities*, 2012.

[59] M.R. Barratt and S.L. Pool, *Principles of Clinical Medicine for Space Flight*, Springer, New York, 2008. p. 67.

[60] NRC, *Managing Space Radiation Risk in the New Era of Space Exploration*, 2008.

RADIATION PROTECTION

FIGURE 4.12 Assessment of radiation-safety systems and capabilities for a human mission to the Mars surface.

medical and life support risks posed by human exploration missions to distant destinations would be greatly mitigated by advanced high-thrust in-space propulsion systems that substantially reduce transit times to and from the destination and thereby reduce exposure to zero-g, space radiation, and psychosocial stress.

For exploration missions, a large portion of the launch mass is the fuel and oxidizer required for in-space propulsion. Therefore, the efficiency of in-space propulsion systems (in terms of I_{sp}) is of key importance. High thrust is also important to reduce transit time. Despite decades of research and study, it remains to be seen what type of advanced in-space propulsion system would provide the best combination of I_{sp} and thrust for future exploration missions. The four technologies of greatest interest are cryogenic propulsion, NEP, NTP, and SEP. Each of these options is discussed below along with fission power systems that could be adapted for operation in space (to power NEP systems) or to provide power on the surface of the Moon or Mars.

As the technologies for in-space propulsion are developed and matured, there will probably need to be a down-selection among the four options because of the high development costs required for each one. SEP and NEP provide the highest I_{sp}, but the megawatts of power needed to provide sufficient thrust for a crewed exploration mission creates high development risk. NTP delivers I_{sp} that is double that of cryogenic systems but comes with a high development risk due in large part to difficulties of safely ground testing an open-cycle nuclear fission system. Cryogenic propulsion is a more mature technology than the other options, but it offers lower performance and poses additional technical challenges, primarily in connection with low-loss, long-term storage and in-space transfer of cryogenic fuels and oxidizers.

4.2.6.1.3.1 Fission Power

Robotic spacecraft typically require only a few hundred watts, whereas larger satellites can require a few kilowatts. The highest-power satellites are geosynchronous communication systems; they require about 20 kW of electric power.

Human space missions require much more power than typical uncrewed spacecraft. The space station has a 100-kW capability, and concepts for long-term missions to the surface of the Moon or Mars typically call for 50 100 kW of installed power. Solar and nuclear fission systems are currently the only viable options for providing long-term power at those levels both on the surface of the Moon or Mars and in space as part of an SEP or NEP system.

Nuclear fission reactors are fundamentally a long-term source of thermal energy. A complete space nuclear power system converts the thermal energy produced by the reactor into electricity. The nuclear reactors that would be used for space applications would be similar in some ways to low-power nuclear fission systems, such as research reactors, currently used on Earth. However, substantial modifications would be required. For example, a space nuclear reactor system must be designed to minimize risk to the public during launch and to operate safely and reliably in the intended environment (in space or on the surface of the Moon or Mars).

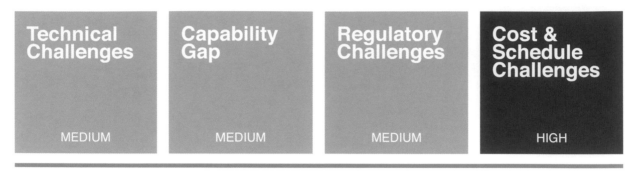

FIGURE 4.13 Assessment of a 100-kW fission power system.

The United States has launched just one fission reactor, the SNAP-10A in 1965. The Soviet Union was the only other country to operate fission reactors in space, and those systems were also launched decades ago. The SP-100 program—which was jointly funded by NASA, the Department of Defense (DOD), and the Department of Energy—was developing fission reactors for a surface power system until it was canceled in 1992. U.S. space reactor designs were based on the use of thermoelectric systems to convert heat into electricity. In the early 1990s, DOD also investigated the feasibility of developing a space reactor with a thermionic energy conversion system that used unfueled components of "Topaz-2" space reactors purchased from Russia, but this effort was terminated without such a system being launched by Russia or the United States.[61] Most of the key people and facilities from prior programs are no longer available. A new program to design, test, and produce space nuclear power systems would be a substantial undertaking. Allowing for increased regulatory complexity, it would cost billions of dollars and take at least a decade to develop a fission power system capable of producing 50-100 kw.[62] Existing federal regulations specify the approval process needed to develop and launch fission reactors. The approval process, however, is very time-consuming and costly, and political opposition to the launch of a nuclear reactor may arise during the approval process.

The assessment of fission power systems needed for a human mission to the Mars surface is summarized in Figure 4.13. Technical challenges are ranked medium because of extensive experience with reactor technologies although some new technologies would be needed to provide reliable, long-term operation in space and on the surface of Mars. The capability gap is ranked medium because, despite past accomplishment in nuclear power technology in general and space nuclear power in particular, it has been almost 50 years since a U.S. space nuclear power program succeeded in conducting a flight test of a fission power reactor. Regulatory challenges are ranked medium because of the difficulty of completing the regulatory process that has been established to obtain launch approval. Cost and schedule challenges are ranked high because of the extraordinary resources and time that would be required to develop an operational reactor system and obtain the necessary launch approvals.

4.2.6.1.3.2 In-Space Cryogenic Propulsion

Given the long and successful history of engines such as the RL-10 and J-2, cryogenic engines for in-space propulsion systems are well defined. In fact, the cryogenic propulsion option for in-space propulsion described in DRA 5.0 proposes the use of existing RL-10-B2 engines for all three of the major in-space propulsion modules: the trans-Mars injection module for the trip to Mars, the Mars orbit insertion module, and the trans-Earth injection

---

[61] NRC, *Priorities in Space Science Enabled by Nuclear Power and Propulsion,* The National Academies Press, Washington, D.C., 2006, p. 114.

[62] V.C. Truscello, *SP-100, The U.S. Space Nuclear Reactor Power Program. Technical Information Report,* Jet Propulsion Laboratory Report 1085, November 1, 1983, Pasadena, Calif., http://www.osti.gov/scitech/servlets/purl/10184691.

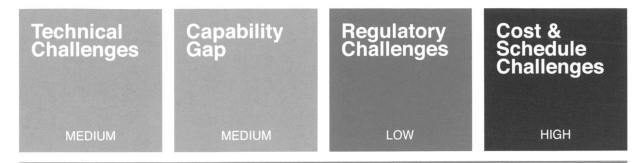

IN-SPACE CRYOGENIC PROPULSION

FIGURE 4.14 Assessment of in-space cryogenic propulsion systems that could be used for a human mission to the Mars surface.

module for the trip home.[63] New technologies are needed, however, to enable the use of these engines for a human mission to Mars. Key requirements include low-loss, long-term storage and in-space transfer of cryogenic fuels and oxidizers. In addition, existing RL-10 engines have operational lifetimes measured in hours. The propulsion modules used for a human Mars mission would need to be stored in LEO for perhaps 4-6 months during vehicle assembly, and the Mars orbit insertion module and trans-Earth injection module would need to operate reliably after being exposed to the space environment for years. Currently, chemical propulsion systems are the only option available for human exploration missions, but chemical propulsion has a lower I_{sp} (450 seconds in vacuum for the shuttle main engine, which uses liquid hydrogen and liquid oxygen) than other in-space candidates. As a result, the use of chemical systems would require large amounts of fuel to be launched from Earth and carried to Mars.

NASA has plans to fly in-space experiments to advance cryogenic propellant storage capabilities with the goal of improving lifetime from hours to months. New technology would still be needed, however, to close the capability gap between the performance of experimental systems now being developed and the performance of operational systems for a Mars mission. Developing the propulsion modules needed for a human mission to the Mars surface and demonstrating the ability to assemble them in space and operate them reliably after the long transit to Mars would also be an expensive and lengthy undertaking.

The assessment of in-space cryogenic propulsion systems that could be used for a human mission to the Mars surface is summarized in Figure 4.14. Technical challenges are ranked medium because, although high-performance in-space cryogenic propulsion systems are already operational, new technologies are needed for in-space fuel-handling and long-term storage. The capability gap is ranked medium because of the improvements needed to extend the in-space storage and operational lifetime of existing systems. Regulatory challenges are ranked low because no regulatory changes are needed. Cost and schedule challenges are ranked high because of the long time that will likely be required to develop the ability to store cryogenic fuel in space for years at a time.

4.2.6.1.3.3 Nuclear Electric Propulsion

Electric propulsion systems are used extensively on Earth-orbiting satellites and on some robotic science missions. Electric propulsion systems accelerate ions to very high velocities. The resulting I_{sp} is in the range of 3,000-6,000 seconds, which is higher than that of any other in-space propulsion technology. As a result, NEP systems would use about 10 percent of the fuel that a cryogenic propulsion system would use to produce an equivalent change in spacecraft velocity. NEP systems would be powered by a fission power system. To date, all electric propulsion systems have been powered by solar energy (see the discussion of SEP systems, below), and that will remain the case until fission power systems for space are developed. Unlike SEP systems, the power available to an NEP system is constant regardless of the distance from the Sun to the spacecraft.

---

[63] NASA, "Human Exploration of Mars Design Reference Architecture 5.0," 2009, http://www.nasa.gov/pdf/373665main_NASA-SP-2009-566.pdf.

NUCLEAR ELECTRIC PROPULSION (MW)

FIGURE 4.15 Assessment of megawatt-class NEP systems for a human mission to the Mars surface.

The major shortcoming of all electric propulsion systems (NEP and SEP) from the standpoint of human transportation is that the governing physics result in accelerations that are very low compared with the alternatives for plausible combinations of system characteristics and power levels. Research to increase the thrust of electric propulsion engines is under way. NASA is actively working on scaling engines to the 60-kW range. However, the technology developed by these efforts will still leave electric propulsion systems far short of the thrust required for crew transport vehicles travelling to Mars or other distant destinations. This would require megawatt-class electric-propulsion technologies, which are not being developed.

SEP and NEP systems both offer the potential to preposition cargo and habitation subsystems. The lack of human passengers would accommodate the low acceleration of these systems. However, the DRA 5.0 architecture for a Mars surface mission uses NTP (see below) for both crewed and cargo missions to reduce the number of separate technology-development programs needed. It would probably not be economical to develop a propulsion system for cargo that could not also be used for crewed vessels.

The assessment of megawatt-class NEP in-space propulsion systems for a human mission to the Mars surface is summarized in Figure 4.15. The cost and schedule challenges, regulatory challenges, and technical challenges are driven largely by the challenges associated with the fission power system (see Figure 4.13) that lies at the heart of the NEP system. A megawatt-class NEP system, however, faces a higher capability gap than the 100-kW fission system discussed above because of the higher power levels that the fission power and electric propulsion engines would be required to meet relative to the state of the art.

4.2.6.1.3.4 Nuclear Thermal Propulsion

NTP systems generate high thrust by using a nuclear fission reactor to heat a propellant (typically hydrogen) to very high temperatures. The propellant gases are then expanded through a nozzle to produce thrust. The resulting I_{sp} (800-1,000 seconds) is higher than that of chemical propulsion systems but lower than the I_{sp} of NEP or SEP systems.

NTP is the baseline in-space propulsion system for the Mars DRA 5.0 mission. However, other than some low-level technology research efforts, NTP technologies are not being developed. A full-scale system, NERVA, was built and ground tested about 40 years ago. The NERVA program built and tested 23 reactors and engines with peak power up to 4,000 MW, system operation up to 1 hour, and an I_{sp} of 850 seconds.[64] Key facilities and personnel from the NERVA program are no longer available, and current environmental regulations are much more stringent. It would be difficult to produce a test facility that could contain the propulsion exhaust of a full-scale NTP system, and the political and regulatory opposition to the construction of such facilities could be a problem. However, a concept for subsurface active filtering of exhaust coupled with adsorption of exhaust in a bed of metal

[64] S.D. Howe, High energy-density propulsion—Reducing the risk to humans in planetary exploration, *Space Policy* 17(4):275-283, 2001, doi:10.1016/S0265-9646(01)00042-X.

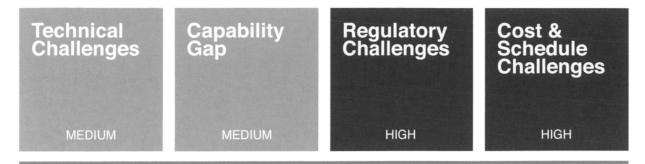

FIGURE 4.16 Assessment of NTP systems for a human mission to the Mars surface.

hydride was viewed as a workable approach during the Strategic Defense Initiative Office's Project Timberwood, which sought to develop NTP well after the NASA NERVA program.[65]

The assessment of NTP systems needed for a human mission to the Mars surface is summarized in Figure 4.16. Technical challenges are ranked medium because the NERVA program developed most of the technologies that would be needed by an operational NTP system. The capability gap is ranked medium; the NERVA program tested a full-scale system, but the state of the art has degraded somewhat during the ensuing 40 years. Regulatory challenges are ranked high because it would be technically and politically difficult to develop test facilities for a large nuclear rocket program. Cost and schedule challenges are ranked high because it would be extraordinarily expensive and time-consuming just to repeat the NERVA work of the past, let alone proceed with development of an operational system.

4.2.6.1.3.5 Solar Electric Propulsion

All electric propulsion systems to date have been powered by solar energy, and that will remain the case until space nuclear power systems are developed, as discussed above. SEP is commonly used in orbital spacecraft, and SEP systems have been used on the Deep Space 1 and Dawn scientific spacecraft, which were launched in 1998 and 2007, respectively. Both spacecraft used 2-kW ion thrusters and SEP systems. Most of the ongoing activity in support of SEP is focused on scaling up the thrust levels. As with an NEP system, thrust levels needed for a nominal 6-month human trip to Mars would require scaling up existing systems to megawatts. This would require tremendous advances in both electric engines (see the discussion of NEP system, above) and solar power systems. As with a megawatt-class NEP system, developing a megawatt-class SEP system would require a long development program supported by substantial resources to overcome the capability gap between required performance levels and the existing state of the art.

The assessment of megawatt-class SEP systems that could be used for a human mission to the Mars surface is summarized in Figure 4.17. Technical challenges are ranked low because SEP systems are well developed and have a long history of operation in space. The capability gap is ranked high because the power level of state-of-the-art systems is far below the power level needed for a crewed spacecraft transiting to and from Mars. Regulatory challenges are ranked low because no regulatory changes are needed. Cost and schedule challenges are ranked high because extraordinary resources and time would be needed to close the capability gap.

An SEP system with a power level of hundreds of kilowatts, which would be suitable for transporting uncrewed cargo vessels to Mars or other distant destinations, would face medium cost and schedule challenges and a medium capability gap. However, as noted above, it would be preferable to have a single in-space propulsion system for both cargo vessels and crewed vehicles.

---

[65] Final Environmental Impact Statement (EIS) for the Space Nuclear Thermal Propulsion (SNTP) Program, Sanitized Version, September 19, 1991, http://oai.dtic.mil/oai/oai?verb=getRecord&metadataPrefix=html&identifier=ADA248408.

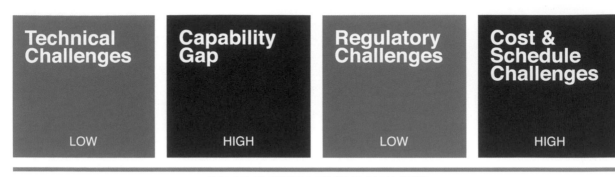

SOLAR ELECTRIC PROPULSION (MW)

FIGURE 4.17 Assessment of megawatt-class SEP systems for a human mission to the Mars surface.

4.2.6.1.4 Heavy-lift Launch Vehicles

Heavy-lift launch systems (that is, launch systems with a payload capability of about 50 MT or more to LEO) would reduce the number of launches required for human exploration missions beyond LEO. Launch vehicles that can accommodate payloads with large mass and volume enable the launch of large mission elements as single units, which reduces or eliminates the cost, time, and technical risk associated with in-orbit assembly. Two heavy-lift launch systems are under development in the United States: the NASA SLS and the SpaceX Falcon Heavy.

4.2.6.1.4.1 Space Launch System

The NASA Authorization Act of 2010 directed NASA to develop the SLS. The system design selected by NASA retains many of the characteristics of the heavy-lift vehicle that was under development as part of the Constellation program, which was canceled in 2010. The Block 1 SLS will have a payload capacity of 70 MT to LEO. Two major upgrades are planned, to increase payload capacity to 105 MT and, subsequently, to 130 MT.

The core stage of the SLS is based on the space-shuttle external tank, stretched and modified to house the main propulsion system at the aft end and an interstage structure at the forward end. The core stage is characterized by NASA personnel as the "long pole" that is pacing SLS development toward the first flight in 2017.

As currently planned, the propulsion system will consist of space-shuttle main engines (RS-25s) left over from the Space Shuttle Program. As with the space shuttle, during the first 2 minutes of flight the RS-25 liquid propulsion system will be augmented by two solid rocket boosters mounted on either side of the core stage. The boosters for the Block I SLS will be modified space-shuttle solid rocket boosters. Boosters for the Block IA and Block II vehicles will have higher performance. They will be procured after a source selection that will be open to both liquid-fuel and solid-fuel systems.

The Block I SLS will use a Delta IV upper stage for the first two missions: the uncrewed Exploration Mission (EM)-1 and the first crewed mission, EM-2. An upper stage developed for the SLS will be used for later flights that require more than the 70-MT payload capacity available with the Block I configuration. The upper stage for Block IB (which will have a payload capacity105 MT) will use four RL-10A-4-2 engines, which are used on the Centaur upper stage for the Atlas V Evolved Expendable Launch Vehicle. Higher-capacity SLS vehicles (130 MT or more) will use J-2X engines, which will be upgraded versions of the upper-stage engines used during the Saturn program.

No technological breakthroughs are required to complete SLS development. In fact, the SLS was intentionally configured so that each key component could be derived from systems with a long heritage of successful flight, such as the external tank, main engine, and solid rocket booster used on the space shuttle and upper-stage engines used for other launch vehicles. Nonetheless, given the physical size of the SLS, its development will be a major undertaking. In fact, the cost of developing SLS and related systems (the Orion Multi-Purpose Crew Vehicle and the ground systems) are so high relative to the budget of the Human Exploration and Operations Mission Director-

FIGURE 4.18 Assessment of the SLS heavy-lift launch vehicle.

ate that the schedule has been stretched to accommodate available funding. As a result, the currently planned time between SLS launches is much greater than in past human spaceflight programs: the first two SLS flights (EM-1 and EM-2) will be launched 4 years apart, in 2017 and 2021.

4.2.6.1.4.2 Falcon Heavy

The Falcon Heavy launch system, now being developed by SpaceX, is a heavy-lift variant of the Falcon 9 launch system. The Falcon 9 system uses Merlin 1D rocket engines fueled by kerosene and liquid oxygen on both its first and second stages (nine engines on the first stage and one on the second). The Falcon Heavy configuration uses a standard Falcon 9 core with two additional Falcon 9 first stages strapped on as boosters. The heaviest-lift variant of the Falcon Heavy will have a payload capacity of 53 MT. Falcon Heavy is designed to tolerate the loss of thrust from several engines and still complete its mission, thus enhancing mission reliability.

The Merlin 1D engine is a higher-thrust version of the flight-tested Merlin 1C engine. Ground testing of the Merlin 1D was completed in June 2012, and SpaceX announced that the engine had achieved flight qualification in March 2013. The successful first launch of a Falcon 9 version 1.1 (with Merlin 1D engines) took place at Vandenberg Air Force Base during September 2013 and was followed by a second successful launch in December 2013.

When launching heavy payloads, the Falcon Heavy uses propellant cross-feed from the side boosters to the center core. As a result, the center core still has most of its fuel after the side boosters separate, and this increases its maximum payload capacity. This unique feature is being implemented through the innovative adaptation of existing technologies.[66]

As with NASA's SLS, no technological breakthroughs are required to complete development of the Falcon Heavy. The first test flight is scheduled for 2015, and the first two operational flights are scheduled for 2015 (for the U.S. Air Force) and 2017 (for Intelsat).

The assessment of the SLS heavy-lift launch vehicles is summarized in Figure 4.18. Technical challenges and the capability gap are ranked low because the SLS was designed to avoid the need for either new technologies or substantial improvements to existing technologies. Regulatory challenges are ranked low because no regulatory changes are needed. Cost and schedule challenges are ranked high because of the very high cost that NASA has projected to complete development and flight testing of the SLS and the long period before the first operational flight.

The assessment of the Falcon Heavy is summarized in Figure 4.19. It is the same as Figure 4.18 for the SLS except that the cost and schedule challenges are ranked low because the Falcon Heavy development program is much closer to completion and a Falcon Heavy launch would cost less than an SLS launch. However, given the smaller payload capacity of the Falcon Heavy, use of the Falcon Heavy for a Mars surface mission might increase cost, technical challenges, and capability gaps for other elements of a Mars program. A smaller launch system would

---

[66] Erik Seedhouse, *SpaceX: Making Commercial Spaceflight a Reality*, Springer-Praxis Books, New York, N.Y., 2013.

FALCON HEAVY-LIFT LAUNCH VEHICLE

FIGURE 4.19 Assessment of the Falcon Heavy heavy-lift launch vehicle.

require more launches, more time in orbit, more docking events, smaller spacecraft modules, and more time for orbital assembly and checkout of the larger number of modules. As a result, mission reliability might be reduced.

As noted above, the Augustine study suggested that human exploration could be accomplished with a 50-MT launch system.[67] The business case for the SLS versus that for multiple launches of smaller rockets depends primarily on the number of missions to be accomplished and the operational tempo. At present, the crossover point at which an SLS-based approach would be more economical than an approach using smaller launch vehicles has yet to be determined. Regardless, the business case for developing the SLS is weakened if the United States is not committed to a robust program of human exploration, large robotic spacecraft, or other high-mass missions (such as large-scale optics). China appears to be examining the tradeoffs among heavy-lift launch vehicles of various capacities. The Long March 9, which is still under study, would have a LEO payload capacity of 130 MT.[68] Russia is developing a new family of launch vehicles. Proposed variants would have LEO payload capacities up to 41.5 MT (the Angara A7V).

4.2.6.1.5 Planetary Ascent Propulsion

The required characteristics of an ascent propulsion system are absolute reliability after a long period of dormancy, high-to-medium thrust levels, and high efficiency. The Mars DRA 5.0 study did not include detailed analysis of the Mars ascent vehicle. However, based on prior studies, the DRA 5.0 ascent propulsion system would be fueled by liquid methane and liquid oxygen. One mission concept would reduce launch mass on Earth and the mass of landed systems on Mars by relying on ISRU systems to produce liquid oxygen on Mars; liquid methane would be brought from Earth. NASA is not working on an ascent propulsion system that is scalable for human transportation.

Development of a planetary ascent propulsion system will be similar to the development of the propulsion system for a new launch vehicle, except that the ascent system will need to operate from a remote location with no ground crew, it must be integrated into a vehicle that descends to the surface of Mars, and it must operate reliably after a long period of dormancy during the transit from Earth and while on the surface of Mars before launch. In addition, new technology will be needed to develop cryogenic propellant storage capabilities that can store cryogens for years with little or no boil-off.

The assessment of planetary ascent systems needed for a human mission to the Mars surface is summarized in Figure 4.20. Technical challenges are ranked medium because experience with lunar ascent engines and existing in-space propulsion systems provide a solid foundation for developing the technologies needed for Mars. In addition, new technologies are needed for long-term storage of cryogenic fuels. The capability gap is ranked medium

---

[67] Review of U.S. Human Space Flight Plans Committee, *Seeking a Human Spaceflight Program Worthy of a Great Nation,* 2009.

[68] B. Perret, Launcher Leap, *Aviation Week & Space Technology,* pp. 22-23, September 22-23 (reporting Chinese mission and launcher concepts from the International Astronautical Congress in September 2013).

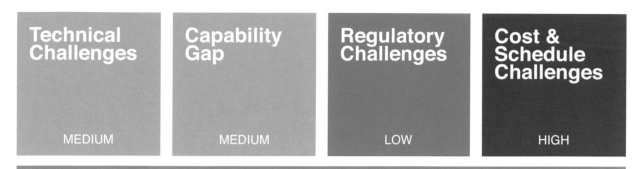

FIGURE 4.20 Assessment of planetary ascent propulsion systems for a human mission to the Mars surface.

because of the improvements needed to advance available technologies enough to provide the power needed for ascent from Mars. Regulatory challenges are ranked low because no regulatory changes are needed. Cost and schedule challenges are ranked high because of the long time that will probably be required to develop the ability to store cryogenic fuel in space for years at a time.

4.2.6.1.6 Environmental Control and Life Support System

A reliable closed-loop ECLSS is needed for spacecraft, surface habitats, and EVA suits to enable long-duration human missions beyond LEO. For missions to Mars and other missions without an early-return abort option, the ECLSS must be highly reliable and easily repairable. The U.S. and Russian ECLSS on the ISS have demonstrated rates of hardware failures that would be unsustainable on a Mars mission.

The ECLSS maintains a safe atmosphere by monitoring and controlling partial pressures of nitrogen, oxygen, carbon dioxide, methane, hydrogen, and water vapor; maintaining total cabin pressure; filtering out particles and microorganisms; and distributing air. The ECLSS also provides potable water and performs habitation functions, such as food preparation and production, hygiene, collection and stabilization of metabolic waste, laundry services, and trash recycling. ECLSS waste management subsystems safeguard crew health, recover resources, and protect planetary surfaces. Key functions include reducing the mass and volume of consumables, including food; controlling odors and the growth of microorganisms; and recovering water, oxygen, other gases, and minerals.[69]

The ECLSS in use today requires constant repair and a large store of spares and uses too many consumables to be practical for missions lasting more than a few weeks. Some progress is being made to improve the reliability and performance of in-space ECLSS, and the ISS is an excellent platform for testing in-space ECLSS subsystems. However, there is still a large gap between current capabilities and the performance that would be needed for long-duration missions in space. It remains to be seen how soon that gap can be closed and what new research capabilities and technologies will be needed. In addition, little effort is being made to develop an ECLSS that is tailored for operation in the partial-gravity environments found on the surface of the Moon or Mars.

The assessment of ECLSS for a human mission to the Mars surface is summarized in Figure 4.21. Technical challenges are ranked medium because ECLSS technologies and systems are already operational. The capability gap is ranked high because of the substantial improvements that are needed to extend the lifetime and increase the reliability of existing technologies and systems. Regulatory challenges are ranked low because no regulatory changes are needed. Cost and schedule challenges are ranked high because extraordinary resources and time would be needed to develop and validate the performance of closed-loop ECLSS that would operate reliably over long periods in space and on the surface of Mars.

---

[69] NRC, *NASA Space Technology Roadmaps and Priorities*, 2012.

ECLSS (ENVIRONMENTAL CONTROL & LIFE SUPPORT SYSTEM)

FIGURE 4.21 Assessment of ECLSS for a human mission to the Mars surface.

4.2.6.1.7 Habitats

All human missions to space require a pressurized and safe environment in which crews can live and work productively. Habitats of interest include short-term in-space habitats, such as the Orion Multipurpose Crew Vehicle; long-term in-space habitats, such as the ISS and the transit habitats for long-duration missions; and surface habitats for missions to the surface of the Moon or Mars. All types of habitats for space exploration have some common requirements and other distinct requirements, based on the environments in which they operate, mission duration, the number of crew, and mission goals.

Conventional habitats are large, complex, and heavy. NASA and private industry have supported development of expandable habitats in recent years that could be used for in-space and surface habitats. NASA is scheduled to attach an expandable habitat developed by Bigelow Aerospace to the ISS in 2015. This will provide an opportunity to characterize its performance under constant loads, the extent of degradation in the space environment, resistance to meteorite impact, and so on.

Key habitat systems include ECLSS and radiation safety systems (discussed above), thermal management, power generation and distribution, and micrometeorite protection. Each of these systems has been constantly improved. However, none has been designed or tested for long-duration missions with no possibility of resupply, with no options for a quick return to Earth in case of mission abort, and with constant exposure to the high-radiation environment of deep space.

The major differences between in-space habitats and surface habitats are the presence of dust and partial gravity (1/6 g on the Moon and 3/8 g on Mars). Lunar dust is highly abrasive and detrimental to mechanical systems, and it could pose a health hazard during long-duration missions if proper isolation cannot be achieved. The martian dust is not as well characterized as lunar dust, but the soil of Mars is known to be toxic because of high concentrations of perchlorates. In addition, the thin atmosphere and high winds on Mars will increase the diffusion of dust over all surface systems. Effective dust mitigation and control technologies and systems for both the Moon and Mars would be essential.

The assessment of habitats that are needed for a human mission to the Mars surface is summarized in Figure 4.22. (This summary pertains to habitat systems other than ECLSS and radiation safety systems, which are addressed separately; see Figures 4.21 and 4.12, respectively.) Technical challenges are ranked medium because NASA has extensive experience in designing and building habitats in LEO, culminating with the ISS. The capability gap is ranked medium because substantial improvements are needed to extend the lifetime and increase the reliability of existing technologies and systems and to assure that habitat systems work as expected in the partial gravity of the Moon or Mars. Regulatory challenges are ranked low because no regulatory changes are needed. Cost and schedule challenges are ranked medium because substantial resources and time would be needed to upgrade and validate the performance of habitat systems that would operate reliably over long periods in space and on the surface of Mars.

FIGURE 4.22 Assessment of habitats for a human mission to the Mars surface exclusive of ECLSS and radiation systems, which are assessed separately (see Figures 4.21 and 4.12, respectively).

4.2.6.1.8 Extravehicular Activity Suits

EVA suits can be viewed as individualized spacecraft. Key performance characteristics of EVA suits include mobility, pressurization, environmental protection (protection from heat, radiation, and micrometeoroids), portable life support (oxygen supply and CO_2 removal), ease of donning and doffing the suit, emergency capabilities, range of sizing, operational reliability, durability, sensory capabilities, data management, adaptability, level of articulation, and the forces and torques that an astronaut must apply to conduct assigned tasks. Given that current EVA suits represent incremental changes to suits that were developed more than 30 years ago, potentially substantial increases in performance are possible. EVA suits are needed for operations in microgravity (during in-space operations) and partial gravity (on the surface of the Moon or Mars). The microgravity environment in LEO is extremely well understood, and there is vast experience in performing EVAs during the past 50 years. However, since the end of the Apollo program, little research has addressed EVA suits for surface operations. EVA suits for the Mars surface will need to accommodate the effects of long-term exposure to the deep-space environment en route to Mars and EVA operations on Mars that will be much more extensive than the Apollo EVA operations on the Moon. Key issues include the effects that the partial gravity on Mars could have on gait, posture, and suit biomechanics; extending suit operational life; reducing suit mass; reducing suit maintenance; and reducing the effects of dust on bearings, seals, and closure mechanisms. It will be important for EVA suits to be designed to integrate easily with the design of rovers, habitats, and robotic assist vehicles during surface operations. It would also be beneficial to improve the mission duration, reliability, and maintainability of the portable life support systems incorporated into EVA suits while reducing system mass.[70]

The assessment of EVA suits needed for a human mission to the Mars surface is summarized in Figure 4.23. Technical challenges are ranked low because there is substantial research and experience with EVA suits in space and, to a lesser extent, on the surface of the Moon. The capability gap is ranked medium because of the advances needed to accommodate the long duration of a human mission to the Mars surface during transit and on the surface. Regulatory challenges are ranked low because no regulatory changes are needed. Cost and schedule challenges are ranked medium because substantial resources and time would be needed to close the capability gap.

4.2.6.1.9 Crew Health

The ability to maintain crew health during long-duration exposure to the space environment is critical for the success of human missions to Mars and other distant destinations. Both physiological and psychosocial issues present medical threats to crew well-being during extended missions.

NASA and the international community have a basic understanding of physiological problems associated with long exposure to microgravity, and they have been executing a methodical plan to reduce the effects of identified

---

[70] NRC, *NASA Space Technology Roadmaps and Priorities*, 2012.

FIGURE 4.23 Assessment of EVA suits for a human mission to the Mars surface.

issues (most notably, bone loss, muscular and cardiovascular deconditioning, and neurosensory decrements) and to screen for as-yet-unidentified problems that may exist. The plausibility of such as-yet-unidentified problems is bolstered by the recent discovery of new problems, such as the ocular impairments experienced by some astronauts. There is considerable individual variability in physiological responses to microgravity, but all astronauts are affected to some degree. One of the universal effects is bone loss, which is caused by a rapid increase in bone resorption and a decrease in bone formation during space missions. Physical countermeasures (specialized exercises) and pharmaceuticals have been studied, but bone loss is not yet manageable for long-duration space missions.[71] Ongoing research plans include testing of astronauts during and after extended stays of up to 12 months on the ISS. However, because of the small number of potential test subjects available on the ISS and the high degree of variation between individuals in both susceptibility and recovery, it takes a long time to accumulate datasets that are large enough to support general conclusions about human health effects.

The extent to which long-term exposure to partial gravity on the Moon or Mars may be a problem remains to be determined. If no adverse effects occur in the 3/8 *g* on Mars, minimal additional efforts (beyond those necessary to enable extended stays on the ISS) would be needed to counteract the effects of weightlessness during the full extent of a human mission to the Mars surface, including the transit times to and from Mars. Managing the effects of weightlessness for a human mission to the Mars surface would be further eased if the partial-gravity environment on the surface of Mars allowed astronauts to recover from at least some of the effects of weightlessness encountered during the transit to Mars.

Apart from the effects of weightlessness, crew physiology would be threatened by other factors, such as space radiation, illness, and injuries. Radiation safety is addressed separately above. Highly capable diagnostic and treatment equipment, including surgical facilities designed for operation in space and on the surface, would reduce the threats posed by injuries and illnesses, but this is a difficult challenge given that (1) the types of injuries and illnesses that might be experienced cannot all be anticipated and (2) the mass and volume of medical facilities on spacecraft and in ground habitats will be limited.

Psychosocial issues could affect the behavior and performance of astronauts on long-duration exploration missions. Studies of personnel in isolated and confined extreme environments (crews of nuclear submarines, groups wintering over in Antarctic research stations, and astronauts) suggest that psychosocial issues can substantially reduce crew performance, health, and well-being. The effects of psychosocial issues on the crews of deep-space missions could be more severe than those documented in the studies above because of reduced prospects for escape and safe return to Earth in case of spacecraft emergencies, the ineffectiveness of real-time two-way communication with friends and family because of time delays, and the longer duration of the mission, which exacerbates all the

---

[71] E.S. Orwoll, R.A. Adler, S. Amin, N. Binkley, E.M. Lewiecki, S.M. Petak, S.A. Shapses, M. Sinaki, N.B. Watts, and J.D. Sibonga, Skeletal health in long-duration astronauts: Nature, assessment, and management recommendations from the NASA bone summit, *Journal of Bone and Mineral Research* 28:1243-1255, 2013.

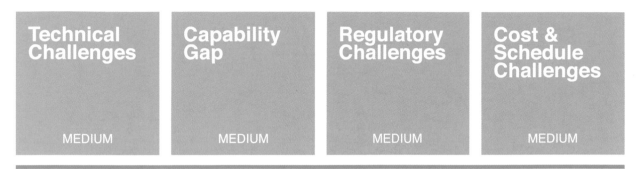

CREW HEALTH

FIGURE 4.24 Assessment of crew health systems for a human mission to the Mars surface exclusive of the effects of space radiation, which are assessed separately (see Figure 4.12).

stresses associated with living and working in a confined space with minimal privacy, barely adequate facilities for personal hygiene, the physiological effects of weightlessness, and so on.[72]

Psychosocial stresses affect individuals and interpersonal relationships. *Individual* effects include changes in personality during and after the mission, anxiety, depression, insomnia, reduced productivity, cognitive impairment, psychosis, and psychosomatic illness. *Interpersonal* effects include a greater desire for privacy, increased tension and conflict, and loss of cohesiveness among the flight crew and between the flight crew and ground personnel. Individual and interpersonal effects tend to increase over the course of a mission. The effects of psychosocial issues could be reduced by research that identifies countermeasures that could be taken before, during, or after long-duration missions. Research of interest includes ground-based simulations of long-duration missions, bed-rest experiments of various durations, and studies conducted on the ISS.[73–79]

The assessment of crew health systems needed for a human mission to the Mars surface is summarized in Figure 4.24. (This summary pertains to crew health systems other than radiation safety systems, which are addressed separately; see Figure 4.12.) Technical challenges are ranked medium because final solutions of many physiological and psychosocial threats to crew health have yet to be identified. The capability gap is ranked medium because solutions to some issues are rather well defined although others still require substantial research. Regulatory challenges are ranked medium because new standards may be needed as research into physiological and psychosocial issues continues, particularly given the results of a recent report on ethical issues associated with human space-flight.[80] Cost and schedule challenges are ranked medium because substantial resources and time would be needed to overcome the technical and regulatory challenges and to close the capability gap.

---

[72] Institute of Medicine, *Health Standards for Long Duration and Exploration Spaceflight: Ethics Principles, Responsibilities, and Decision Framework*, The National Academies Press, Washington, D.C., 2014.

[73] G.G. De La Torre, B. van Baarsen, F. Ferlazzo, N. Kanas, K. Weiss, S. Schneider, and I. Whiteley, Future perspectives on space psychology: Recommendations on psychosocial and neurobehavioural aspects of human spaceflight, *Acta Astronautica* 81(2):587-599, 2012.

[74] N. Kanas, Psychological, psychiatric, and interpersonal aspects of long-duration space missions, *Journal of Spacecraft and Rockets* 27(5):457-463, 1990.

[75] N. Kanas, From Earth's orbit to the outer planets and beyond: Psychological issues in space, *Acta Astronautica* 68(5-6):576-581, 2011.

[76] N. Kanas, G. Sandal, J.E. Boyd, V.I. Gushin, D. Manzey, R. North, G.R. Leon, et al., Psychology and culture during long-duration space missions, *Acta Astronautica* 64:659-677, 2009; N. Kanas et al., Erratum to "Psychology and culture during long-duration space missions," *Acta Astronautica* 66(1-2):331, 2010.

[77] L.A. Palinkas, Psychosocial issues in long-term space flight: Overview, *Gravitational and Space Biology Bulletin,* 14(2):25-33, 2001.

[78] M.P. Paulus, A neuroscience approach to optimizing brain resources for human performance in extreme environments, *Neuroscience and Biobehavioral Reviews* 33(7):1080-1088, 2009.

[79] Institute of Medicine, *Safe Passage: Astronaut Care for Exploration Missions*, The National Academies Press, Washington, D.C., 2001, Chapter 5.

[80] Institute of Medicine, *Health Standards for Long Duration and Exploration Spaceflight*, 2014.

FIGURE 4.25 Assessment of ISRU systems designed to produce consumables from the Mars atmosphere.

4.2.6.1.10 In Situ Resource Utilization (Mars Atmosphere)

ISRU refers to the use of natural resources found on the Moon, Mars, or an asteroid to support space science or exploration missions. Resources of interest include water, volatile substances implanted by solar wind in surface rocks, metals, minerals, and the atmosphere (for missions to Mars). ISRU systems can potentially transform these resources into materials needed for life support, propellant, manufacturing, and construction. To the extent that the mass of the materials produced by ISRU systems exceeds the mass of the ISRU system itself, ISRU capabilities offer the potential to reduce the launch mass and cost of space missions.

Without an ISRU capability, a human mission to the Mars surface would need to carry all the propellant, air, food, water, radiation shielding, and so on from Earth. This would increase the launch mass from Earth by about 10-15 percent and the landed mass on Mars by about 25-30 percent compared to a mission scenario that includes ISRU capability.

Highly advanced ISRU systems could conceivably use lunar regolith or the soil on Mars to produce a wide variety of materials. As a first step, however, the ISRU system specified for the Mars DRA 5.0 mission would use only the atmosphere of Mars as its raw material. The primary output of this system would be oxygen, which would be used for life support and as the oxidizer for the ascent propulsion system (in the form of liquid oxygen). The ISRU system would convert the CO_2 in the Mars atmosphere into oxygen and carbon monoxide and then vent the carbon monoxide back into the atmosphere. The ISRU plant would also separate and collect nitrogen and argon from the Mars atmosphere for use as buffer gases for crew breathing. In addition, the ISRU system could be designed to produce water by reacting hydrogen brought from Earth with oxygen produced on Mars. This water would be used to replace water lost during crew and EVA operations.[81]

The assessment of ISRU systems designed to produce consumables from the Mars atmosphere is summarized in Figure 4.25. Technical challenges are ranked low because technologies to achieve the ISRU capabilities described above have been demonstrated on Earth. The capability gap is ranked high because there is a large gap between the capabilities of the small-scale experiments that have been completed and the development of a full-scale operational system capable of reliable operation during long-term exposure to the partial gravity, dust, atmosphere, and radiation environment on the surface of Mars. Regulatory challenges are ranked low because no regulatory changes are needed. Cost and schedule challenges are ranked medium because substantial resources and time would be needed to close the capability gap.

---

[81] NASA, "Human Exploration of Mars Design Reference Architecture 5.0," 2009, http://www.nasa.gov/pdf/373665main_NASA-SP-2009-566.pdf, p. 40.

4.2.6.2 Additional Capabilities

In addition to the high-priority capabilities described above, advances in many other capabilities will be essential for a human mission to the Mars surface or to some of the other DRMs that appear in the pathways. Examples of the additional essential capabilities include the following:

- Autonomous systems.
- EDL systems for return to Earth.
- In-space operations, including assembly of large structures and propellant storage and transfer.
- ISRU systems capable of using lunar regolith or the soil of Mars to produce materials for manufacturing, construction, or repair of systems.
- Mission operations and communication.
- Planetary protection (to minimize the biological contamination of explored environments and to protect Earth from biological contamination in case life is encountered by human exploration missions).
- Surface mobility.
- Surface operations.

Capabilities such as these are not included in the list of high-priority capabilities. Advances in these areas are not urgent because they are not needed for early missions in any of the pathways and/or they can be achieved more quickly and with fewer resources than the high-priority capabilities. For example, the state of the art for EDL systems for Earth return, for mission operations and communication, and for autonomous systems is sufficient for missions in cislunar space; surface mobility systems for lunar sorties could be procured without having to overcome any major technical challenges; crewed surface mobility systems for Mars could be developed in time for a mission to Mars in the 2030s or 2040s even without substantial research and technology development in the near term; ISRU systems that can process lunar regolith are not essential for lunar surface missions; and ISRU systems that can process Mars soil are not essential for the first generation of Mars surface missions.

4.2.6.3 Summary of Challenges in Developing High-Priority Capabilities

The four parameters used to assess each of the high-priority technical capabilities discussed above in section 4.2.6.1 ("High Priority Capabilities") are summarized in Figure 4.26. Without belaboring the point, the relative paucity of green in this summary highlights the difficulty and cumulative scale of technology development required to achieve the horizon goal of a human mission to the Mars surface, whatever the intermediate destinations along the pathway. This technology development challenge bears directly on the next major section of this chapter, which addresses the affordability of a human spaceflight program over the decades required to extend human presence beyond LEO and make meaningful progress in addressing the enduring questions (see Chapters 1 and 2).

4.2.7 Affordability

The biggest challenge in implementing pathways of human exploration beyond LEO may be financial rather than technical. A pathway is affordable if the costs of all flight programs and associated development fit within the resources available. Any proposed pathway to Mars must be affordable by the U.S. taxpayer (and international partners) to be sustainable. Recent history has shown that budget constraints have dictated the pace of development of exploration systems, and while it is thought that it is feasible to overcome the technical and development challenges to landing humans on Mars, the technical challenges are daunting, and substantial development effort is required.

4.2.7.1 Potential Budget Available to Human Spaceflight Beyond LEO

Total annual spending by NASA on human spaceflight programs, adjusted for inflation to 2013 dollars, has fluctuated over the past 30 years, but the trend has been flat with an annual budget around the current level of

FIGURE 4.26 Summary of the assessments of the high-priority capabilities. High, medium, and low are defined for each of the assessment areas in Figure 4.8.

approximately $8 billion in FY 2013 dollars (see Figure 4.27).[82] The most recent presidential budget request[83] shows $7.9 billion for the Human Exploration and Operations Mission Directorate (HEOMD) for FY 2014. The budget request proposes continuation of $7.9 billion for FY 2015 and then annual increases of 1 percent through the budgeting horizon of FY 2019, which represents a decreasing budget in constant dollars. This report uses the NASA FY 2014 presidential budget request, which projects NASA's budget through FY 2018, as a departure point for projections beyond 2018. Future developments in NASA's human spaceflight program will also attempt to leverage work done by the Space Technology Mission Directorate (STMD), which was funded at about $0.6 billion in FY 2014 and is proposed to increase to $0.7 billion in FY 2015.

NASA's human spaceflight program has four main areas: operations, research, support, and development, broken out by approximate percentage of the proposed annual FY 2018 budget in Figure 4.28. The operations budget, 41 percent, is dominated by the ISS, including its transportation costs. Research, at 11 percent, includes the STMD's Exploration Technology Development Program (ETDP) and lays the foundation for future developments by advancing technologies and reducing knowledge gaps. Research funding is spread across many competing technologies with the goal of developing enhanced capabilities that are relevant to a variety of potential missions but without a generally accepted guiding roadmap for what is specifically required for future human spaceflight beyond LEO. The development of some critical capabilities, such as in-space transportation and EDL, are funded at low levels across broad trade spaces; decisions as to where to focus efforts have not been made. This spreading of resources poses a serious challenge to progress in human spaceflight. Differential investment is one of the few tools that program leadership can use in highly constrained situations, but differential investment is extremely

---

[82] Figure 4.26 shows the NASA funding approved specifically for human spaceflight programs does not fully account for funding provided to cross-agency elements that indirectly support human spaceflight.

[83] NASA, "FY 2015 President's Budget Request Summary," http://www.nasa.gov/news/budget/index.html.

NASA HUMAN SPACEFLIGHT BUDGET (FY13 $)

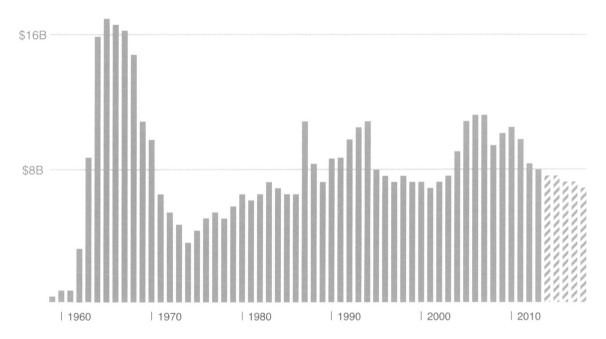

FIGURE 4.27 Historical funding of NASA human spaceflight programs in constant FY 2013 dollars. SOURCE: NASA FY 2014 President's Budget Request Summary, http://www.nasa.gov/pdf/750614main_NASA_ FY_2014_Budget_Estimates-508.pdf, accessed January 24, 2014; NASA Historical Data Books, SP-4012, Volumes 2-7, http://history.nasa.gov/SP-4012/cover.html.

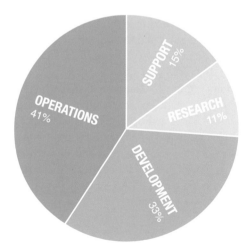

FIGURE 4.28 Approximate distribution of NASA's FY 2018 proposed human spaceflight budget, which is *used as the basis for projecting the cost of current human spaceflight programs beyond 2018.* NOTE: The research funding depicted in this figure is a lower bound. STMD funds several research and technology programs. One, the Exploration Technology Development Program, is focused on exploration, and its budget is included above. The funding for other STMD programs that support all three NASA mission directorates is not included above, because it is not clear how much of this funding supports the HEOMD as opposed to the Science Mission Directorate and the Aeronautics Research Mission Directorate. SOURCE: President's Budget Request for FY 2014.

difficult to implement in highly bureaucratic organizations in which level-of-effort funding practices are the status quo. Support costs, 15 percent, are needed to maintain the required infrastructure, such as manufacturing facilities, and other supporting efforts, such as the deep-space communication networks for future missions. Development, 33 percent, funds the design and testing of next-generation systems. Major systems now being developed are the commercial cargo and crew systems, the Orion spacecraft, the Block 1 SLS, and their associated ground systems. When completed, the latter three systems will provide the basic transportation systems and infrastructure needed for some missions in cislunar space. The Orion capsule as currently planned limits mission duration to about 21 days for a crew of four. To proceed with longer-duration missions beyond LEO, substantial development will be required in longer-duration in-space systems.

Projecting beyond the current budget horizon provides insight into the availability of funds for future development and flight operations. For this study, a few key assumptions were made to facilitate understanding of potential future scenarios. Human spaceflight budgets through FY 2018 are assumed to be those specified in the NASA FY 2014 presidential budget request. For budgets beyond FY 2018, this analysis assumes a lower bound of a flat budget for human spaceflight beyond FY 2018 (with no increase for inflation) and an upper bound of a human spaceflight budget increasing with inflation, projected to be approximately 2.5 percent per year in NASA's 2013 new start inflation index.[84] Figure 4.29 is a projection of the current human spaceflight program of record relative to these upper and lower bounds (see also Box 4.1). Budget uncertainty is indicated with a solid black line to represent the lower bound (a flat budget) and a dashed line to represent the upper-bound budget (increasing with inflation). The projection includes funding for ISS operations to its previous baseline decommissioning date of 2020 and a possible extension of operations through the current engineering estimated limit of 2028.[85] Also included is funding for two performance upgrades of the SLS (the latter occurring in the late 2030s), which would increase payload capacity to LEO to 130 MT. The light and dark turquoise areas under the upper and lower budget bounds indicate funds available for new projects beyond LEO.

Using the budget uncertainty projection bounded by flat and inflation-adjusted growth, the total available funding for development and operations for new projects of potential future human spaceflight systems and pathways can be projected. Figure 4.30 illustrates the cumulative funds available from FY 2015 to FY 2065 for new development and flight operations beyond that required for fixed infrastructure, ISS operations, and development of Orion and the 130-MT SLS. Previous studies have provided a wide range of estimates for the costs of developing and operating the critical elements for a Mars surface mission, most of which are in the hundreds of billions of dollars.[86,87] The four available funding cases shown in Figure 4.30 include those with (turquoise) and without (purple) an extension of ISS operations from 2020 to 2028 for both a flat budget projection (solid) and a budget increasing with inflation (dashed) at a projected annual rate of 2.5 percent. Two observations can be drawn from this figure:

- The lower bound of the budget uncertainty, or flat human spaceflight budget projected from 2018 guidance in then-year dollars (not adjusted for inflation), accumulates a total of less than $100 billion in then-year dollars that can be used to develop, build, and operate the new critical elements—such as in-space habitats, new in-space propulsion systems, and landers—that will be required to venture and explore beyond cislunar space. Although the exact effects of the fixed support and infrastructure costs are unknown and must be evaluated in more detail, the combination of a flat budget for the extended future and the prerequisite large infrastructure costs associated with the current NASA plan would prevent NASA from proceeding very far down any human exploration pathway to Mars.

---

[84] NASA, "2013 NASA New Start Inflation Index for FY14," http://www.nasa.gov/sites/default/files/files/2013_NNSI_FY14(1).xlsx, accessed March 11, 2014.

[85] In January 2014, the administration expressed a commitment to continue to operate the ISS to 2024. At the time of writing of the present report, the international partners and Congress had not committed to operations beyond 2020. In this chapter, the analysis considers the bounding cases of 2020 and 2028.

[86] M. Reichert and W. Seboldt, "What Does the First Manned Mars Mission Cost? Scientific, Technical and Economic Issues of a Manned Mars Mission," DLR–German Aerospace Research Establishment. Cologne, Germany, presented at the 48th International Astronautical Congress, October 6-10, 1997, Turin, Italy, IAA-97-IAA.3.1.06.

[87] M. Humboldt, Jr., and K. Cyr. "Lunar Mars Cost Review with NASA Administrator, NASA Exploration Initiative," October 31, 1989.

NASA HUMAN SPACEFLIGHT PROJECTED AVAILABLE BUDGET (THEN-YEAR $)

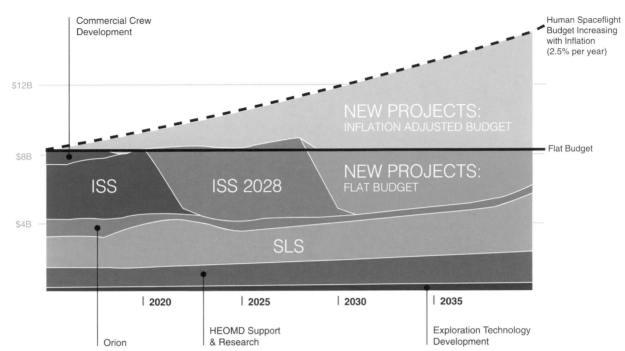

FIGURE 4.29 Projected available budget and costs of the currently planned human spaceflight program.

BOX 4.1
Sand Charts

The figures in this report that show notional projections of annual costs and available funding for human spaceflight as a function of time are commonly referred to as sand charts. The sand chart that shows projected available budget and costs of the currently planned human spaceflight program (Figure 4.29) was derived as follows: Near-term costs are based on the NASA FY 2014 presidential budget request,[a] which projects budget requests through FY 2018. For years after 2018, the costs of operations, support, and research projects are held at their proposed 2018 funding levels adjusted for inflation (using the NASA new start index). The exceptions to this are ISS costs, which are held at proposed FY 2018 funding levels until the year of termination and then followed by 2 years of ramping down budgets to cover termination-related costs. By 2018, SLS and Commercial Crew will have reached their initial operation capability (IOC) leaving the Orion spacecraft as the only unfinished currently funded major development project. Orion's costs after 2018 are ramped down until a steady state is reached in the 2022 timeframe corresponding to the predicted IOC. Planned upgrades of the SLS would increase its capability beyond that provided at IOC and are included in the projected SLS fixed costs.

The sand charts that project the costs of the pathways (Figures 4.31a, 4.32a, and 4.33a) were derived as described below in section 4.2.7.2 ("Pathway Cost Range Methodology").

Projections of the annual budget available for human spaceflight in all the sand charts are shown for two scenarios: a flat budget at FY 2015 levels and a budget that increases with inflation.

---

[a] NASA FY 2014 President's Budget Request Summary, http://www.nasa.gov/pdf/740512main_FY2014%20CJ%20 for%20Online.pdf, retrieved January 24, 2014.

CUMULATIVE THEN-YEAR $ FOR NEW PROJECTS

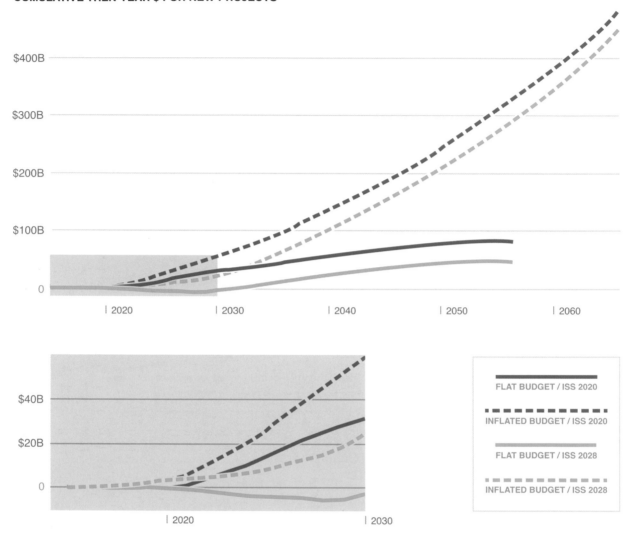

FIGURE 4.30 Projected cumulative then-year dollars available for new projects for human spaceflight beyond LEO, with inset showing detail through 2030.

- For any given cost estimate for a pathway to a Mars surface mission, Figure 4.30 indicates when, for a budget increasing with inflation, a landing could be achieved. For example, a $400 billion cost for a pathway to Mars cannot be achieved before roughly 2060. An alternative way to look at this information is that to achieve a landing before 2050 and still be affordable, the pathway to Mars would have to be technically feasible and cost less than about $220 billion.

The scale of the government investment required to send humans to Mars is, to a rough order of magnitude, equivalent to:

- The cost of perhaps 75-150 "flagship class" robotic exploration spacecraft (assuming an average cost of $1 to $2 billion each).

- Twice as much as the National Science Foundation budget over the corresponding period.
- Two to four times the U.S. investment in the ISS, which amounted to roughly $150 billion, including launch costs.

Barring unforeseen changes such as significant budget increases, substantial increases in operational efficiency, or technological game changers, progress toward deep-space destinations will be measured on time scales of *decades*. Policy goals that state shorter time horizons cannot change this reality. Thus, it would be well to note that 20 years before Apollo 11 landed on the Moon, it would have been difficult to anticipate the technological progress in many area—from computing to guided missile technology—that would enable Apollo. Many, perhaps most, of those technology advances resulted from research outside NASA. The architectures imagined for travel to the Moon by visionaries of the late 1940s bore little resemblance to the path ultimately followed. Thus, the committee treads cautiously in noting the difficulties associated with human space exploration beyond LEO, based primarily on the reference architectures developed by NASA.

4.2.7.2 Pathway Cost Range Methodology

Now that the range of available resources, or budget, for human spaceflight has been established, affordability can be assessed. The first step is to determine the likely cost range of the three representative pathways. Projecting the cost of the development, production, and operations of the exploration elements required to land humans on Mars over the next several decades involved a high degree of uncertainty, so all cost assessments in this report are notional.

New developments and operations beyond those currently funded are projected using a combination of historical analogies (such as the space shuttle and Apollo) and results of previous exploration studies (such as the Space Exploration Initiative[88] and Mars DRM 3[89]). The use of historical analogies is preferred, previous studies are used when there are few or no valid historical analogies. These sources lead to cost projections that align with NASA's traditional ways of doing business in human spaceflight and assume no significant international cost savings. New ways of doing business are not assumed except where explicitly noted (such as for the Commercial Crew and Cargo program). The funding needed to advance required technologies to the point where project development can begin is not included, because it is assumed that technology will be advanced to sufficient levels through technology incubation programs, such as NASA's ETDP and Exploration Research and Development program, which are continued at their proposed FY 2018 funding levels (adjusted for inflation). Notional cost projections are generated for development of new systems and for their production and operation. Development costs are adjusted to account for potential heritage from previous developments. For example, the projected cost of a deep-space habitat to support a Mars Moons DRM is reduced if a deep-space habitat was developed for the Asteroid in Native Orbit DRM. Projected development costs are spread over development times (nominally 8-10 years) using a beta curve distribution. Cost for production and operations are split into fixed and variable costs for systems that are sustained over extended periods, and the variable costs associated with each mission are spread over 2- to 3-year procurement periods.

Many variables affect the mission rates and development schedules for the various DRMs, pathways, and affordability scenarios. In all cases, the missions are scheduled using an optimistic assumption of no catastrophic development or mission failures. The crewed mission timelines assumed full operational capability in the first flights. The missions are planned to minimize the time before the first human mission to the Mars surface subject to available budget or imposed restrictions on mission rate. Each new DRM in a pathway requires a substantial amount of development time and funding and is phased to meet the budgetary guidelines of the specific scenario being considered. Specific mission operations are placed between the peaks in development funding, and there is an attempt to add as many missions as possible given the budget available in each scenario. The projected costs

---

[88] NASA, *Report of the 90-day Study on Human Exploration of the Moon and Mars,* November 1989, available in NASA Historical Reference Collection, History Office, Washington, D.C., http://history.nasa.gov/90_day_study.pdf.

[89] NASA, "Reference Mission Version 3.0 Addendum to the Human Exploration of Mars: The Reference Mission of the NASA Mars Exploration Study Team," NASA/SP-6107-ADD, 1998, http://ston.jsc.nasa.gov/collections/trs/_techrep/SP-6107-ADD.pdf.

of predeployed cargo missions (such as prepositioning surface outposts) are included, but precursor technology demonstrators are not, under the previously noted assumption that those costs are wrapped into either the development costs or the technology incubation programs. Human missions to Mars are constrained by orbital mechanics to occur only every 2 years or so, depending on mission type. In the actual figures, the costs for each DRM for a particular pathway are time-phased and stacked in order above the projections of fixed infrastructure costs (gray area) to create the sand chart that indicates the annual cost projections until the first human landing on Mars. The SLS and Orion marginal mission costs for both crew and cargo are included with the specific DRM; their fixed infrastructure and upgrade costs are held in the gray area because they support all the DRMs.

Uncertainty in the projection of funding requirements is evident in all cost modeling. Trying to project the costs of systems that will not be designed for decades adds to the uncertainty. As already noted, many technological advances will be required just to initiate the design of some mission elements. For each mission element, a range of possible costs was generated to account for various levels of margin added to the historically derived numbers as well as variable growth rates in fixed infrastructure costs. The margin accounts for incomplete knowledge of the system requirements; low to minimal design maturity; uncertainties about the ability to scale up the capabilities of current technologies, such as SEP, by orders of magnitude; and historical cost growth of NASA projects. In general, 50 percent margins were used for developmental efforts, and 25 percent margins for production and operations.[90] Although these margins are reasonably consistent with historical patterns at NASA, averaged over many programs, some programs experience cost overruns that greatly exceed historical averages. In addition, the degree of challenge for some of the technologies needed for travel to Mars is much greater than that faced by NASA in recent decades. Thus, if anything, the cost projections generated for this study are optimistic. *This effort is intended both to produce a notional cost range for each pathway so that this range can be compared with future funding levels and to evaluate affordability and sustainability in terms of flight frequency and the required time span to land humans on Mars.* More detailed cost analysis that could be used for budget planning will have to be carried out by NASA and other independent groups once decisions are made to pursue particular goals.

4.2.7.3 Cost Implications of International Collaboration

An additional uncertainty to consider is the effect of international collaboration. It is generally accepted that landing humans on Mars will require substantial international cooperation for many reasons. With respect to affordability, it is unclear how the goals of the various spacefaring nations will translate to budget commitments and how potential hardware contributions will effectively and efficiently contribute to overall pathway design. The ISS would not exist today without strong international cooperation and participation. However, only 17 percent of the total ISS cost has been covered by non-U.S. stakeholders.[91] The ISS is also an example of the typical increases in complexity and inefficiency associated with multinational endeavors. In fact, independent cost models often add an international complexity factor that increases projected costs for international programs. Considering increased complexity and likely additional cost elements to satisfy specific and unique political goals of an international partner, international partnerships with cost-sharing on the scale required for human spaceflight tend to be cost-neutral relative to the cost of the program without international partners. If large enough, international contributions could overcome the cost increases associated with management complexity, but that was not the case with the ISS. To make the Mars pathways affordable to the United States, international contributions would need to be large enough to cover the costs of increased complexity and additional cost elements introduced by the partnership with enough excess to cover the gap between the projected human spaceflight budget and the projected costs of each pathway, which are described in the sections that follow.

International partnerships can be highly effective when interfaces are clearly defined and each partner has specific responsibilities for providing one or more particular mission elements. However, the amounts and types of contributions from partner nations for exploration efforts beyond LEO are unknown, and potential international

---

[90] D.L. Emmons, M. Lobbia, T. Radcliffe, and R.E. Bitten, Affordability Assessments to Support Strategic Planning and Decisions at NASA, Proceedings of IEEE Aerospace Conference, 2010.

[91] Claude LaFleur, Costs of U.S. piloted programs, *The Space Review*, March 8, 2010.

partners with the technical capability and budget commitment may also have views that differ from those of the United States when it comes to selecting a pathway to Mars. A variety of possible international contributions—such as cost support, innovation, and stability—are outlined in the *Global Exploration Roadmap*,[92] and those contributions may be important. Nonetheless, NASA is still likely to carry most of the financial burden, especially if the United States wants to have a strong influence on pathway and technology direction.

The pathway cost profiles in this report do not reflect the value of international contributions to human space exploration, nor do they include any increase in costs associated with the international complexity factor.

4.2.7.4 Schedule-Driven Affordability Scenarios

To understand the more detailed aspects of the three representative pathways, it is helpful first to determine what an unconstrained, NASA-only program might look like if it is desirable to land on Mars as early as possible. The three previously described representative pathways have been modeled using optimistic development times and flight rates with minimal gaps between major pathway developments. In the history of U.S. human spaceflight, the longest flight gap was the transition from the Apollo to the Space Shuttle Program. That gap was initially expected to be 4 years, but it stretched to 6 years because of unplanned delays in space shuttle development. The Apollo Program launched 11 crewed missions at an average of one launch every 4 months. The Space Shuttle Program sustained an average flight rate of one launch every 3 months over a span of 3 decades. The time between shuttle flights ranged from 17 days to 32 months (after the loss of *Challenger*). The time between the first two shuttle flights was 7 months. In contrast, the planned time between the first two SLS launches is 4 years.

In some cases, in particular the Moon-to-Mars and Enhanced Exploration pathways with an ISS retirement date of 2020, the flight rates modeled in this scenario are representative of sustainable, historical flight rates. However, as noted above, an optimistic bias with respect to program margin is included in all the affordability analyses. In addition, it is assumed that each developmental system is available when it is needed for operations and that there are no mission failures (with associated inquiries and program delays, as happened for Apollo 1 and in the loss of *Challenger* and *Columbia*). Also ignored are programmatically necessary robotic precursor missions. For example, the SLS will be launched just once with no crew before the first human flight (EM-2). One can scarcely imagine that a Mars EDL system will not have been extensively tested robotically in the atmospheres of Earth and Mars before a human crew depends on it for survival. Thus, for all affordability analyses shown below, estimated dates of first landing on Mars are highly optimistic.

As a representative example, the cost profiles for each of the stepping-stone DRMs of the Enhanced Exploration pathway are presented in Figure 4.31a. This figure breaks out the cost profiles associated with the development and operation of the payloads for the individual DRMs, as well as basic infrastructure and support. The costs shown are representative of medium point cost projections of new systems and mission operations (with margins based on historical cost growth) on top of the baseline costs. Figure 4.31a presents the total human spaceflight then-year annual budget required where the fixed infrastructure costs and the development costs for Orion and SLS have been grayed out and the required pathway developmental and operational costs (including marginal SLS and Orion costs) for each DRM are shown in the colors indicated. The extension of the ISS to 2028 is also shown. Because of the notional nature of the cost projections in this study, the vertical cost axes in Figure 4.31a and similar figures are not marked with dollar values. Even so, the committee is confident that the cost projections that are summarized in these figures provide a sound basis for making relative comparisons among the pathways and between the pathways and budget projections.

Figure 4.31a shows that a large increase in the human spaceflight budget would be required to use the Enhanced Exploration pathway to land on Mars prior to 2040. The schedule-driven results for all three pathways with and without the ISS extension to 2028 are shown in Figure 4.31b. The total annual cost to NASA for the Enhanced Exploration pathway and the extension of the ISS to 2028 (Figure 4.31a) is represented as a thick purple line and is compared with the other pathway schedule-driven scenarios. In these schedule-driven scenarios, the estimated dates of the Mars surface landing are driven primarily by technology development timelines and the associated

---

[92] ISECG, *The Global Exploration Roadmap,* 2013.

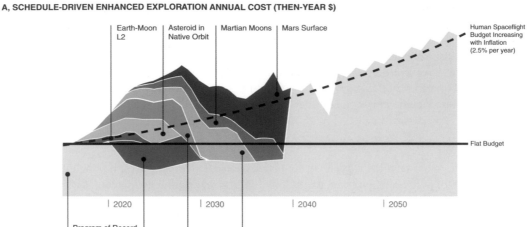

A, SCHEDULE-DRIVEN ENHANCED EXPLORATION ANNUAL COST (THEN-YEAR $)

B, SCHEDULE-DRIVEN PATHWAYS ANNUAL COST (THEN-YEAR $)

C, SCHEDULE-DRIVEN PATHWAYS CREWED FLIGHTS

FIGURE 4.31 (a) Schedule-driven cost profile of the Enhanced Exploration Pathway in then-year dollars. (b) Schedule-driven cost profile comparison in then-year dollars of three representative pathways with and without the ISS extended to 2028. (c) Schedule-driven crewed mission timeline assumptions for three representative pathways with and without the ISS extended to 2028.

increase in funding to achieve the assumed dates. Figure 4.31c shows the planned crewed launch dates for each scenario with lunar surface missions occurring twice a year, cislunar and asteroid missions occurring once a year, and Mars-based missions occurring every 26 months because of planetary launch constraints. In all those cases, the final destination of the pathway is reached in the 2030s, but, as seen in Figure 4.31b, the projected costs for all three pathways extend well above the projected human spaceflight budgets and are not considered affordable. To fund such programs fully, the human spaceflight budget would have to increase at a rate 2-4 times that of inflation for the next 15 years, depending on the specific pathway and the status of the ISS.

The schedules shown here are considered optimistic in that they assume a fully funded and success-oriented program, and a landing on Mars may be delayed, depending on the successful demonstration of the required technologies.

4.2.7.5 Budget-Driven Affordability Scenarios

Budget-driven affordability scenarios are based on the assumption that the three representative pathways to Mars are constrained by the human spaceflight budget increasing with inflation. The lower bound of the budget uncertainty, or flat budget, was not considered, because this condition cannot sustain any pathway to land humans on Mars. To achieve pathway cost scenarios constrained to trend with the human spaceflight budget increasing with inflation, the year in which humans first land on Mars must slip to the right, flight gaps between DRMs must increase, and crewed flight rates will have to be lowered to below historical rates. Figure 4.32a indicates that each DRM of the Enhanced Exploration pathway can be delayed until sufficient budget is available to proceed after the extension of the ISS to 2028 and that a Mars landing would occur in about 2054. However, to achieve this budget-constrained scenario, the number of crewed flights must be greatly reduced and the mission rate lowered to just two crewed flights per 5-year period, as shown in Figure 4.32c.

Figures 4.32b and 4.32c compare the three representative pathways with respect to the budget-driven scenarios. With respect to timeline, as expected the ARM-to-Mars pathway potentially could lead to a human mission to the Mars surface as early as 2037-2046, depending on the ISS extension. The Moon-to-Mars pathway could yield a landing between 2043 and 2050. All of these landing dates are likely optimistic in that delays will inevitably occur as developmental challenges and potential failures delay the specific pathway schedule. Such issues will be exacerbated by tight budget constraints and the limited ability of project managers to react to unexpected issues and concerns. These budget-constrained scenarios result in unrealistic and unsustainable mission rates well below any historical precedent. For the ARM-to-Mars pathway, there are only five crewed missions in the 18 years between ISS retirement in 2028 and landing on Mars. For the Moon–to-Mars pathway, there are only six crewed missions to the Moon for both the Lunar Sortie and Lunar Outpost DRMs with planned periods of up to 7 years with no flights. Only once missions to the vicinity of Mars are undertaken do the flight rates for SLS increase substantially. For each of the pathways, the number of SLS launches in any year prior to 2030 varies from zero to four. Conversely, missions to Mars (either the moons of Mars or the Mars surface) will require salvos of SLS launches for each mission. (In both cases, some of the launches will take place in the launch opportunity prior to the launch of the crew to preposition resources for their use and sustainment.) Even when SLS launch rates are in a sustainable range,[93] the mission rates in the budget-driven scenario are much lower than in previous experience in the U.S. human spaceflight program.

The serious potential programmatic and operational risks attendant on the low operational tempo of the budget-driven scenarios led the committee to consider augmentation of the mission rate beyond what is strictly necessary to effect each of the pathways, based primarily on the professional experience of the Technical Panel and those members of the committee experienced in spaceflight programs and associated engineering. This is discussed in the following section.

---

[93] According to NASA leadership in a recent presentation to the NASA Advisory Council's Human Exploration and Operations Committee, that rate is once a year.

A, BUDGET-DRIVEN ENHANCED EXPLORATION ANNUAL COST (THEN-YEAR $)

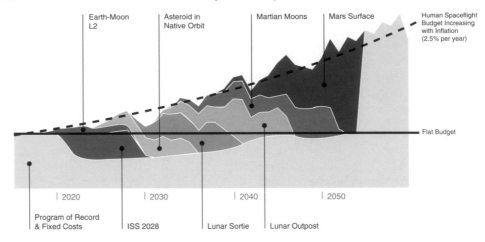

B, BUDGET-DRIVEN PATHWAYS ANNUAL COST (THEN-YEAR $)

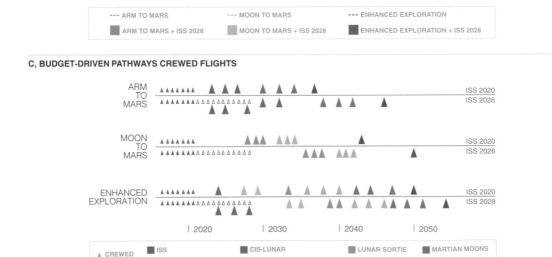

C, BUDGET-DRIVEN PATHWAYS CREWED FLIGHTS

FIGURE 4.32 (a) Budget-driven cost profile of the Enhanced Exploration Pathway in then-year dollars. (b) Budget-driven cost profile comparison in then-year dollars of three representative pathways with and without the ISS extended to 2028. (c) Budget-driven crewed mission timeline assumptions for three representative pathways with and without the ISS extended to 2028.

4.2.7.6 Operationally Viable Scenarios

Examination of the schedule-driven and budget-driven affordability scenarios for each pathway indicates, independently of the ISS extension, that *the pathways using historical mission rates are not affordable,* and *affordable pathways based on a human spaceflight budget increasing with inflation are not sustainable.* For each pathway, the purpose of this section is to illustrate optimistic but potentially sustainable mission rates with the minimum budget possible. The average time between crewed missions during the period from ISS retirement until the first Mars mission would be 19-28 months, depending on the combination of pathway and ISS retirement date. Some gaps in crewed missions would occur between major mission operations to allow hardware development and predeployment of mission assets. Additionally, robotic missions to test some key systems, such as Mars EDL, would need to be completed before crews could be committed to a Mars surface mission.

Figures 4.33a-c indicate the budget required for the three representative pathways to achieve an operationally viable mission rate. Assuming that the ISS is extended to 2028 and that the human spaceflight budget is increased by 5 percent per year (twice the rate of inflation), the earliest that a crewed surface mission to Mars is likely to occur is about 2040-2050. Again, these dates are probably optimistic in that delays would inevitably occur as developmental challenges and failures require design modifications and schedule delays. If the exploration budget grows at 5 percent per year, the benefit of terminating the ISS in 2020 is not that great from an affordability perspective, in that a human landing on Mars may be advanced by just 2-4 years, depending on the pathway and the associated risk.

4.2.8 Assessment of Pathways Against Desirable Pathway Properties

Many possible pathways can be conceived, even given the relative paucity of feasible destinations for humans, because of the combinatoric complexity of ordering destinations and because of the large number of DRMs that have accumulated for each of them. However, it is possible to articulate desirable properties of pathways that can guide choices for the nation's human spaceflight program. Six such desirable properties follow:

1. The horizon and intermediate destinations have profound scientific, cultural, economic, inspirational, and/or geopolitical benefits that justify public investment.
2. The sequence of missions and destinations permits stakeholders, including taxpayers, to see progress and develop confidence in NASA's ability to execute the pathway.
3. The pathway is characterized by logical feed-forward of technical capabilities.
4. The pathway minimizes the use of dead-end mission elements that do not contribute to later destinations on the pathway.
5. The pathway is affordable without incurring unacceptable development risk.
6. The pathway supports, in the context of available budget, an operational tempo that ensures retention of critical technical capability, proficiency of operators, and effective use of infrastructure.

The committee is not recommending any particular pathway, but the pathways outlined above are assessed in the following sections in terms of these desirable properties.

4.2.8.1 Significance of the Pathway Destinations

Any of the pathways, if executed to completion, would rate highly in terms of the significance of the pathway destination because they all include a human mission to the surface of Mars. This is the most challenging destination for human exploration in the context of foreseeable technology. In addition, Mars has high scientific value in the context of the evolution of terrestrial planets and the possible origins of life.

The Moon-to-Mars and Enhanced Exploration pathways arguably rate higher on this property because they include the Moon and near-Earth asteroids large enough to be scientifically interesting. Although some have dismissed the Moon as no longer interesting because humans have visited it before, this is similar to considering

A, OPERATIONALLY VIABLE ENHANCED EXPLORATION ANNUAL COST (THEN-YEAR $)

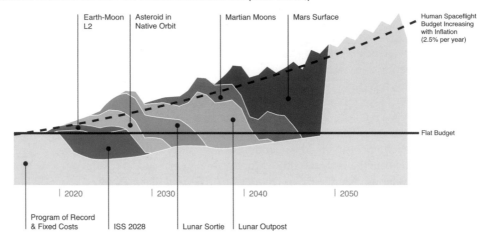

B, OPERATIONALLY VIABLE PATHWAYS ANNUAL COST (THEN-YEAR $)

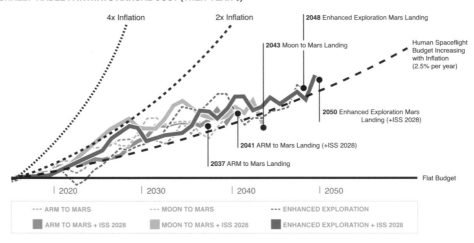

C, OPERATIONALLY VIABLE PATHWAYS CREWED FLIGHTS

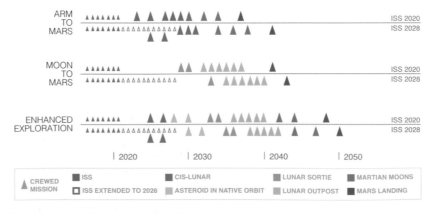

FIGURE 4.33 (a) Operationally viable cost profile of Enhanced Exploration Pathway in then-year dollars. (b) Operationally viable cost profile comparison in then-year dollars of three representative pathways with and without the ISS extended to 2028. (c) Operationally viable crewed mission timeline assumptions for three representative pathways with and without the ISS extended to 2028.

the New World to have been adequately explored after the first four voyages of Columbus, whereas the continued exploration and exploitation of the New World had profound cultural, economic, and geopolitical impact on the Old World. The New World in Columbus's time was discovered to be verdant and populated. Clearly, neither adjective is relevant to the Moon, but some of the most significant effects of the New World on the Old were as a result of the systematic extraction of mineral wealth, principally in the form of gold, which had been mined by the indigenous peoples and later by European colonists. That mineral wealth was not apparent after the initial reconnaissance of the New World by Columbus. It was discovered later during more complete exploration. Similarly, as a result of further exploration (in this case, robotic exploration) of the Moon, it is thought that the Moon probably contains "mineral wealth" worth much more than its weight in gold in the form of water in permanently shadowed craters. That water could be a critically enabling resource for human exploration and perhaps a space-based economy as a source of oxygen, fuel, and potable water that would not require expensive transport from the bottom of Earth's gravity well. The characterization of near-Earth asteroids, although almost certainly most cost-effectively explored robotically, is also important because of the implications for planetary defense and perhaps ultimately their potential for exploitation in a space-based economy.

4.2.8.2 Sequence Shows Progress with Intermediate Destinations

As has been noted earlier, human spaceflight beyond LEO would take place over the course of many decades. Given current budget realities, it is tempting to suggest that the "best" pathway to Mars is the shortest and least expensive, that is, the ARM-to-Mars pathway. However, even with the decommissioning of the ISS in 2020, a budget-constrained schedule means that the earliest conceivable human presence in the vicinity of Mars would occur in the late 2030s, assuming that all technological challenges are met without setbacks. Until then, the only human spaceflight missions would be trips to the ISS (until 2020) and voyages in cislunar space without landings on any naturally occurring celestial body. Thus, without a considerable increase in human spaceflight funding for NASA, the ARM-to-Mars pathway presents the prospect of a long period of technology development during which NASA's stakeholders do not see human exploration missions taking place. This problem poses one of the most serious challenges to program sustainability that the study's Technical Panel identified.

ARM has failed to engender substantial enthusiasm either in Congress or in the scientific community. Support for such voyages and the extensive and expensive technology development needed to take the next step in this pathway (to the moons of Mars) may not be sustainable in the context of such distant goals.

In contrast, the Moon-to-Mars and Enhanced Exploration pathways would allow Congress and the public at large to see an expanding horizon of human activity with intermediate milestones that are distinctly different. Such pathways may be more sustainable, even though they would cost more than the ARM-to-Mars pathway.

4.2.8.3 Logical Technological Feed-Forward

The Moon-to-Mars and Enhanced Exploration pathways have a relatively steady pace of enabling system development. They lack the "cliff" present in Figure 4.7 for the ARM-to-Mars pathway, in which only a few primary mission elements are developed in the near term to practice human spaceflight in cislunar space, and more than half of the total number of primary mission elements must be developed to take the last step from the moons of Mars to the surface of Mars. This need for such a large increment of capability poses a very high development risk. One of the best motivators for successful technology development is to put that technology development into a program that requires it in a specific time frame. This tends to prevent technology development from falling into a mode of self-perpetuation that may ultimately have little or no value. As noted earlier, NTP was developed over a period of almost 20 years with multiple variants. Although significant progress was made technologically, the lack of a program that defined actual requirements ultimately doomed these efforts, and all the resources dedicated to the effort led to essentially no payoff.

4.2.8.4 Minimizing Dead-End Mission Elements

The ARM-to-Mars pathway has four dead-end mission elements: the asteroid capture vehicle and three elements needed to support human crews for long periods on or near the moons of Mars. The Enhanced Exploration pathway also has four dead-end mission elements: the three needed for the Mars Moons mission and one (a large storage stage) for missions to an asteroid in its native orbit. The Moon-to-Mars pathway has only a single dead-end mission element, the disposable descent stage for lunar sorties. This analysis shows how one could improve pathway affordability. Deleting the moons of Mars from the Enhanced Exploration pathway would substantially reduce the number of dead-end mission elements while still allowing exploration of an asteroid in its native orbit. Such tradeoffs must take place in the context of the value of the Mars Moons missions in a program of human exploration and the effect that eliminating these missions would have on other pathway characteristics, such as program development risk. Of course, the nation is unlikely to adopt any of the pathways exactly as they are presented in this report. The goals of "minimizing dead-end mission elements" are to maximize the logical feed-forward of systems and to make the best use of constrained resources. Only in the context of a government-consensus pathway can one ultimately determine whether a given mission element is a dead end.

4.2.8.5 Affordability and Development Risk

Table 4.4 summarizes the affordability, development risk, and operational tempo for every combination of pathway and affordability scenarios examined in this report. The table also shows the number of crewed flights (including the first Mars landing) and the earliest year possible for the Mars landing for both ISS scenarios, that is, ISS decommissioning in either 2020 or 2028.

The total number of crewed flights and the earliest possible year for a human mission to the Mars surface vary widely for the various combinations of pathways and affordability scenarios. The ARM-to-Mars pathway generally has the fewest crewed flight missions (no more than nine). It may be difficult to engage the public's interest with so few missions, but if it is able to overcome the exceedingly high development risk, this pathway would yield the earliest possible landing on Mars. The Enhanced Exploration pathway tends to have the largest number of crewed flights, and it includes missions to the largest number of destinations. It would probably enhance public interest, but it tends to delay the first human mission to the Mars surface relative to the other pathways.

Affordability and operational tempo are determined by the choice of affordability scenario. The schedule-driven scenarios are not affordable, because the cost of executing them far exceeds any feasible increases in the NASA human spaceflight budget. (These scenarios would require the human spaceflight budget to increase at 4 times the rate of inflation for at least 15 years.) The budget-driven scenarios are the most affordable, but they have unacceptably low operational tempos. The operationally viable scenarios are all marginal both in affordability (they would require the human spaceflight budget to increase at twice the rate of inflation) and in operational tempo (the mission rate would still be well below historical precedents, although not as low as in the budget-driven scenarios).

The development risk is determined by the choice of pathway. As discussed above in section 4.2.5 ("Contribution of Key Mission Elements to the Pathways"), the ARM-to-Mars pathway has three destinations, of which the first two (the ARM mission and Mars Moons mission) develop just 5 of the 11 primary missions elements needed for a mission to the Mars surface (see Figure 4.7). In addition, missions to the first two destinations include minimal time at the destinations (about 10 days for the ARM mission and about 60 days for the Mars Moons mission), and neither would provide any data on human health or system performance in an environment that has substantial partial gravity. During the Mars surface mission, however, astronauts would be on the surface for approximately 500 days. A tremendous development effort and major advances in capabilities would be needed for the subsequent Mars surface mission. Thus, the ARM-to-Mars pathway has exceedingly high developmental risk.

The Moon-to-Mars pathway develops 7 of the 11 primary mission elements during lunar missions, which have greater mission-abort capability than missions beyond cislunar space. Compared to the ARM-to-Mars pathway, the Moon-to-Mars pathway features a smoother progression of developing mission elements required for the Mars surface mission. In addition, several transitional mission elements are developed during the lunar missions, which would further reduce development risk for the Mars surface mission. Even so, it would be a big step to go from lunar missions to a Mars surface mission, and the Moon-to-Mars pathway carries very high development risk.

TABLE 4.4 Summary of Pathway Affordability, Development Risk, and Operational Viability

	ARM TO MARS		MOON TO MARS		ENHANCED EXPLORATION	
	ISS 2020	ISS 2028	ISS 2020	ISS 2028	ISS 2020	ISS 2028
TOTAL # CREWED FLIGHTS	Schedule driven: **9**	**9**	Schedule driven: **17**	**17**	Schedule driven: **20**	**20**
	Budget driven: **7**	**9**	Budget driven: **7**	**7**	Budget driven: **11**	**14**
	Operationally viable: **9**	**9**	Operationally viable: **9**	**8**	Operationally viable: **14**	**14**
	ISS 2020	ISS 2028	ISS 2020	ISS 2028	ISS 2020	ISS 2028
EARLIEST POSSIBLE YEAR MARS LANDING	Schedule driven: **2033**	**2033**	Schedule driven: **2035**	**2035**	Schedule driven: **2039**	**2039**
	Budget driven: **2037**	**2046**	Budget driven: **2043**	**2050**	Budget driven: **2050**	**2054**
	Operationally viable: **2037**	**2041**	Operationally viable: **2041**	**2043**	Operationally viable: **2048**	**2050**
OPERATIONAL TEMPO SATISFACTORY	Schedule driven: **NO**		Schedule driven: **YES**		Schedule driven: **YES**	
	Budget driven: **NO**		Budget driven: **NO**		Budget driven: **NO**	
	Operationally viable: MARGINAL		Operationally viable: MARGINAL		Operationally viable: MARGINAL	
AFFORDABLE[a]	Schedule driven: **NO**		Schedule driven: **NO**		Schedule driven: **NO**	
	Budget driven: **YES**		Budget driven: **YES**		Budget driven: **YES**	
	Operationally viable: MARGINAL		Operationally viable: MARGINAL		Operationally viable: MARGINAL	
DEVELOPMENT RISK	Schedule driven: **EXCEEDINGLY HIGH**		Schedule driven: **VERY HIGH**		Schedule driven: **HIGH**	
	Budget driven: **EXCEEDINGLY HIGH**		Budget driven: **VERY HIGH**		Budget driven: **HIGH**	
	Operationally viable: **EXCEEDINGLY HIGH**		Operationally viable: **VERY HIGH**		Operationally viable: **HIGH**	

[a] In the "Affordable" column, "Yes" means that the scenario can be executed with budget increases equal to inflation, "Marginal" means that the scenario requires human spaceflight budgets to increase at twice the rate of inflation (that is, about 5 percent per year), and "No" means that the human spaceflight budget would need to increase at 4 times the rate of inflation (that is, about 10 percent per year).

The Enhanced Exploration pathway shows a long incremental growth of capability (see Figure 4.7) while avoiding multiple concurrent developments of major mission elements. Transitional elements are used to advance technology gaps. This incremental growth in capabilities implies that the Enhanced Exploration pathway has lower development risk than either the ARM-to-Mars or Moon-to-Mars pathway. However, given the level of technological advances required to develop the 11 primary mission elements, supporting systems, and associated capabilities, the Enhanced Exploration pathway still has high development risk.

4.2.8.6 Operational Tempo

Operational tempo is assessed by examining crewed mission rates and SLS flight rates in the context of historical norms for successful programs. The first two SLS flights are scheduled for 2017 and 2021. Table 4.5 shows average time between SLS launches from 2022 until the first launch of an SLS for a human mission to the moons of Mars or the Mars surface. Once Mars missions are under way, the timing of launches will be driven by the synodic period of Mars, with launch windows opening an average of once every 26 months.

The rate of crewed launches will be driven by the ISS until its retirement. Table 4.5 shows the average time between crewed missions from the retirement of the ISS until the first launch of a crewed mission to the moons of Mars or the Mars surface. As with SLS launches, once Mars missions begin, the timing of launches would be driven by the synodic period of Mars.

TABLE 4.5 Operational Tempo: SLS Launches and Crewed Missions

Pathway	ARM-to-Mars		Moon-to-Mars		Enhanced Exploration	
Year of ISS Retirement	2020	2028	2020	2028	2020	2028
Average time between SLS flights from 2022 until first launch for Mars Moons or Mars surface mission (months)						
• Schedule-driven scenario	16	16	4	4	3	3
• Budget-driven scenario	24	32	9	12	9	8
• Operationally viable scenario	18	29	6	7	5	5
Average time between crewed missions from ISS retirement through first Mars moons or Mars surface mission (months)						
• Schedule-driven scenario	17	12	11	7	11	7
• Budget-driven scenario	30	40	46	44	33	29
• Operationally viable scenario	16	18	28	23	22	19

After the ISS is retired, crewed mission rates would be equal to or lower than SLS flight rates for the same period, so crewed mission rates are the key factor in assessing operational tempo. The lower the operational tempo, the more difficult it is to retain critical technical capabilities, to retain operator proficiency, and to use personnel and infrastructure effectively. Action needed to overcome these risks would tend to increase costs. As the disparity between historical and proposed launch rates increases, past experience becomes less relevant as a basis for planning future programs, and at some point the increased cost and risk make pathways unsustainable. Accordingly, it is unrealistic to assume that very low SLS flight rates or crewed mission rates are compatible with an exploration pathway that would span decades.

The operational tempo of each of the pathways depends on whether the pathway is executed in a schedule-driven, budget-driven, or operationally viable scenario and on the retirement date of the ISS. Because of budget constraints, all the pathways in the budget-driven scenario present mission rates much lower than previous successful U.S. programs, which is a significant program risk. Operational tempo in the schedule-driven scenario is in general excellent except for the ARM-to-Mars pathway, where the technology development "cliff" imposes a long delay in advancing from the ARM mission to the Mars Moons mission. The operationally viable scenario constitutes a marginal case in that the crewed mission rate is lower than previous U.S. experience, but it might be achievable.

4.3 TECHNOLOGY PROGRAMS

This section describes ongoing efforts to develop new technologies and capabilities to support future human spaceflight endeavors. The discussion is a summary of key exploration technology programs being conducted by NASA, industry, DOD, and foreign governments. This is not intended to be a rigorous survey or assessment, but it does indicate the breadth of potentially relevant programs and the magnitude of investment, in NASA and elsewhere, that is available to address human spaceflight challenges.

The new capabilities that enable future human spaceflight will be drawn from a wide variety of sources. Foremost among them are the technology programs managed by NASA that respond directly to the needs of NASA's human spaceflight program. These technology programs bring to bear the results of NASA-funded research to inform related engineering efforts. Similarly, the mission requirements should guide researchers in focusing on mission-enabling capabilities. Important new capabilities may also be derived from commercial activities that are pursuing analogous or related goals for their own purposes, including traditional aerospace activities, such as launch vehicle or satellite development, as well as nascent industries, such as space tourism. Other government agencies, principally DOD, also produce new technologies that can be leveraged into the human spaceflight program. Finally, many activities in non-U.S. space programs are producing capabilities of interest to NASA, and international partnerships would allow the United States to draw on these capabilities.

4.3.1 NASA Technology Programs

NASA's development of new and advanced exploration systems and technologies is funded by the exploration budget of the Human Exploration and Operations Mission Directorate (HEOMD) and STMD. HEOMD development activities are organized into three theme areas:

- Exploration System Development
  — Orion Multi-Purpose Crew Vehicle (MPCV) Program
  — SLS Program
- Exploration Ground Systems Program
  — Exploration Research and Development
  — Human Research Program
- Advanced Exploration Systems Program
  — Commercial Spaceflight
  — Commercial Crew Program

STMD manages one program, the Exploration Technology Development Program, which is wholly dedicated to the development of exploration technology, and several other programs that support the development of technology for exploration as well as other applications.

The status of HEOMD and STMD exploration technology programs is summarized below.

4.3.2 Human Exploration and Operations Mission Directorate

4.3.2.1 Exploration System Development

The SLS heavy-lift launch vehicle, Orion MPCV, and related ground systems are being developed with the goal of restoring the ability of the United States to conduct human spaceflight beyond LEO. The development of these systems is not directed at any particular destination. Rather, they are part of NASA's effort to enhance human spaceflight capabilities that could support missions to a variety of destinations in anticipation of a future commitment to initiate a flight program to one or more specific destinations.

The first flight of an SLS launch vehicle (EM-1) is scheduled for 2017, with a second flight (EM-2) to follow in 2021.[94] Both of these vehicles will have a payload capacity of 70 MT to LEO. Upgrades to the SLS could increase the payload capacity to 130 MT or more. Additional information on the SLS program appears above in section 4.2.6.1.4 ("Heavy-lift launch Vehicles").

The main components of the Orion MPCV are the launch abort system, the crew module, and the service module. Orion will have a crew capacity of four, and it will serve as the crew vehicle for launch and return to Earth. Its capabilities will include rendezvous, docking, and EVA. Orion's maximum mission length will depend on the capabilities of the service module, which will provide consumables needed to sustain the crew. The first uncrewed test flight is scheduled to take place in 2014 using a Boeing Delta IV Heavy launch vehicle. The second flight will take place in 2017 as part of the first SLS flight, EM-1. The first crewed flight of the Orion will take place in 2021 on SLS flight EM-2. The service module for the first Orion flight will be provided by NASA. The service module for the second Orion flight (EM-1) will be provided by the European Space Agency (ESA) as part of its cost-sharing obligation to the United States as an ISS partner. The design of the ESA service module will be based on the design of the Automated Transfer Vehicle that has flown to the ISS several times.

The Exploration Ground Systems Program is developing new and refurbished launch site infrastructure to support integration, launch, and recovery operations for the SLS and Orion MPCV flight systems. NASA's Kennedy Space Center has established a single office to manage the Exploration Ground Systems Program and the 21st Century Ground Systems Program. The latter is intended to improve NASA's ability to support non-NASA launch customers in government and industry.

---

[94] These dates appeared to be in flux as this report was in its final preparation but are those reported by NASA in January 2014.

4.3.2.2 Exploration Research and Development

NASA's Exploration Research and Development Program consists of the Human Research Program and the Advanced Exploration Systems Program.

The Human Research Program is concerned with investigating, understanding, preventing, and mitigating risks to human health and performance that are associated with long-duration space missions. Areas of particular interest include exposure to space radiation, emergency medical care in space, psychosocial issues associated with the effects of confinement and isolation, and the effects of microgravity on the human body. This program makes extensive use of the ISS to conduct experiments and collect data on astronaut health in the microgravity environment. The Space Radiation NASA Research Announcement engages the external scientific community to provide better understanding and risk reduction of the space radiation hazard faced by spaceflight crews on exploration missions. This research announcement and associated NASA research are critically aligned to U.S. deep-space exploration objectives.

The Advanced Exploration Systems Program is developing new capabilities and operational concepts to improve safety, reduce risk, and lower the cost of human spaceflight missions beyond LEO. The scope of this program includes crew mobility, habitat systems, vehicle systems, and operations robotic technology for precursor missions. Activities include the demonstration and evaluation of prototype systems during test on the ground and in space on the ISS. The program is also cooperating with the Science Mission Directorate to develop instruments and concepts for robotic precursor missions that would collect data on potential destinations for human exploration missions.

4.3.2.3 Commercial Spaceflight

Anticipating the termination of space shuttle operations in 2011, in 2006 NASA began the process of partnering with industry to foster the development of commercial systems to transport cargo and crew to and from the ISS. The development of commercial cargo launch systems is complete, as evidenced by successful demonstration and operational flights to the ISS by SpaceX and Orbital Sciences Corporation. SpaceX and Orbital are under contract to provide a total of 20 cargo flights to the ISS, and funding for commercial cargo activities has been transferred from exploration to operations in the NASA HEOMD budget. As of April 2014, SpaceX had completed three ISS resupply missions, using its Falcon 9 launch vehicles and Dragon capsules. Orbital completed its first resupply mission in January 2014, using its Antares launch vehicle and Cygnus capsule. NASA is continuing to support the development of commercial crew capabilities (to transport astronauts to the ISS and return them to Earth) with three industry partners: Boeing, SpaceX, and Sierra Nevada. NASA has financial agreements with these companies to provide a total of up to $1.1 billion for all three companies over a 2-year period. Subsequent agreements are expected to result in crewed flights to the ISS starting no earlier than 2017. The main effect of the commercial crew program on NASA is to mitigate the United States' complete dependence on Russia for crew transport. It is too early to know if this approach will reduce costs. What is certain, however, is that the U.S. industrial base capable of supporting human spaceflight has been stabilized and expanded by the commercial cargo and commercial crew programs. This will most probably provide NASA with additional capabilities in the future. In addition, it may enable the establishment of a space-based economy that is not completely dependent on NASA, much as the commercial imaging and communication satellite industries are no longer dependent on NASA. It is well to note, however, that establishment of a commercial space-based economy with human spaceflight as a major component is highly speculative.

4.3.2.4 Space Technology Mission Directorate

The Exploration Technology Development Program is developing technologies to support human exploration beyond Earth orbit with a focus on advanced technologies that have a long development time. The scope of this program includes advanced technologies for SEP, ECLSS, ISRU, EDL, storage and transfer of cryogenic fluids in space, and robotic systems to improve crew safety and effectiveness, and it consumes about one-third of STMD's budget.

In addition to the Exploration Technology Development Program, STMD also funds three other programs:

- Crosscutting Space Technology Development Program.
- Small Business Innovation Research/Small Business Technology Transfer Program.
- Partnership Development and Strategic Integration Program.

Each of these three programs supports HEOMD, the Science Mission Directorate, and the Aeronautics Research Mission Directorate.

The Crosscutting Space Technology Development Program develops technology that is broadly applicable to the needs of future NASA science and exploration missions as well as the needs of other federal agencies and industry. Various elements of this program investigate technologies across a wide range of maturity levels, from conceptual studies through flight demonstrations. Areas of interest include advanced manufacturing technologies, nanotechnology, and synthetic biology. This program also conducts technology demonstration missions to validate selected technologies that have successfully completed ground testing. These missions are currently being developed for low-density supersonic decelerators, laser communications, deep-space atomic clocks, and solar sails.

The Small Business Innovation Research/Small Business Technology Transfer Program is intended to increase the participation of small businesses in NASA research and technology development and to facilitate the commercial application of NASA research results. The 2012 solicitation for this program had a broad scope, encompassing topics such as the reduction of airframe noise and drag; the development and use of launch vehicles, in-space propulsion systems, and space habitats; and advanced space telescopes for astrophysics and Earth science.

The Partnership Development and Strategic Integration Program coordinates technology development activities throughout NASA and takes the lead in NASA's technology transfer and commercialization activities.

4.3.2.5 NASA Infrastructure

The United States had a fairly rudimentary space program when the Apollo program started. As a result, in addition to developing the required technology, the supporting infrastructure for testing, assembly, launch, and so on had to be created ab initio. Some of this infrastructure, such as the Vehicle Assembly Building and the Pad 39 Complex (A and B) at Kennedy Space Center, was repurposed after Apollo. Even so, NASA carries a very large set of aging infrastructure. NASA's Office of the Inspector General (OIG), the Government Accountability Office, and the National Research Council have each reported on NASA infrastructure and offered worrisome observations.[95] NASA's OIG determined that about 80 percent of NASA facilities are more than 40 years old, that maintenance costs for these facilities amount to more than $24 million a year, and that continuing shortfalls in maintenance are adding to an already substantial backlog in deferred maintenance.[96] Carrying such costs constitute latent threats to the development of newer infrastructure that will be needed to support human exploration beyond LEO. For example, some of the infrastructure needed to develop and test fission surface power and NEP systems will undoubtedly be extremely challenging to plan, permit, fund, and construct.

4.3.3 Commercial Programs

A wide range of corporations are developing advanced launch vehicles, capsules, and space habitats to support traditional customers, such as the commercial satellite industry, NASA, and DOD. However, some of the new initiatives to provide orbital and suborbital flight capabilities are focused on new markets such as space tourism or transportation to a privately developed space station, should one be constructed. Some of the new transportation systems are being developed by aerospace companies, such as Boeing and Orbital Sciences Corporation, which

---

[95] A.D. McNaull, House Space Subcommittee discusses aging NASA infrastructure, *FYI: The AIP Bulletin of Science Policy News*, Number 145, October 4, 2013.

[96] NASA Office of Inspector General, "NASA's Efforts to Reduce Unneeded Infrastructure and Facilities," February 12, 2013, http://oig.nasa.gov/audits/reports/FY13/IG-13-008.pdf.

have substantial spaceflight experience. Others are being developed by relatively new ventures such as Bigelow Aerospace, Blue Origin, SpaceX, Virgin Galactic, and XCOR Aerospace. In some cases, the development of new vehicles has been aided by NASA (to support the development of commercial cargo and crew space transportation systems, for example). In other cases, development programs have relied entirely on private funding. In all cases, however, these companies have invested heavily in making commercial human spaceflight profitable, and market-based evidence indicates that there is a sustainable path forward for at least some of these initiatives. In fact, many of the companies are making good progress in developing new systems; as noted above, the Orbital Sciences Antares launch vehicle and the SpaceX Falcon 9 launch vehicle have entered commercial cargo service, and a space habitat developed by Bigelow Aerospace will be docked to the ISS in 2015 for a 2-year evaluation period.

Looking beyond LEO, some new ventures have been established with the goal of launching privately funded crewed missions to the Moon or Mars and uncrewed missions to asteroids. However, none of these more ambitious endeavors has evidenced substantial funding or vehicle development activities. This is not surprising because exploration missions beyond LEO would be extremely expensive, they would need to overcome substantial technical risks, and the time to produce a positive return on investment would be well beyond the normal time horizon of corporate business plans.

The largest of the new commercial vehicles under development, the SpaceX Falcon Heavy, would be able to launch robotic missions to Mars, but it would have limited capabilities to support human spaceflight beyond LEO unless the United States committed itself to in-orbit assembly, replenishment, fuel depots, and so on. Given the congressional mandate to develop the SLS, that seems unlikely. Smaller launch vehicles now in service or under development would be less suitable. This is not surprising given the small market for such capabilities and the fact that NASA is developing its own launch vehicle, the SLS, for human spaceflight beyond LEO.

4.3.4 Department of Defense

Research and development of new technology and capabilities in government agencies other than NASA could benefit future human spaceflight programs. Most prominently, DOD is highly invested in improving the capabilities and reducing the costs of its space-related activities. Launch costs are of particular interest. The U.S. Air Force recently announced that it will allow new launch vehicle providers, such as Space X, to pursue certification to launch national security payloads. In contrast, there has been little overt DOD interest in the SLS. DOD research agencies such as the Defense Advanced Research Projects Agency (DARPA), the Air Force Research Laboratory, and the Naval Research Laboratory are also advancing space technologies at various levels of maturity, from basic building blocks, such as advanced materials and power generation technologies, to complex in-space demonstrations of systems such as automated rendezvous and docking. Specific examples include the DARPA Fast Access Spacecraft Testbed, which facilitated the development of solar arrays that have higher power and less mass; the DARPA Orbital Express missions, which demonstrated automated rendezvous and docking as well as the transfer of noncryogenic fluids between two spacecraft; and the Air Force Research Laboratory's research into advanced thermal protection systems.

Research focused on terrestrial applications can also contribute to future human spaceflight missions. For example, some of the medical and behavioral advances to improve the performance and safety of soldiers on the battlefield may help assure the health of astronauts on long-duration missions, and advances in robotics technology for military applications may contribute to the development of advanced robotics by NASA. In addition, technology developed for the Army's Human Universal Load Carrier could enable NASA to develop advanced spacesuits that reduce the forces and torques that a wearer would need to exert during EVA operations.

4.3.5 International Activities

A growing number of nations have substantial human exploration programs, including system or technology development programs that could possibly contribute directly to future NASA human space activities. The programs of greatest note are being conducted by Russia, ESA, Japan, and China. In addition, Canada is an active partner in the ISS program, and India has a long-term plan for human space exploration. Altogether, 38 nations have had their astronauts fly in space. As noted in Chapter 2, human space exploration enhances international stature and national pride, and this seems to be a key motivation for emerging space powers to enter and expand human spaceflight activities.[97]

Like the United States, Russia has focused its current human spaceflight program on supporting the ISS. Russia has a long-term space policy, and, as discussed in the mass media, it has the capabilities to develop technologies and systems in support of human exploration programs beyond LEO, particularly with regard to nuclear propulsion, power systems, and human exploration. Russian nuclear development facilities, however, would likely require major refurbishment before they could support a new system development effort. As in the United States, nearly all Russia's work in the development of space nuclear systems took place decades ago.

ESA's human spaceflight program is focused on supporting the ISS. ESA is conducting cargo flights to the ISS and exploring options for a crewed spacecraft, and it is also committed to developing a service module for the Orion MPCV. The service module will incorporate technology from ESA's Automated Transfer Vehicle, which has been used for ISS resupply.

Chinese astronauts (also known as yǔhángyuán or taikonauts) are viewed domestically as symbols of Chinese cultural and technological prominence. China's human spaceflight program is focused on LEO missions. China intends to develop a modular space station and a crewed lunar exploration program, although details of the technologies and systems being developed for these missions are not publicly available. NASA is prohibited by law from bilateral cooperation with the Chinese, so the U.S. and Chinese space programs are proceeding independently. This policy, driven by congressional sentiment, denies the United States the option to partner with a nation that will probably be capable of making truly significant contributions to international collaborative missions. It may be time to reexamine whether this policy serves the long-term interests of the United States.

India and Japan are separately considering options for developing a crewed spacecraft to be launched on indigenous rockets.[98]

The 2013 *The Global Exploration Roadmap*[99] authored by the International Space Exploration Coordination Group (ISECG) includes a single reference mission scenario that leads to a human mission to the Mars surface. This scenario "reflects the importance of a stepwise evolution of critical capabilities which are necessary for executing increasingly complex missions to multiple destinations. . . . The roadmap demonstrates how initial capabilities can enable a variety of missions in the lunar vicinity, responding to individual and common goals and objectives, while contributing to building the partnerships required for sustainable human space exploration."[100] The 12 space agencies and countries that created the ISECG *The Global Exploration Roadmap* continue to support human exploration beyond LEO.

4.3.6 Robotic Systems

4.3.6.1 Robotic Science and Exploration

Robotic space exploration, which is driven by science objectives, has greatly expanded knowledge of the solar system, including its origin and evolution. Robotic missions involving orbiters, landers, and rovers have discovered extensive evidence of past warm and wet conditions on Mars. The evidence includes lake systems, alkali flats, and volcanic vents.[101] The possibility that robotic and/or human missions may discover evidence of past life on Mars is a driving scientific reason for continued interest in exploring and understanding the red planet. Other targets of interest for human and/or robotic expeditions include asteroids, Earth's Moon, and the two moons of Mars, Phobos and Deimos. Analysis of lunar material brought back by Apollo and by robotic surface missions launched by the Soviet Union, along with the results of subsequent robotic orbital missions to the Moon, indicate that it

---

[97] D.A. Mindell, S.A. Uebelhart, A.A. Siddiqi, and S. Gerovitch, *The Future of Human Spaceflight: Objectives and Policy Implications in a Global Context*, American Academy of Arts and Sciences, Cambridge, Mass., 2009.

[98] Ibid.

[99] ISECG, *The Global Exploration Roadmap*, 2013.

[100] Ibid.

[101] J.P. Grotzinger, D. Y. Sumner, L.C. Kah, K. Stack, S. Gupta, L. Edgar, D. Rubin, et al., A habitable fluvio-lacustrine environment at Yellowknife Bay, Gale Crater, Mars, *Science* 343(6169), doi:10.1126/science.1242777.

was created when a Mars-size planetesimal collided with Earth early in its history.[102] Earth-based observations and robotic exploration of asteroids show a plethora of evolutionary states, from primitive bodies to remnants of iron-nickel cores of impact-disrupted objects.[103] Phobos and Deimos are particularly interesting because these two small bodies appear to be composed of very primitive materials that are quite different than the inferred average composition of Mars.[104] Given the high scientific interest in Mars, the Moon, asteroids, and the moons of Mars, there will be continued interest in additional robotic missions to the Moon and beyond.

Robotic science missions provide information about the environmental, terrain, and surface properties and other information that is essential for the safety and effectiveness of crewed missions. For example, radiation data collected by the MSL mission en route to Mars and on the ground (by the *Curiosity* rover) are providing NASA with more accurate information about the nature and magnitude of radiation that astronauts would encounter during a mission to the Mars surface.

Evolutionary improvements in the capabilities of robotics would enable robotic servicing of spacecraft and robotic precursor missions to conduct ground operations at a landing site before a crewed lander arrives. Experiments on the ISS are testing the ability of robotic systems to refuel and repair spacecraft. Surface operations on the Moon and Mars could include exploration of potentially hazardous terrains and environments near landing sites, and a robotically operated ISRU system on the surface of Mars could generate needed supplies of oxygen, some of which could be converted to water using hydrogen brought from Earth. After the crew arrives on the surface, robotic systems could relieve the crew of menial tasks, such as housekeeping. In addition, surface exploration could be conducted as a cooperative effort involving humans and robotic systems. For example, robots capable of traversing hazardous terrain might be used to explore the safest paths for humans to sample strata on the walls of canyons and crater rims, or they could be commanded to conduct reconnaissance and sampling if the terrain proved to be too difficult for humans to explore.

4.3.6.2 A Game Changing Vision of Robotics

As observed elsewhere in this chapter, visionaries in the late 1940s correctly presumed that humans would eventually travel to the Moon, but the approaches that they envisioned (a single huge rocket that would make the round trip) was not the approach ultimately used. That was in part because more practical approaches were developed in the constrained resource environment of a real program and in part because there had been tremendous technological advances unforeseen by the earlier visionaries, particularly with regard to digital computers.

Although the fundamental technology of chemical propulsion has hardly advanced in the past 50 years, the capability of computers has continued to advance at an exponential pace with no end in sight. The capabilities of sensor technologies have also increased dramatically. Inexpensive consumer devices now have more computing power, sensor resolution, and programmability than were available to NASA in the midst of Apollo.

Robotics is a manifestation of the exponential increase in computational capability, sensor quality, and mechatronics (which combines mechanical, electric, and control technologies). Industrial robots are commonplace. Self-driving cars, which in the recent past were the subject of highly speculative cutting-edge experiments, are now being developed commercially. Robots are seriously contemplated as an answer to demographic crises in Western societies, where eldercare will require highly autonomous systems that are capable of complex interactions with humans.

The most common response to the criticism that human spaceflight is extremely expensive is that humans' capacity for contextual reasoning and problem-solving is indispensable in exploration scenarios. Nonetheless, much more has been learned about our solar system by relatively primitive robots than by human explorers. Similarly, oceanographic research, until recently in decline because of the prohibitive costs of ship-based oceanography, is undergoing a revolution through the use of autonomous systems that can gather data in larger quantities and much less expensively than conventional ocean expeditions. Even in the culturally resistant domain of warfare, where

---

[102] D.J. Stevenson, Origin of the Moon—The collision hypothesis, *Annual Review of Earth and Planetary Sciences* 15(1):271-315, 1987.

[103] H.Y. McSween, Jr., *Meteorites and Their Parent Planets* (2nd ed.), Oxford University Press, 1999.

[104] A. Fraeman, S.L. Murchie, R.E. Arvidson, R.N. Clark, R.V. Morris, A.S. Rivkin, and F. Vilas, Spectral absorptions on Phobos and Deimos in the visible/near infrared wavelengths and their compositional constraints, *Icarus* 229:196-205, doi:10.1016/j.icarus.2013.11.021.

it has been assumed that human presence is mandatory, uncrewed aircraft, uncrewed submersibles (to detect and destroy mines), and other robotic systems are increasingly taking the place of humans for dangerous missions. In coming decades, advances in robotic capabilities may likewise present new options for space exploration and open new pathways that are technologically achievable and can be implemented affordably without unacceptable developmental risk.

4.4 KEY RESULTS FROM THE PANEL'S TECHNICAL ANALYSIS AND AFFORDABILITY ASSESSMENT

The Technical Panel's technical analysis and affordability assessment yielded the following key results.

1. **Feasible Destinations for Human Exploration.** For the foreseeable future, the only feasible destinations for human space exploration are the Moon, asteroids, Mars, and the moons of Mars. A human mission to the Mars surface is feasible, although doing so will require overcoming unprecedented technical risk, fiscal risk, and programmatic challenges. Mars is humanity's horizon destination.
2. **Pace and Cost of Human Exploration.** Progress in human exploration beyond LEO will be measured in decades with costs measured in hundreds of billions of dollars and significant risk to human life.
3. **Human Spaceflight Budget Projections.** With current flat or even inflation-adjusted budget projections for human spaceflight, there are no viable pathways to Mars.
   a. A continuation of flat budgets for human spaceflight is insufficient for NASA to execute any pathway to Mars and limits human spaceflight to LEO until after the end of the ISS program.
   b. Even with a NASA human spaceflight budget adjusted for inflation, technical and operational risks do not permit a viable pathway to Mars.
   c. The currently planned crewed flight rate is far below the flight rate of past human spaceflight programs.[105]
   d. Increasing NASA's budget to allow increasing the human spaceflight budget by 5 percent per year would enable pathways with potentially viable mission rates, greatly reducing technical, cost, and schedule risk.
4. **Potential Cost Reductions.** The decadal timescales reflected above are based on traditional NASA acquisition. Acceleration *might* be possible with substantial cost reductions resulting from
   a. More extensive use of broadly applicable commercial products and practices.
   b. Robust international cost sharing (that is, cost sharing that greatly exceeds the level of cost sharing with the ISS).
   c. Unforeseen significant technological advances in the high-priority capabilities.
5. **Highest-Priority Capabilities.** The highest-priority capabilities that are needed to enable human surface exploration of Mars are related to
   a. Entry, descent, and landing for Mars.
   b. Radiation safety.
   c. In-space propulsion and power.
6. **Continuity of Goals.** Frequent changes in the goals for U.S. human space exploration (in the context of the *decades* that will be required to accomplish them) dissipate resources and impede progress.
7. **Maintaining Forward Progress.** Within the current budget scenario, there are only a few actions, all of which would be difficult to take, that can ensure that the U.S. human exploration program keeps moving forward during the next decade following the pathways laid out in this report. These options include the following:

---

[105] Late in the production of this report, NASA leadership told the NASA Advisory Council's Human Exploration and Operations Committee that inasmuch as a "repetitive cadence is necessary" for a viable SLS program, the SLS would launch every year. Although that statement is consistent with the findings of the present committee, the "assumed long-term SLS manifest" shown to the Human Exploration and Operations Committee did not identify payloads past the EM2 mission currently planned for 2021. It has been suggested that military, commercial, or science missions will take up the slack, but no commitments have been announced. The SLS heavy-lift capacity suggests that the cost of payloads appropriate for its use will provide a substantial barrier to increasing the SLS launch rate to once per year.

    a. Aggressively divesting NASA of nonessential facilities and personnel to free up resources for future-oriented programs.[106]

    b. Focusing the use of the ISS to support future exploration beyond LEO, including the use of the ISS as a testbed for technologies and human health. (However, for technical reasons and because the operation of nuclear fission systems in LEO is prohibited,[107] the ISS cannot be used to support development of any of the highest-priority capabilities identified in key result 5.)

    c. Making substantial research and technology investments in the highest-priority capabilities.

    d. Maintaining a robust planetary and space science program to engage and maintain the interest of the public in deep-space exploration, to advance technological capabilities relevant to human exploration operations in space and on the surface of the Moon and/or Mars, and to better understand space and surface environments.

    e. Investing in people: developing a workforce with hands-on experience in flight programs.

    f. Continuing to leverage commercial products and practices to strengthen the industrial base and increase NASA's efficiency.

    g. Strengthening international relationships and partnerships with the goal of reducing duplication, exploiting worldwide expertise, and substantially reducing the cost of the total program to the United States.

Human spaceflight is an extraordinarily challenging endeavor that is fraught with daunting technical, political, and programmatic hurdles and carries a high degree of physical risk to the explorers who push forward the boundaries of human presence. The scale of U.S. investment in this endeavor is substantial and has remained so for decades despite periodic financial strictures and without broad and deeply committed public support. The frequently cited rationales for human spaceflight are self-evident to some and unconvincing to others, and there are unlikely to be novel rationales that are more potent.

The United States and a group of international partners have succeeded in building and operating the most ambitious space engineering project ever: the ISS. This required a substantial investment by multiple countries, and it is natural to attempt to amortize that investment and retain the capability. Programmatic and budgetary factors led to the completion of the ISS just as the main launch system that enabled its construction was retired from service, and Congress mandated the development of a new heavy-lift launcher, the SLS. In addition, national leadership has called for the United States to venture beyond LEO, where more distant destinations beckon. The SLS is but one part of the technology that needs to be developed to enable even a return to the Moon, let alone human visits to near-Earth asteroids or to the vicinity or surface of Mars. Although national leadership has sustained operations in LEO, it seems disinclined to increase the level of investment in human spaceflight substantially. Thus, the human spaceflight program faces a dilemma. Maintaining the ISS and developing the SLS leave precious little budgetary maneuvering room to plan the next steps beyond LEO. However, the affordability analysis summarized above has shown that with currently projected human spaceflight budgets it will take decades to achieve the next significant human spaceflight milestone even with optimistic assumptions about costs, technology development, and programmatic stability.

Probably the most significant factors in progressing beyond LEO are the development of a strong national (and international) consensus about the pathway to be undertaken and sustained discipline to maintain course over many administrations and Congresses. Without that consensus and discipline, it is all too likely that the potential of the SLS will be wasted, human spaceflight to LEO will be increasingly routine (although still with risk to life), and the horizons of human existence will not be expanded—at least not by the United States. With such a consensus, however, and with strict adherence to the pathways approach and principles outlined in this report, the United States could maintain its historical position of leadership in space exploration and embark on a program of human spaceflight beyond LEO that, perhaps for the first time in the more than half-century of human spaceflight, would be sustainable.

---

[106] The political difficulty of taking this action suggests that it may require an extraordinary process, akin to NASA's "real property assessment" of 2004 but using an independent, bipartisan assessment commission and later an up-or-down vote by Congress.

[107] United Nations Principles Relevant to the Use of Nuclear Power Sources in Outer Space, UN A/RES/47/68, 1992, available at http://www.un.org/documents/ga/res/47/a47r068.htm.

Appendixes

A

Statement of Task

In accordance with Section 204 of the NASA Authorization Act 2010, the National Research Council (NRC) will appoint an ad hoc committee to undertake a study to review the long-term goals, core capabilities, and direction of the U.S. human spaceflight program and make recommendations to enable a sustainable U.S. human spaceflight program.

The committee will:

1. Consider the goals for the human spaceflight program as set forth in (a) the National Aeronautics and Space Act of 1958, (b) the National Aeronautics and Space Administration Authorization Acts of 2005, 2008, and 2010, and (c) the National Space Policy of the United States (2010), and any existing statement of space policy issued by the president of the United States.

2. Solicit broadly-based, but directed, public and stakeholder input to understand better the motivations, goals, and possible evolution of human spaceflight—that is, the foundations of a rationale for a compelling and sustainable U.S. human spaceflight program—and to characterize its value to the public and other stakeholders.

3. Describe the expected value and value proposition of NASA's human spaceflight activities in the context of national goals—including the needs of government, industry, the economy, and the public good—and in the context of the priorities and programs of current and potential international partners in the spaceflight program.

4. Identify a set of high-priority enduring questions that describe the rationale for and value of human exploration in a national and international context. The questions should motivate a sustainable direction for the long-term exploration of space by humans. The enduring questions may include scientific, engineering, economic, cultural, and social science questions to be addressed by human space exploration and questions on improving the overall human condition.

5. Consider prior studies examining human space exploration, and NASA's work with international partners, to understand possible exploration pathways (including key technical pursuits and destinations) and the appropriate balance between the "technology push" and "requirements pull". Consideration should include the analysis completed by NASA's Human Exploration Framework Team, NASA's Human Spaceflight Architecture Team, the Review of U.S. Human Spaceflight Plans (Augustine Commission), previous NRC reports, and relevant reports identified by the committee.

6. Examine the relationship of national goals to foundational capabilities, robotic activities, technologies, and missions authorized by the NASA Authorization Act of 2010 by assessing them with respect to the set of enduring questions.

7. Provide findings, rationale, prioritized recommendations, and decision rules that could enable and guide future planning for U.S. human space exploration. The recommendations will describe a high-level strategic approach to ensuring the sustainable pursuit of national goals enabled by human space exploration, answering enduring questions, and delivering value to the nation over the fiscal year (FY) period of FY2014 through FY2023, while considering the program's likely evolution in 2015-2030.

B

Methodological Notes About the Public Opinion Data

RESEARCH ORGANIZATION/SPONSOR NAME ABBREVIATIONS

ANES	American National Election Studies
AP	Associated Press
AP/Ipsos	Ipsos Associated Press Polls
CBS	CBS News Poll
CBS/NYT	CBS News/*New York Times* Poll
Gallup	The Gallup Organization
GSS	National Opinion Research Center General Social Survey
HI	Harris Interactive
LSAY	Longitudinal Study of American Youth
NBC/AP	NBC News/Associated Press Poll
NBC/WSJ	NBC News/*Wall Street Journal* Poll
NSF Surveys of Public Attitudes	National Science Foundation Surveys of Public Attitudes Toward and Understanding of Science and Technology
ORC	ORC Macro
Pew	Pew Research Center for People and the Press

SURVEY METHODS AND QUESTION WORDING

The survey questions referred to in Chapter 3 are listed in this section by topic and in the order in which they appear in the chapter. Unless otherwise noted, the data cited are available online from one or more of the following sources: the Roper Center for Public Opinion Research at the University of Connecticut, the Inter-university Consortium for Political and Social Research at the University of Michigan, the Odum Institute at the University of North Carolina and the Pew Research Center.

Interest and the Attentive Public

1981-2000: NSF Surveys of Public Attitudes
2003-2007: Science News Study
2008: ANES Panel

The NSF Surveys of Public Attitudes were conducted by phone by the Public Opinion Laboratory at Northern Illinois University. The 2003-2007 Science News Studies and 2008 ANES Panel were conducted by Knowledge Networks using online national probability samples. Analysis of the Science News Study data was provided to the panel by the Principal Investigator, Jon Miller.

Interest in Space

> There are a lot of issues in the news, and it is hard to keep up with every area. I'm going to read you a short list of issues, and for each one I would like you to tell me if you are very interested, moderately interested, or not at all interested.
>
> [Space exploration]

Being Well Informed about Space

> For each issue I'd like you to tell me if you are very well informed, moderately well informed, or poorly informed.
>
> [Space exploration]

Attentiveness to Space

The measure of attentiveness was created based on (1) a high level of interest in space exploration, (2) a sense of being well informed about space exploration, and (3) a minimal regular pattern of national news consumption. A minimal pattern of news consumption was defined as doing any one of the following: reading a newspaper in print or online "a few times a week" or "everyday"; reading a news magazine "regularly, that is most issues"; reading a science magazine "regularly, that is most issues"; reading a health magazine "regularly, that is most issues"; watching network or cable television news three or more days a week; reading news online three or more days a week. The computations in the 1980s did not include the online option for newspapers or online news.

Other Interest Questions

GSS 2012. In-person interview, n=1,974.

> There are a lot of issues in the news, and it is hard to keep up with every area. I'm going to read you a short list of issues, and for each one—as I read it—I would like you to tell me if you are very interested, moderately interested, or not.
>
> [Issues about space exploration].

Support for Spending

Pew 10/13. Telephone interview, conducted by Abt SRBI, n=1,504.

> Now I would like to ask you about some parts of the government. Is your overall opinion of [The National Aeronautic and Space Administration, NASA] very favorable, mostly favorable, mostly unfavorable, or very unfavorable?

Gallup 8/06. Survey conducted on behalf of the Space Foundation, telephone interview, n=1,000. Data from Gallup Organization report, Public Opinion Regarding America's Space Program, submitted to the Space Foundation.

> NASA'S budget request this year is under one percent of the federal budget which would amount to approximately $58 per year for the average citizen. Do you think the nation should continue to fund space exploration at this current level, a slightly increased level, at a significantly increased level, at a slightly decreased level, at a significantly decreased level, or not fund at all?

GSS, annually between 1972-1978, 1980, 1982-1991, 1993; every other year 1994-2012. In-person interview, telephone interview in later years, Sample size varied.

> We are faced with many problems in this country, none of which can be solved easily or inexpensively. I'm going to name some of these problems, and for each one I'd like you to name some of these problems, and for each one I'd like you to tell me whether you think we're spending too much money on it, too little money, or about the right amount. First [the space exploration program] . . . are we spending too much, too little, or about the right amount on [the space exploration program]?

Pew 1/04. Telephone interview, n=1,503.

> I'd like to ask you some questions about priorities for President (George W.) Bush and Congress this year (2004). As I read from a list, tell me if you think the item I read should be a top priority, important but lower priority, not too important, or should it not be done? Should [expanding America's space program] be a top priority, important but lower priority, not too important, or should it not be done?

The Apollo Program

Gallup 5/61. In-person interview, n=3,449.

> It has been estimated that it would cost the United States 40 billion dollars—or an average of about $225 per person— to send a man to the moon. Would you like to see this amount spent for this purpose, or not?

HI 4/67. Telephone interview, n=1,250.

> It could cost the United States $4 billion a year for the next few years to finally put a man on the moon and to explore other planets and outer space. All in all, do you feel the space program is worth spending that amount of money or do you feel it isn't worth it?

Gallup 2/67. Telephone interview, n=2,344.

> In your opinion, do you think it is important or is not important to try to send a man to the moon before Russia does?

CBS/NYT 7/79, CBS/NYT 6/94, CBS 7/97, CBS 8/99, CBS 7/09. Telephone interview, 1979: n=1,192; 1994: n=978; 1997: n=1,042; 1999: n=1,165; 2009: n=944.

> (*X number of*) years ago, the United States spent a great deal of time, effort and money to land men on the moon. Looking back now, do you think that effort was worth it, or not?

International Space Station and Space Shuttle

NSF Surveys of Public Attitudes, 1988, 1992, 1997, 1999. Telephone interview conducted by ORC Macro, 1988: n=2,041; 1992: n=2,001; 1997: n=2,000; 1999: n=1,882.

> The American space program should build a space station large enough to house scientific and manufacturing experiments. Do you strongly agree, agree, disagree, or strongly disagree?

CBS/NYT 1/87, CBS/NYT 1/88, CBS/NYT 10/88, CBS 12/93, CBS 8/99, CBS 7/05. Telephone interview, 1987: n=1,590; 1988: n=1,663; 1988 n=1,530; 1993: n=892; 1999: n=1,165; 2005: n=1,222.

Given the costs and risks involved in space exploration, do you think the space shuttle is worth continuing, or not?

NBC/AP 10/81, NBC/AP 11/82, NBC/WSJ 10/85, NBC/WSJ 6/86, Pew 6/11. Telephone interview, 1981: n=1,598; 1982: n=1,583; 1985: n=1,573; 1986: n=1,599; 2011: n=1,502.

Do you think the space shuttle program has been a good investment for this country, or don't you think so?

(Between 1981-1986 the question read: Do you think the space shuttle program in a good investment . . .)

Mars

CBS/NYT 6/94, CBS 7/97, CBS 8/99, CBS 1/04, CBS 7/09. Telephone interview, 1994: n=978; 1997: n=1,042; 1999: n=1,165; 2004: n=1,022; 2009 n=944.

Would you favor or oppose the United States sending astronauts to explore Mars?

Gallup 7/69, Gallup 7/99, Gallup/CNN/USA Today 6/05. Telephone interview, 1969: n=1,555; 1999: n=1,061; 2005 n=1,009.

There has been much discussion about attempting to land an astronaut on the planet Mars. How would you feel about such an attempt—would you favor or oppose the United States setting aside money for such a project?

AP/Ipsos 1/04. Telephone interview, n=1,000.

Version 1: As you may have heard, **the United States** is considering expanding the space program by building a permanent space station on the Moon with a plan to eventually send astronauts to Mars. Considering all the potential costs and benefits, do you favor or oppose expanding the space program this way or do you oppose it?

Version 2: As you may have heard, **the Bush administration** is considering expanding the space program by building a permanent space station on the Moon with a plan to eventually send astronauts to Mars. Considering all the potential costs and benefits, do you favor expanding the space program this way or do you oppose it?

Science News Study, 2007. Web interview, n=1,407. Analysis of the data was provided to the panel by the principal investigator, Jon Miller.

The United States should begin planning for a manned mission to Mars in the next 25 years. (strongly agree, agree, not sure, disagree, strongly disagree)

NSF Surveys of Public Attitudes, 1988. Telephone interview conducted by ORC Macro, n=2,041.

The American space program should try to land astronauts on Mars within the next 25 years. Do you strongly agree, agree, disagree, or strongly disagree?

Human Versus Robotic Missions

Gallup/CNN/USA Today 2/03. Telephone interview, n=1,000.

Some people feel the U.S. space program should concentrate on unmanned missions like Voyager 2, which send back information from space. Others say the U.S. should concentrate on maintaining a manned space program like the space shuttle. Which comes closer to your view?

AP/Ipsos 1/04. Telephone interview, n=1,000.

> Some have suggested that space exploration on the Moon and Mars would be more affordable using robots than sending humans to do the exploration. Would you rather see exploration of the Moon and Mars done by robots or with human astronauts?

> In light of the space shuttle accident last February, in which seven astronauts were killed, do you think the United States should or should not continue to send humans into space?

NASA's Role, International Collaboration, and Commercial Firms

Pew 6/11. Telephone interview, n=1,502.

> In your view, is it essential or not essential that the United States continue to be a world leader in space exploration?

Time/Yankelovich 1/88. Telephone interview, n= 957.

> How important do you think it is for this country to be the leading nation in space exploration—very important, somewhat important, or not important at all?

AP/Ipsos1/04. Telephone interview, n= 1,000.

> How important do you think it is for this country to be the leading nation in space exploration—very important, somewhat important, or not important at all?

CNN/ORC 7/11. Telephone interview, n=1,009.

> How important do you think it is for the United States to be ahead of Russia in space exploration—very important, fairly important, or not too important?

Gallup 6/61. In person interview, n=1,625.

> How important do you think it is for the United States to be ahead of Russia in space exploration—very important, fairly important, or not too important?

Gallup 3/06. Telephone interview, n=1,001.

> A number of Asian and European countries now have space programs of their own or have announced plans for space activities and exploration. As more countries embark on space programs, how concerned are you that the US will lose its leadership in space?

Gallup 4/08. Survey conducted on behalf of the Coalition for Space Exploration, telephone interview, n=1,002. Data from Gallup Organization report, Public Opinion Regarding America's Space Program, submitted to the Coalition for Space Exploration.

> Both China and the US have announced plans to send astronauts to the moon. China has announced plans to send astronauts to the moon by 2017 and the US has announced plans to send astronauts to the moon by 2018, a year later. To what extent, if any, are you concerned that China would become the new leader in space exploration or take the lead over the US?

Time/Yankelovich, 1/88. Telephone interview, n= 957.

> Do you think it would be a good idea or a bad idea for the United States and the Soviet Union to undertake cooperative efforts in space—such as going to Mars?

HI 7/97. Telephone interview, n=1,002.

(Would you favor or oppose the following?) . . . Joint space missions involving Americans, Russians and people from other countries.

CBS 7/97. Telephone interview, n=1,042.

In general, do you think the US should work with Russia on space missions, or not?

Gallup 4/08. Survey conducted on behalf of the Coalition for Space Exploration, telephone interview, n=1,002. Data from Gallup Organization report, Public Opinion Regarding America's Space Program, submitted to the Coalition for Space Exploration.

As you may know the US space shuttles will retire in 2010. The first launch of the Constellation Program is scheduled for 2015, leaving a five-year gap between 2010, the space shuttle's last scheduled mission, and the first schedule launch of the Constellation Program. During the five-year period America will need to access the International Space Station through the Russian Space Agency who would ferry crew members to the Station. How concerned are you that the US will not have direct access to the Space Station during these five years?

Yankelovich 12/97. Survey conducted on behalf of the The Boeing Company, telephone interview, n=1,510. Data from Yankelovich report, Public Opinion and the Space Program: Understanding Americans' Attitudes and Developing a Communications Strategy for Space, Third Year Tracking, submitted to The Boeing Company.

Some people believe that the space program should be funded and manned by private business, while others believe it should be funded and managed by the federal government. Which of these opinions do you agree with more?

CNN/ORC 7/11. Telephone interview, n=1,009.

In general, do you think the US (United States) should rely more on the government or more on private companies to run the country's manned space missions in the future?

Rationales for Space Exploration

CBS/NYT 6/94. Telephone interview, n=978.

What do you think is the best reason for exploring space? (open ended question)

Gallup 6/04. Survey conducted on behalf of the Space Foundation, telephone interview, n=1,000. Data from Gallup Organization report, Public Opinion Regarding America's Space Program, submitted to the Space Foundation.

What do you think is the main reason why America continues to explore space?: It is human nature to explore, to maintain our status as an international leader in space, to provide benefits on earth, to keep our nation safe, it inspires us and motivates our children, or some other reason?

Pew 6/11. Telephone interview, n=1,502.

Thinking about the space program more generally, how much does the U.S. space program contribute to: Scientific advances that all Americans can use? This country's national pride and patriotism? Encouraging people's interest in science and technology?

Gallup 6/04. Survey conducted on behalf of the Space Foundation, telephone interview, n=1,000. Data from Gallup Organization report, Public Opinion Regarding America's Space Program, submitted to the Space Foundation.

How much do you agree or disagree with this statement: The quality of our daily lives has benefited from the knowledge and technology that have come from our nation's space program.

Gallup 5/05. Survey conducted on behalf of the Space Foundation, telephone interview, n=1,001. Data from Gallup Organization report, Public Opinion Regarding America's Space Program, submitted to the Space Foundation.

To what extent do you agree or disagree with the following statements:

America's space program helps give America the scientific and technological edge it needs to compete with other nations in the international marketplace.

America's space program benefits the nation's economy by inspiring students to pursue careers in technical fields.

Gallup 4/08. Survey conducted on behalf of the Coalition for Space Exploration, telephone interview, n=1,000. Data from Gallup Organization report, Public Opinion Regarding America's Space Program, submitted to the Coalition for Space Exploration.

To what extent do you agree or disagree that the scientific, technical and other benefits of space exploration are worth the risks of human space flight?

To what extent do you believe America's space program inspires young people to consider an education in science, technology, math, or engineering fields?

Correlates of Support for Space Exploration

LSAY, 1987, 1988, 1989, 2008, 2011. Self-administered questionnaire, 1987: n=4,491, 1988: n=3,708, 1989: n=3,191; 2008: n=2,568; 2011: n=3,154.

There are a lot of issues discussed in the news and it is hard to keep up with every area. For each issue area listed below, please indicate how interested and informed you are about that issue.

[Space exploration]

C

Stakeholder Survey Methods

To obtain stakeholder input, the Human Spaceflight Public and Stakeholder Opinion Panel conducted a survey of several key stakeholder groups. The survey was conducted by NORC at the University of Chicago. This appendix describes the methods used to conduct the survey.

SAMPLING FRAME

After initial informal exploratory discussions with a variety of experts and stakeholders, the panel, in consultation with the committee, developed a list of stakeholders customized to meet the needs of this project (Table C.1). To build a sampling frame for the survey, the panel identified leadership positions within each of the stakeholder groups of interest, and then identified the individuals occupying these positions. For example, in the case of the industry group, this included CEOs and Presidents of corporations that are members of the Aerospace Industries Association, Commercial Spaceflight Federation, and the American Institute of Aeronautics and Astronautics. In the case of space scientists, the sampling frame included members of the National Academy of Sciences, Institute of Medicine, and National Academy of Engineering, with an interest in space, as well as officers and Board Directors of professional associations, such as the American Astronautical Society and American Astronomical Society.

Because the committee's charge was to make recommendations for a sustainable program spanning the next couple of decades, the views of the younger generation were particularly important to capture. To develop a sampling frame of young space scientists, the panel assembled lists of American Institute of Aeronautics and Astronautics early career and graduate student award winners, National Science Foundation postdoctoral fellowship winners, and NASA Aeronautics Graduate Scholarship Program winners from the past 3 years.

A description of the lists used to generate the sampling frames for each of the stakeholder groups is provided in Table C.1.

The approach described yielded an overall sampling frame of over 10,000 individuals. Within each of the groups, a systematic random sample was drawn, for an overall sample of 2,054 cases. Duplicate records within each group were removed before sampling. Those individuals who appeared in the final sample for more than one group were flagged, for the purposes of the analyses, as members of each of the groups they were sampled from, but they did not receive duplicate requests to complete the survey. Because "NASA's stakeholders" are not a clearly defined population and because we selected the stakeholders using sampling frames that were reasonable and convenient rather than comprehensive, the results from this survey cannot be generalized to all stakeholders.

TABLE C.1 Lists Used to Generate Sampling Frame for the Stakeholder Survey

Stakeholders	Description
Economic/ industry	Aerospace Industries Association Commercial Spaceflight Federation American Institute of Aeronautics and Astronautics Aerospace States Association House Committee on Science, Space and Technology, Subcommittee on Space individuals with an industry affiliation who testified during the past 3 years Senate Commerce, Science and Transportation Committee, Science and Space Subcommittee individuals with an industry affiliation who testified during the past 3 years
Space scientists and engineers	National Academy of Sciences members from relevant fields National Academy of Engineering members from relevant fields Institute of Medicine members who indicated an interest in space-related research NASA Advisory Council and Committees Aerospace Safety Advisory Panel American Astronautical Society American Geophysical Union American Astronomical Society American Society for Gravitational and Space Research House Committee on Science, Space and Technology, Subcommittee on Space, space scientists who testified during the past 3 years Senate Commerce, Science and Transportation Committee, Science and Space Subcommittee, space scientists who testified during the past 3 years
Young space scientists and engineers	American Institute of Aeronautics and Astronautics winners of early career and student awards, past 3 years National Science Foundation postdoctoral fellowship winners in relevant fields, past 3 years NASA Aeronautics Graduate Scholarship Program winners, past 3 years
Other scientists and engineers	National Academy of Sciences members from non-space-related fields National Academy of Engineering members from non-space-related fields Institute of Medicine members who did not indicate an interest in space-related research National Science Board American Association for the Advancement of Science *Science* magazine editorial board House Committee on Science, Space and Technology, Subcommittee on Space non-space scientists who testified during the past 3 years Senate Commerce, Science and Transportation Committee, Science and Space Subcommittee non-space scientists who testified during the past 3 years
Higher education	Deans and heads of graduate departments in relevant fields from research universities and doctoral/ research universities (Carnegie codes 15, 16, 17)
Security/defense/ foreign policy	Department of Defense, Defense Policy Board National Academy of Sciences Air Force Studies Board National Academy of Sciences Board on Army Science and Technology National Academy of Sciences Naval Studies Board Department of Defense Federally Funded Research and Development Centers House Committee on Science, Space and Technology, Subcommittee on Space individuals with a defense background who testified during the past 3 years Senate Commerce, Science and Transportation Committee, Science and Space Subcommittee individuals with a defense background who testified during the past 3 years

continued

TABLE C.1 Continued

Stakeholders	Description
Space writers and science popularizers	Planetarium directors
	National Association of Science Writers members who indicated "astronomy/space" as one of their areas of expertise
	Individuals with NASA social media credentials from @NASASocial Twitter lists
	House Committee on Science, Space and Technology, Subcommittee on Space testifiers who are space writers or science popularizers, past 3 years
	Senate Commerce, Science and Transportation Committee, Science and Space Subcommittee testifiers who are space writers or science popularizers, past 3 years
Space advocates	Explore Mars
	Mars Society
	National Space Society
	Planetary Society
	Space Foundation
	Space Frontier Foundation
	Space Generation
	Students for the Exploration and Development of Space
	100 Year Starship
	SpaceUp
	House Committee on Science, Space and Technology, Subcommittee on Space, testifiers with a space advocacy background, past 3 years
	Senate Commerce, Science and Transportation Committee, Science and Space Subcommittee testifiers with a space advocacy background, past 3 years

In addition, we have not made any attempt to weight the data to compensate for differences in the sizes of the various groups or overlaps in their composition. As a result, combining the responses from each group does not represent the universe of all of NASA's stakeholders. Nevertheless, the sample from each frame was a probability sample and we believe this methodology provides a broader and more diverse perspective on stakeholder views than a nonprobability sample would.

SURVEY QUESTIONNAIRE

The objective of the stakeholder survey was to provide input to the committee on stakeholder views of human spaceflight, and specifically:

- The rationales traditionally provided for human spaceflight and any new or emerging rationales
- The importance of human spaceflight in the context of tradeoffs and alternatives
- The consequences of discontinuing NASA's human spaceflight program
- The characteristics and goals of a worthwhile and feasible program for the near future (e.g., next two decades)

The survey instrument was developed by the panel in consultation with the committee. The goal was to develop standardized questions that could be both self-administered (via a paper questionnaire or web) and administered by a trained interviewer (via phone). The final questionnaire is included in Appendix D. To minimize response order effects (respondent's potential tendency to favor answer options appearing toward the beginning of self-administered lists and toward the end of lists administered by phone), two versions of the questionnaire were produced, reversing the order of the response items in 3, 4, 5, and 11. Respondents were randomly assigned to receive either version A or version B of the questionnaire, consistently across administration modes. (The version included in Appendix D is version A of the mail questionnaire).

DATA COLLECTION

To carry out the data collection, the panel selected NORC at the University of Chicago. Data collection began on September 17, 2013, and concluded on November 6, 2013.

To maximize response rates, the survey was conducted in three administration modes—mail, web and telephone—all monitored by a case management system that tracked the status and outcomes of all sampled stakeholders across modes. This ensured that sample members were not contacted once they had completed the survey. It also controlled the number of emails received by sample members who had not completed the survey so as to limit unnecessary burden.

As a first step, a paper questionnaire was sent to all sample members where a mailing address was available. The questionnaire was accompanied by a cover letter signed by the Chair of the NRC, along with a postage paid envelope for returning the completed questionnaire.

Follow-up mailings were sent to cases where the only contact information after several rounds of locating attempts was a mailing address. In addition, repeat mailings were sent to a small number of sample members who requested another mailed questionnaire when contacted by a telephone interviewer. Near the end of data collection, reminder letters were mailed via FedEx to the remaining sample members who had not yet responded within the Industry and Young Space Scientist strata to help boost participation among these groups. These letters included the link to the web survey, individual PINs, and the toll-free number that sample members could use to complete the survey by phone.

Receipt control and data entry systems were used to input responses received via mail with 100 percent data entry verification.

Following the initial mailings and emails, all non-responding sample members with a located telephone number were contacted by phone. Telephone interviewers were selected and trained from the existing group employed by NORC's telephone center in Chicago, IL. Prior to interviewing, all interviewers passed a certification test to demonstrate that they were well-versed in the project's purpose and survey administration. A system was also implemented to receive inbound calls via a toll-free number.

A web link to this mode of the survey was sent to all sample members with a known email address. Sample members accessed the web survey using an individualized PIN assigned at random during the compilation of the sample. A link to the web survey was also included in the initial paper mailing, through regularly scheduled email blasts, and by request.

Telephoning began on October 1, 2013. Follow-up calls were made on a regular basis to all non-completers. Exemptions were made for those who refused to participate. NORC's telephone center supervisors monitored interviewers throughout the data collection process.

The research staff continuously monitored results for signs of any complications with the data collection process. The research staff also monitored the progress of the survey on a daily basis, tallying the number of completed surveys by mode and by strata, as well as the total number of completed surveys. Careful monitoring of the transition of cases between modes of contact was critical throughout. Any cases where it was determined that the sample member lived outside the United States, was deceased or incapacitated/unable to complete the survey were considered not eligible.

Six survey questions requested open-ended responses that required coding by the research staff. The answers were coded into categories developed by the panel. Two coders independently coded 100 cases completed early in data collection and inter-coder reliability was calculated. The resulting kappa statistics are shown in Table C.2 and indicating the coding was adequately reliable. These cases were selected proportionally from the eight stakeholder groups. NORC survey managers and the coding supervisor debriefed the panel on this initial coding process. Then, upon the completion of data collection, all open-ended responses were coded in compliance with guidelines delineated during the initial batch of open-ends.

By the end of the data collection period 1,104 individuals, or 54 percent of the initial sample, completed the survey. The AAPOR Response Rate 3, which adjusts for ineligible cases, was 55.4 percent (Table C.3). In an effort to reduce the differences in response rates among the groups, at the end of the field period an additional mailing was sent via FedEx to nonresponders in the groups with the lowest response rates. To some extent, the differences

TABLE C.2 Inter-Coder Reliability Calculated Based on the First 100 Cases Coded

Question Number	Kappa Statistic
Q1A	0.82
Q1B	0.73
Q2A	0.87
Q2B	0.75
Q6	0.73
Q18	0.88

TABLE C.3 Number of Completed Cases and Response Rates

Stratum Name	Original Sample	Number of Complete Cases	Simple Response Rate (%)	AAPOR Response Rate 3 (%)
Economic/industry	384	104	27	28.6
Space scientists and engineers	395	261	66	67.1
Young space scientists and engineers	195	90	46	49.7
Other scientists and engineers	396	201	51	51.3
Higher education	399	294	74	74.1
Security/defense/foreign policy	110	71	65	66.4
Space writers and science popularizers	99	53	54	56.4
Space advocates	96	46	48	51.7
TOTAL	2,054	1,104	54	55.4

in the final response rates reflect differences in the contact information available for the different groups, including whether efforts to reach the sample member were likely to be screened by a gatekeeper, which was especially likely in the case of the industry group.

D

Stakeholder Survey Mail Questionnaire (Version A)

Version A of the stakeholder survey mail questionnaire is reprinted in this appendix.

NATIONAL RESEARCH COUNCIL
OF THE NATIONAL ACADEMIES

Human Spaceflight Study

This survey is being conducted by the National Research Council of the National Academies as part of a request from Congress to review the U.S. human spaceflight program. The purpose of this study is to gather input from stakeholders and members of the scientific community. Your input will help the committee develop recommendations about the long-term goals and direction of NASA's programs.

Before continuing, please know that your participation is voluntary. You may choose to skip any question or end the survey at any point. We will take all possible steps to protect your privacy and we can use your answers only for statistical research. This means that no individual will be identified in any of the analyses or reports from this study. The survey will take less than 20 minutes to complete.

NORC at the University of Chicago is conducting this study on behalf of the NRC. If you have any questions about your rights as a study participant, you may call the NORC Institutional Review Board, toll free, at 866-309-0542. Any other questions can be sent to the study's email address: HumanSpaceflightStudy@norc.org.

Human Spaceflight Study

The first question is about space exploration *in general* and the next question is about *human* space exploration.

1. What do you consider to be the main reasons for and against space exploration *in general?*

   FOR: _____

   _____

   _____

   _____

   _____

   _____

   _____

   _____

   _____

   _____

   _____

   _____

   AGAINST: _____

   _____

   _____

   _____

   _____

   _____

   _____

   _____

   _____

   _____

   _____

   _____

   _____

2. What do you consider to be the main reasons for and against *human* space exploration?

   FOR: _____

   _____

   _____

   _____

   _____

   _____

   _____

   _____

   _____

   _____

   _____

   _____

   AGAINST: _____

   _____

   _____

   _____

   _____

   _____

   _____

   _____

   _____

   _____

   _____

   _____

2

3. **Below are some reasons commonly given for space exploration. For each, please indicate whether you think it is a very important, somewhat important, not too important, or not at all important reason for space exploration *in general*, and for *human* spaceflight in particular.**

	SPACE EXPLORATION IN GENERAL				HUMAN SPACEFLIGHT			
	Very important	Somewhat important	Not too important	Not important at all	Very important	Somewhat important	Not too important	Not important at all
Expanding knowledge and scientific understanding	1 ☐	2 ☐	3 ☐	4 ☐	1 ☐	2 ☐	3 ☐	4 ☐
Driving technological advances	1 ☐	2 ☐	3 ☐	4 ☐	1 ☐	2 ☐	3 ☐	4 ☐
Extending human economic activity beyond Earth	1 ☐	2 ☐	3 ☐	4 ☐	1 ☐	2 ☐	3 ☐	4 ☐
Paving the way for future settlements in space	1 ☐	2 ☐	3 ☐	4 ☐	1 ☐	2 ☐	3 ☐	4 ☐
Paving the way for commercial space travel	1 ☐	2 ☐	3 ☐	4 ☐	1 ☐	2 ☐	3 ☐	4 ☐
Creating opportunities for international cooperation	1 ☐	2 ☐	3 ☐	4 ☐	1 ☐	2 ☐	3 ☐	4 ☐
Maintaining our national security	1 ☐	2 ☐	3 ☐	4 ☐	1 ☐	2 ☐	3 ☐	4 ☐
Enhancing U.S. prestige	1 ☐	2 ☐	3 ☐	4 ☐	1 ☐	2 ☐	3 ☐	4 ☐
Inspiring young people to pursue careers in science, technology, math and engineering	1 ☐	2 ☐	3 ☐	4 ☐	1 ☐	2 ☐	3 ☐	4 ☐
Satisfying a basic human drive to explore new frontiers	1 ☐	2 ☐	3 ☐	4 ☐	1 ☐	2 ☐	3 ☐	4 ☐

If in question 3 you marked two or more reasons as *very important* for space exploration *in general,* please answer question 4. Otherwise go to question 5.

4. Enter a "1" in the box next to the reason you consider to be the *most important* for space exploration *in general,* and a "2" in the box next to the reason you consider to be the *next most important* for space exploration *in general.*

   1. ☐ Expanding knowledge and scientific understanding
   2. ☐ Driving technological advances
   3. ☐ Extending human economic activity beyond Earth
   4. ☐ Paving the way for future settlements in space
   5. ☐ Paving the way for commercial space travel
   6. ☐ Creating opportunities for international cooperation
   7. ☐ Maintaining our national security
   8. ☐ Enhancing U.S. prestige
   9. ☐ Inspiring young people to pursue careers in science, technology, math and engineering
   10. ☐ Satisfying a basic human drive to explore new frontiers

If in question 3 you marked two or more reasons as *very important* for *human* spaceflight, please answer question 5. Otherwise go to question 6.

5. Enter a "1" in the box next to the reason you consider to be the *most important* for *human* spaceflight, and a "2" in the box next to the reason you consider to be the *next most important* for *human* spaceflight.

   1. ☐ Expanding knowledge and scientific understanding
   2. ☐ Driving technological advances
   3. ☐ Extending human economic activity beyond Earth
   4. ☐ Paving the way for future settlements in space
   5. ☐ Paving the way for commercial space travel
   6. ☐ Creating opportunities for international cooperation
   7. ☐ Maintaining our national security
   8. ☐ Enhancing U.S. prestige
   9. ☐ Inspiring young people to pursue careers in science, technology, math and engineering
   10. ☐ Satisfying a basic human drive to explore new frontiers

6. **In your opinion, if NASA's *human* spaceflight program was terminated, what, if anything, would be lost?**

_____

_____

_____

_____

_____

_____

_____

_____

_____

_____

_____

_____

_____

_____

7. **Do you think that NASA should focus mainly or exclusively on human space flight, mainly or exclusively on robotic space exploration, or a combination of both?**

   1 ☐ Mainly or exclusively human spaceflight
   2 ☐ Mainly or exclusively robotic space exploration
   3 ☐ A combination of both human spaceflight and robotic space exploration

8. **Do you think that NASA or the private sector should take the lead on each of the following activities over the next 20 years?**

	NASA	Private sector	Neither
Space exploration for scientific research	1 ☐	2 ☐	3 ☐
Extending human economic activity beyond Earth	1 ☐	2 ☐	3 ☐
Space travel by private citizens	1 ☐	2 ☐	3 ☐
Establishing an off-planet human presence	1 ☐	2 ☐	3 ☐

9. **Which of the following best describes your views on NASA human space exploration missions beyond Low Earth Orbit:**

   1 ☐ NASA should not conduct human space exploration missions beyond Low Earth Orbit
   2 ☐ NASA should conduct human space exploration missions beyond Low Earth Orbit mainly or exclusively as U.S.-only missions
   3 ☐ NASA should conduct human space exploration missions beyond Low Earth Orbit mainly or exclusively in collaboration with current international partners (such as ISS partners)
   4 ☐ NASA should conduct human space exploration missions beyond Low Earth Orbit mainly or exclusively as part of an international collaboration that includes current partners as well as new and emerging space powers

10. **Looking beyond the very near term, consider what goals a worthwhile and feasible U.S. human space exploration program might work toward over the next 20 years. How strongly do you favor or oppose the following options for NASA, bearing in mind that these are multi-year projects and the costs given are approximate overall costs.**

	Strongly favor	Somewhat favor	Somewhat oppose	Strongly oppose
LEAST EXPENSIVE (Tens of Billions)				
Continue with Low Earth Orbit flights to the International Space Station until 2020	1 ☐	2 ☐	3 ☐	4 ☐
Extend the International Space Station to 2028	1 ☐	2 ☐	3 ☐	4 ☐
Send humans to a Near-Earth asteroid in its native orbit	1 ☐	2 ☐	3 ☐	4 ☐
MORE EXPENSIVE (Hundreds of Billions)				
Return to the Moon and explore more of it with short visits	1 ☐	2 ☐	3 ☐	4 ☐
Establish outposts on the Moon	1 ☐	2 ☐	3 ☐	4 ☐
Conduct orbital missions to Mars to teleoperate robots on the surface	1 ☐	2 ☐	3 ☐	4 ☐
Land humans on Mars	1 ☐	2 ☐	3 ☐	4 ☐
MOST EXPENSIVE (Trillions)				
Establish a human presence (base) on Mars	1 ☐	2 ☐	3 ☐	4 ☐

11. **How important is it for NASA to do each of the following over the next 20 years?**

	Very important	Somewhat important	Not too important	Not important at all
Maintain the International Space Station as a laboratory for scientific research	1 ☐	2 ☐	3 ☐	4 ☐
Make the investments necessary to sustain a vigorous program of *human* space exploration	1 ☐	2 ☐	3 ☐	4 ☐
Make the investments necessary to sustain a vigorous program of *robotic* space exploration	1 ☐	2 ☐	3 ☐	4 ☐
Limit human space exploration to Earth-orbit missions while maintaining robotic missions for exploring in and beyond the solar system	1 ☐	2 ☐	3 ☐	4 ☐
Maintain world leadership in human space exploration	1 ☐	2 ☐	3 ☐	4 ☐
Improve orbital technologies such as weather and communication satellites	1 ☐	2 ☐	3 ☐	4 ☐
Plan for a manned mission to Mars	1 ☐	2 ☐	3 ☐	4 ☐
Expand space exploration collaborations with other countries	1 ☐	2 ☐	3 ☐	4 ☐

Now, we would like to ask a few questions about yourself.

12. Currently, how involved are you in space-related work?

1 ☐ Very involved

2 ☐ Somewhat involved

3 ☐ Not involved ➝ *Go to question 14*

13. How involved are you in work related to *human* spaceflight?

1 ☐ Very involved

2 ☐ Somewhat involved

3 ☐ Not involved

14. What is the highest degree you completed?

1 ☐ High school or some college

2 ☐ Bachelor's degree

3 ☐ Master's degree

4 ☐ Professional degree

5 ☐ Doctorate

15. In what year were you born?

☐☐☐☐

16. Are you male or female?

1 ☐ Male

2 ☐ Female

17. Are you currently employed?

1 ☐ Yes

2 ☐ No ➝ *Go to question 19*

18. Which of the following best describes your current job?

1 ☐ Postsecondary educator

2 ☐ Scientist in a non-teaching position

3 ☐ Engineer

4 ☐ Managerial or professional

5 ☐ Other, please specify

_____

19. Please use the space below for any further thoughts you might want to share on space exploration.

_____

_____

_____

_____

_____

_____

_____

_____

_____

_____

_____

_____

_____

Thank you for your time.
Please return the questionnaire in the postage-paid envelope.

E

Frequency Distributions of Responses to the Stakeholder Survey by Respondent Group

Stakeholder groups:

1 = Economic/Industry 4 = Other scientists and engineers 7 = Space writers and science popularizers
2 = Space scientists and engineers 5 = Higher education 8 = Space advocates
3 = Young space scientists and engineers 6 = Security/defense/foreign policy

What do you consider to be the main reasons for space exploration in general?

	Stakeholder group																		
	1	%	2	%	3	%	4	%	5	%	6	%	7	%	8	%	Total	%
Knowledge and scientific understanding	64	64	157	81.8	68	78.2	185	77.7	222	78.2	57	82.6	39	75	32	72.7	814	77.5
Technological advances	45	45	61	31.8	39	44.8	63	26.5	103	36.3	20	29	17	32.7	24	54.6	369	35.1
Human economic activity beyond Earth	16	16	15	7.81	17	19.5	16	6.72	38	13.4	8	11.6	5	9.62	4	9.09	116	11
Future settlements in space	11	11	13	6.77	16	18.4	9	3.78	23	8.1	4	5.8	6	11.5	6	13.6	86	8.18
Commercial space travel	0	0	0	0	1	1.15	1	0.42	0	0	0	0	0	0	0	0	2	0.19
International cooperation	6	6	12	6.25	3	3.45	2	0.84	5	1.76	1	1.45	2	3.85	1	2.27	30	2.85
National security	7	7	12	6.25	2	2.3	14	5.88	6	2.11	3	4.35	3	5.77	2	4.55	49	4.66
U.S. prestige	15	15	17	8.85	3	3.45	14	5.88	13	4.58	5	7.25	3	5.77	4	9.09	72	6.85
Careers in science, technology, math and engineering	8	8	13	6.77	7	8.05	8	3.36	10	3.52	7	10.1	3	5.77	1	2.27	56	5.33
Basic human drive to explore new frontiers	36	36	63	32.8	31	35.6	63	26.5	92	32.4	20	29	16	30.8	25	56.8	337	32.1
Search for signs of life	3	3	7	3.65	11	12.6	6	2.52	14	4.93	5	7.25	3	5.77	3	6.82	52	4.95
Prevent threats from space	1	1	5	2.6	4	4.6	6	2.52	9	3.17	3	4.35	0	0	3	6.82	31	2.95
Other	12	12	22	11.5	7	8.05	14	5.88	20	7.04	5	7.25	1	1.92	7	15.9	86	8.18
None/No compelling reason for human space exploration	0	0	1	0.52	0	0	0	0	0	0	0	0	0	0	1	2.27	2	0.19

What do you consider to be the main reason against space exploration *in general*?

	Stakeholder group																	Total	%
	1	%	2	%	3	%	4	%	5	%	6	%	7	%	8	%		
Cost in absolute sense	37	38.5	72	42.1	45	52.3	93	41	141	52.4	31	45.6	24	47.1	7	17.5	442	44.5
Cost in a relative sense	36	37.5	64	37.4	34	39.5	97	42.7	88	32.7	29	42.7	13	25.5	14	35	369	37.2
Lack of clarity about goals or benefits	8	8.33	10	5.85	8	9.3	5	2.2	9	3.35	5	7.35	3	5.88	3	7.5	50	5.04
Risks	6	6.25	5	2.92	10	11.6	15	6.61	15	5.58	1	1.47	3	5.88	1	2.5	56	5.64
Private sector could do it better	3	3.13	2	1.17	0	0	0	0	0	0	0	0	1	1.96	0	0	6	0.6
Other	7	7.29	9	5.26	7	8.14	15	6.61	15	5.58	7	10.3	4	7.84	8	20	71	7.15
None/No compelling reason against human space exploration	15	15.6	28	16.4	9	10.5	30	13.2	36	13.4	8	11.8	13	25.5	12	30	149	15

What do you consider to be the main reasons for human space exploration?

Stakeholder group

	1	%	2	%	3	%	4	%	5	%	6	%	7	%	8	%	Total	%
Knowledge and scientific understanding	30	32.3	53	28.8	27	31	67	29.4	65	24.1	23	34.3	14	26.4	11	24.4	283	28
Technological advances	20	21.5	24	13	17	19.5	19	8.33	38	14.1	9	13.4	2	3.77	10	22.2	137	13.5
Human economic activity beyond Earth	5	5.38	8	4.35	9	10.3	6	2.63	8	2.96	4	5.97	5	9.43	5	11.1	50	4.94
Future settlements in space	13	14	36	19.6	26	29.9	25	11	45	16.7	15	22.4	10	18.9	17	37.8	183	18.1
Commercial space travel	0	0	2	1.09	0	0	1	0.44	0	0	0	0	0	0	0	0	3	0.3
International cooperation	3	3.23	8	4.35	2	2.3	2	0.88	4	1.48	3	4.48	0	0	0	0	20	1.98
National security	2	2.15	5	2.72	1	1.15	4	1.75	1	0.37	1	1.49	1	1.89	1	2.22	16	1.58
U.S. prestige	11	11.8	27	14.7	5	5.75	11	4.82	15	5.56	7	10.5	1	1.89	3	6.67	78	7.71
Careers in science, technology, math and engineering	11	11.8	16	8.7	2	2.3	4	1.75	10	3.7	8	11.9	1	1.89	7	15.6	58	5.73
Basic human drive to explore new frontiers	25	26.9	63	34.2	31	35.6	54	23.7	84	31.1	24	35.8	14	26.4	16	35.6	305	30.1
Search for signs of life	1	1.08	1	0.54	3	3.45	1	0.44	5	1.85	1	1.49	1	1.89	2	4.44	15	1.48
Prevent threats from space	1	1.08	0	0	0	0	0	0	1	0.37	1	1.49	0	0	0	0	3	0.3
Humans can accomplish more than robots	30	32.3	46	25	41	47.1	52	22.8	100	37	22	32.8	25	47.2	14	31.1	327	32.3
Public support	9	9.68	17	9.24	7	8.05	20	8.77	33	12.2	5	7.46	2	3.77	3	6.67	94	9.29
Other	12	12.9	15	8.15	7	8.05	16	7.02	10	7.02	1	1.49	4	7.55	2	4.44	66	6.52
None/No compelling reason for human space exploration	1	1.08	13	7.07	0	0	26	11.4	18	6.67	2	2.99	1	1.89	1	2.22	61	6.03

What do you consider to be the main reasons against *human* space exploration?

	Stakeholder group																		
	1	%	2	%	3	%	4	%	5	%	6	%	7	%	8	%	Total	%
Cost in absolute sense	33	35.5	76	42	39	45.9	93	38.9	139	52.1	39	56.5	18	34	10	25.6	441	43.5
Cost in a relative sense	20	21.5	36	19.9	11	12.9	54	22.6	31	11.6	11	15.9	6	11.3	4	10.3	170	16.8
Lack of clarity about goals or benefits	5	5.38	17	9.39	3	3.53	9	3.77	8	3	4	5.8	1	1.89	0	0	45	4.43
Risks	29	31.2	47	26	56	65.9	89	37.2	113	42.3	25	36.2	26	49.1	10	25.6	391	38.5
Private sector could do it better	2	2.15	1	0.55	0	0	0	0	0	0	0	0	0	0	0	0	3	0.3
Robots could do it better	21	22.6	62	34.3	23	27.1	58	24.3	79	29.6	27	39.1	11	20.8	9	23.1	287	28.3
Other	18	19.4	38	21	17	20	26	10.9	37	13.9	13	18.8	6	11.3	11	28.2	165	16.3
None/No compelling reason against human space exploration	11	11.8	7	3.87	2	2.35	12	5.02	12	4.49	6	8.7	6	11.3	8	20.5	64	6.31

Below are some reasons commonly given for space exploration. For each, please indicate whether you think it is a very important, somewhat important, not too important, or not at all important reason for space exploration in general:

Expanding knowledge and scientific understanding

	Stakeholder group																		
	1	%	2	%	3	%	4	%	5	%	6	%	7	%	8	%	Total	%
Very important	87	83.7	172	86.9	73	81.1	203	76.9	259	88.1	58	81.7	50	94.3	40	87	931	84.3
Somewhat important	15	14.4	23	11.6	16	17.8	50	18.9	26	8.84	12	16.9	3	5.66	4	8.7	145	13.1
Not too important	1	0.96	0	0	0	0	3	1.14	6	2.04	0	0	0	0	1	2.17	11	1
Not important at all	0	0	1	0.51	0	0	3	1.14	2	0.68	0	0	0	0	0	0	6	0.54
DON'T KNOW	0	0	0	0	0	0	0	0	0	0	0	0	0	0	0	0	0	0
REFUSED	1	0.96	2	1.01	1	1.11	5	1.89	1	0.34	1	1.41	0	0	1	2.17	11	1

Driving technological advances

	Stakeholder group																		
	1	%	2	%	3	%	4	%	5	%	6	%	7	%	8	%	Total	%
Very important	78	75	116	58.6	70	77.8	149	56.4	211	71.8	37	52.1	45	84.9	35	76.1	731	66.2
Somewhat important	23	22.1	59	29.8	15	16.7	88	33.3	61	20.8	27	38	7	13.2	9	19.6	284	25.7
Not too important	0	0	14	7.07	4	4.44	15	5.68	16	5.44	5	7.04	0	0	0	0	54	4.89
Not important at all	1	0.96	4	2.02	0	0	4	1.52	2	0.68	0	0	0	0	1	2.17	11	1
DON'T KNOW	0	0	0	0	0	0	0	0	0	0	0	0	0	0	0	0	0	0
REFUSED	2	1.92	5	2.53	1	1.11	8	3.03	4	1.36	2	2.82	1	1.89	1	2.17	24	2.17

Extending human economic activity beyond Earth

	Stakeholder group																		
	1	%	2	%	3	%	4	%	5	%	6	%	7	%	8	%	Total	%
Very important	27	26	38	19.2	21	23.3	34	12.9	57	19.4	10	14.1	24	45.3	24	52.2	233	21.1
Somewhat important	41	39.4	54	27.3	35	38.9	84	31.8	101	34.4	23	32.4	15	28.3	15	32.6	362	32.8
Not too important	22	21.2	58	29.3	24	26.7	72	27.3	78	26.5	25	35.2	11	20.8	4	8.7	290	26.3
Not important at all	12	11.5	43	21.7	7	7.78	61	23.1	51	17.4	11	15.5	3	5.66	1	2.17	185	16.8
DON'T KNOW	0	0	1	0.51	1	1.11	3	1.14	2	0.68	0	0	0	0	0	0	7	0.63
REFUSED	2	1.92	4	2.02	2	2.22	10	3.79	5	1.7	2	2.82	0	0	2	4.35	27	2.45

Paving the way for future settlements in space

	Stakeholder group																		
	1	%	2	%	3	%	4	%	5	%	6	%	7	%	8	%	Total	%
Very important	38	36.5	39	19.7	29	32.2	19	7.2	60	20.4	10	14.1	21	39.6	26	56.5	239	21.7
Somewhat important	29	27.9	47	23.7	35	38.9	69	26.1	99	33.7	23	32.4	20	37.7	14	30.4	332	30.1
Not too important	18	17.3	58	29.3	18	20	80	30.3	74	25.2	23	32.4	7	13.2	4	8.7	279	25.3
Not important at all	16	15.4	48	24.2	7	7.78	84	31.8	55	18.7	13	18.3	4	7.55	0	0	222	20.1
DON'T KNOW	0	0	0	0	0	0	2	0.76	1	0.34	0	0	0	0	0	0	3	0.27
REFUSED	3	2.88	6	3.03	1	1.11	10	3.79	5	1.7	2	2.82	1	1.89	2	4.35	29	2.63

Paving the way for commercial space travel

Stakeholder group

	1	%	2	%	3	%	4	%	5	%	6	%	7	%	8	%	Total	%
Very important	24	23.1	28	14.1	24	26.7	19	7.2	50	17	6	8.45	22	41.5	16	34.8	186	16.9
Somewhat important	39	37.5	53	26.8	33	36.7	71	26.9	97	33	28	39.4	19	35.9	20	43.5	355	32.2
Not too important	22	21.2	70	35.4	23	25.6	97	36.7	98	33.3	22	31	5	9.43	7	15.2	339	30.7
Not important at all	18	17.3	40	20.2	9	10	64	24.2	42	14.3	13	18.3	7	13.2	2	4.35	192	17.4
DON'T KNOW	0	0	0	0	0	0	2	0.76	2	0.68	0	0	0	0	0	0	4	0.36
REFUSED	1	0.96	7	3.54	1	1.11	11	4.17	5	1.7	2	2.82	0	0	1	2.17	28	2.54

Creating opportunities for international cooperation

Stakeholder group

	1	%	2	%	3	%	4	%	5	%	6	%	7	%	8	%	Total	%
Very important	27	26	71	35.9	30	33.3	69	26.1	99	33.7	20	28.2	34	64.2	20	43.5	365	33.1
Somewhat important	47	45.2	89	45	40	44.4	124	47	135	45.9	23	32.4	15	28.3	16	34.8	482	43.7
Not too important	23	22.1	30	15.2	16	17.8	49	18.6	52	17.7	23	32.4	4	7.55	8	17.4	203	18.4
Not important at all	6	5.77	7	3.54	3	3.33	14	5.3	6	2.04	3	4.23	0	0	1	2.17	38	3.44
DON'T KNOW	0	0	0	0	0	0	0	0	0	0	0	0	0	0	0	0	0	0
REFUSED	1	0.96	1	0.51	1	1.11	8	3.03	2	0.68	2	2.82	0	0	1	2.17	16	1.45

Maintaining our national security

Stakeholder group

	1	%	2	%	3	%	4	%	5	%	6	%	7	%	8	%	Total	%
Very important	53	51	79	39.9	28	31.1	92	34.9	126	42.9	30	42.3	27	50.9	17	37	448	40.6
Somewhat important	34	32.7	66	33.3	30	33.3	99	37.5	94	32	26	36.6	15	28.3	17	37	377	34.2
Not too important	9	8.65	30	15.2	24	26.7	51	19.3	52	17.7	10	14.1	8	15.1	7	15.2	186	16.9
Not important at all	3	2.88	19	9.6	5	5.56	12	4.55	19	6.46	4	5.63	3	5.66	3	6.52	65	5.89
DON'T KNOW	0	0	0	0	0	0	4	1.52	1	0.34	0	0	0	0	0	0	5	0.45
REFUSED	5	4.81	4	2.02	3	3.33	6	2.27	2	0.68	1	1.41	0	0	2	4.35	23	2.08

Enhancing U.S. prestige

Stakeholder group

	1	%	2	%	3	%	4	%	5	%	6	%	7	%	8	%	Total	%
Very important	40	38.5	70	35.4	20	22.2	60	22.7	92	31.3	19	26.8	19	35.9	10	21.7	325	29.4
Somewhat important	43	41.4	86	43.4	37	41.1	116	43.9	116	39.5	30	42.3	20	37.7	22	47.8	464	42
Not too important	18	17.3	28	14.1	22	24.4	61	23.1	64	21.8	18	25.4	9	17	8	17.4	225	20.4
Not important at all	2	1.92	7	3.54	10	11.1	19	7.2	20	6.8	2	2.82	5	9.43	4	8.7	68	6.16
DON'T KNOW	0	0	0	0	0	0	1	0.38	0	0	0	0	0	0	0	0	1	0.09
REFUSED	1	0.96	7	3.54	1	1.11	7	2.65	2	0.68	2	2.82	0	0	2	4.35	21	1.9

Inspiring young people to pursue careers in science, technology, math and engineering

Stakeholder group

	1	%	2	%	3	%	4	%	5	%	6	%	7	%	8	%	Total	%
Very important	65	62.5	127	64.1	62	68.9	133	50.4	199	67.7	38	53.5	42	79.3	29	63	685	62.1
Somewhat important	32	30.8	61	30.8	21	23.3	90	34.1	79	26.9	25	35.2	7	13.2	15	32.6	326	29.5
Not too important	5	4.81	4	2.02	5	5.56	28	10.6	12	4.08	6	8.45	4	7.55	1	2.17	64	5.8
Not important at all	0	0	3	1.52	1	1.11	7	2.65	1	0.34	1	1.41	0	0	0	0	13	1.18
DON'T KNOW	0	0	0	0	0	0	0	0	0	0	0	0	0	0	0	0	0	0
REFUSED	2	1.92	3	1.52	1	1.11	6	2.27	3	1.02	1	1.41	0	0	1	2.17	16	1.45

Satisfying a basic human drive to explore new frontiers

Stakeholder group

	1	%	2	%	3	%	4	%	5	%	6	%	7	%	8	%	Total	%
Very important	56	53.9	140	70.7	50	55.6	134	50.8	188	64	32	45.1	41	77.4	32	69.6	662	60
Somewhat important	27	26	42	21.2	30	33.3	93	35.2	88	29.9	27	38	12	22.6	11	23.9	327	29.6
Not too important	17	16.4	10	5.05	6	6.67	22	8.33	15	5.1	10	14.1	0	0	0	0	78	7.07
Not important at all	3	2.88	4	2.02	2	2.22	7	2.65	2	0.68	2	2.82	0	0	2	4.35	22	1.99
DON'T KNOW	0	0	0	0	0	0	1	0.38	0	0	0	0	0	0	0	0	1	0.09
REFUSED	1	0.96	2	1.01	2	2.22	7	2.65	1	0.34	0	0	0	0	1	2.17	14	1.27

Below are some reasons commonly given for space exploration. For each, please indicate whether you think it is a very important, somewhat important, not too important, or not at all important reason for human spaceflight:

Expanding knowledge and scientific understanding

	Stakeholder group																		
	1	%	2	%	3	%	4	%	5	%	6	%	7	%	8	%	Total	%
Very important	54	51.9	51	25.8	51	56.7	69	26.1	112	38.1	25	35.2	36	67.9	26	56.5	419	38
Somewhat important	28	26.9	62	31.3	22	24.4	84	31.8	94	32	24	33.8	10	18.9	14	30.4	332	30.1
Not too important	12	11.5	46	23.2	11	12.2	62	23.5	49	16.7	14	19.7	4	7.55	2	4.35	196	17.8
Not important at all	4	3.85	29	14.7	2	2.22	31	11.7	22	7.48	5	7.04	1	1.89	1	2.17	94	8.51
DON'T KNOW	0	0	0	0	1	1.11	1	0.38	0	0	0	0	0	0	0	0	2	0.18
REFUSED	6	5.77	10	5.05	3	3.33	17	6.44	17	5.78	3	4.23	2	3.77	3	6.52	61	5.53

Driving technological advances

Q3B2 ([Importance for human space exploration:] Driving technological advances)	Stakeholder group																		
	1	%	2	%	3	%	4	%	5	%	6	%	7	%	8	%	Total	%
Very important	58	55.8	58	29.3	53	58.9	66	25	117	39.8	27	38	38	71.7	31	67.4	442	40
Somewhat important	25	24	73	36.9	22	24.4	94	35.6	96	32.7	26	36.6	9	17	8	17.4	347	31.4
Not too important	9	8.65	33	16.7	9	10	58	22	44	15	13	18.3	4	7.55	2	4.35	170	15.4
Not important at all	4	3.85	19	9.6	3	3.33	23	8.71	15	5.1	2	2.82	0	0	1	2.17	65	5.89
DON'T KNOW	0	0	0	0	0	0	0	0	0	0	0	0	0	0	0	0	0	0
REFUSED	8	7.69	15	7.58	3	3.33	23	8.71	22	7.48	3	4.23	2	3.77	4	8.7	80	7.25

Extending human economic activity beyond Earth

	Stakeholder group																Total	%
	1	%	2	%	3	%	4	%	5	%	6	%	7	%	8	%		
Very important	32	30.8	30	15.2	20	22.2	21	7.95	40	13.6	11	15.5	22	41.5	26	56.5	200	18.1
Somewhat important	25	24	51	25.8	31	34.4	58	22	84	28.6	20	28.2	15	28.3	9	19.6	289	26.2
Not too important	29	27.9	43	21.7	26	28.9	83	31.4	85	28.9	20	28.2	10	18.9	6	13	296	26.8
Not important at all	12	11.5	59	29.8	9	10	79	29.9	63	21.4	16	22.5	4	7.55	2	4.35	240	21.7
DON'T KNOW	0	0	1	0.51	1	1.11	3	1.14	3	1.02	0	0	0	0	0	0	8	0.72
REFUSED	6	5.77	14	7.07	3	3.33	20	7.58	19	6.46	4	5.63	2	3.77	3	6.52	71	6.43

Paving the way for future settlements in space

	Stakeholder group																Total	%
	1	%	2	%	3	%	4	%	5	%	6	%	7	%	8	%		
Very important	41	39.4	54	27.3	48	53.3	30	11.4	93	31.6	22	31	29	54.7	28	60.9	341	30.9
Somewhat important	28	26.9	40	20.2	18	20	55	20.8	72	24.5	20	28.2	12	22.6	11	23.9	252	22.8
Not too important	15	14.4	47	23.7	14	15.6	73	27.7	56	19.1	18	25.4	7	13.2	1	2.17	227	20.6
Not important at all	12	11.5	44	22.2	7	7.78	86	32.6	53	18	10	14.1	3	5.66	1	2.17	212	19.2
DON'T KNOW	0	0	0	0	0	0	1	0.38	1	0.34	0	0	0	0	0	0	2	0.18
REFUSED	8	7.69	13	6.57	3	3.33	19	7.2	19	6.46	1	1.41	2	3.77	5	10.9	70	6.34

Paving the way for commercial space travel

	Stakeholder group																Total	%
	1	%	2	%	3	%	4	%	5	%	6	%	7	%	8	%		
Very important	31	29.8	28	14.1	39	43.3	24	9.09	59	20.1	12	16.9	20	37.7	25	54.4	235	21.3
Somewhat important	34	32.7	54	27.3	26	28.9	60	22.7	80	27.2	23	32.4	22	41.5	11	23.9	307	27.8
Not too important	20	19.2	55	27.8	15	16.7	76	28.8	82	27.9	22	31	4	7.55	5	10.9	273	24.7
Not important at all	13	12.5	45	22.7	7	7.78	81	30.7	45	15.3	11	15.5	5	9.43	1	2.17	205	18.6
DON'T KNOW	0	0	0	0	0	0	1	0.38	3	1.02	0	0	0	0	0	0	4	0.36
REFUSED	6	5.77	16	8.08	3	3.33	22	8.33	25	8.5	3	4.23	2	3.77	4	8.7	80	7.25

Creating opportunities for international cooperation

Stakeholder group

	1	%	2	%	3	%	4	%	5	%	6	%	7	%	8	%	Total	%
Very important	30	28.9	47	23.7	33	36.7	47	17.8	68	23.1	16	22.5	28	52.8	18	39.1	285	25.8
Somewhat important	43	41.4	76	38.4	32	35.6	102	38.6	134	45.6	26	36.6	15	28.3	20	43.5	439	39.8
Not too important	18	17.3	40	20.2	18	20	69	26.1	57	19.4	19	26.8	8	15.1	2	4.35	229	20.7
Not important at all	6	5.77	24	12.1	4	4.44	27	10.2	18	6.12	6	8.45	0	0	2	4.35	84	7.61
DON'T KNOW	0	0	0	0	0	0	0	0	0	0	0	0	0	0	0	0	0	0
REFUSED	7	6.73	11	5.56	3	3.33	19	7.2	17	5.78	4	5.63	2	3.77	4	8.7	67	6.07

Maintaining our national security

Stakeholder group

	1	%	2	%	3	%	4	%	5	%	6	%	7	%	8	%	Total	%
Very important	31	29.8	28	14.1	14	15.6	27	10.2	49	16.7	16	22.5	18	34	8	17.4	190	17.2
Somewhat important	36	34.6	48	24.2	22	24.4	64	24.2	69	23.5	16	22.5	15	28.3	17	37	281	25.5
Not too important	17	16.4	46	23.2	31	34.4	94	35.6	95	32.3	27	38	12	22.6	12	26.1	328	29.7
Not important at all	10	9.62	63	31.8	18	20	54	20.5	64	21.8	11	15.5	6	11.3	7	15.2	230	20.8
DON'T KNOW	0	0	1	0.51	0	0	4	1.52	0	0	0	0	0	0	0	0	5	0.45
REFUSED	10	9.62	12	6.06	5	5.56	21	7.95	17	5.78	1	1.41	2	3.77	2	4.35	70	6.34

Enhancing U.S. prestige

Stakeholder group

	1	%	2	%	3	%	4	%	5	%	6	%	7	%	8	%	Total	%
Very important	40	38.5	66	33.3	29	32.2	35	13.3	78	26.5	15	21.1	17	32.1	19	41.3	293	26.5
Somewhat important	42	40.4	64	32.3	27	30	91	34.5	102	34.7	32	45.1	21	39.6	13	28.3	386	35
Not too important	13	12.5	35	17.7	21	23.3	80	30.3	71	24.2	18	25.4	7	13.2	5	10.9	247	22.4
Not important at all	3	2.88	19	9.6	10	11.1	34	12.9	25	8.5	4	5.63	6	11.3	5	10.9	105	9.51
DON'T KNOW	0	0	0	0	0	0	1	0.38	0	0	0	0	0	0	0	0	1	0.09
REFUSED	6	5.77	14	7.07	3	3.33	23	8.71	18	6.12	2	2.82	2	3.77	4	8.7	72	6.52

Inspiring young people to pursue careers in science, technology, math and engineering

	Stakeholder group																	Total	
	1	%	2	%	3	%	4	%	5	%	6	%	7	%	8	%	Total	%
Very important	64	61.5	78	39.4	59	65.6	76	28.8	147	50	33	46.5	36	67.9	33	71.7	516	46.7
Somewhat important	29	27.9	73	36.9	22	24.4	97	36.7	95	32.3	20	28.2	11	20.8	9	19.6	353	32
Not too important	4	3.85	23	11.6	4	4.44	53	20.1	29	9.86	9	12.7	4	7.55	1	2.17	126	11.4
Not important at all	1	0.96	14	7.07	2	2.22	19	7.2	5	1.7	6	8.45	0	0	0	0	45	4.08
DON'T KNOW	0	0	0	0	0	0	0	0	0	0	0	0	0	0	0	0	0	0
REFUSED	6	5.77	10	5.05	3	3.33	19	7.2	18	6.12	3	4.23	2	3.77	3	6.52	64	5.8

Satisfying a basic human drive to explore new frontiers

	Stakeholder group																	Total	
	1	%	2	%	3	%	4	%	5	%	6	%	7	%	8	%	Total	%
Very important	58	55.8	84	42.4	59	65.6	72	27.3	130	44.2	31	43.7	36	67.9	36	78.3	496	44.9
Somewhat important	25	24	56	28.3	21	23.3	98	37.1	96	32.7	18	25.4	13	24.5	5	10.9	330	29.9
Not too important	11	10.6	29	14.7	3	3.33	48	18.2	39	13.3	14	19.7	2	3.77	0	0	143	13
Not important at all	4	3.85	18	9.09	4	4.44	27	10.2	12	4.08	6	8.45	0	0	2	4.35	72	6.52
DON'T KNOW	0	0	0	0	0	0	0	0	0	0	0	0	0	0	0	0	0	0
REFUSED	6	5.77	11	5.56	3	3.33	19	7.2	17	5.78	2	2.82	2	3.77	3	6.52	63	5.71

Most important reason for space exploration in general

	Stakeholder group																	
	1	%	2	%	3	%	4	%	5	%	6	%	7	%	8	%	Total	%
Expanding knowledge and scientific understanding	32	34.4	117	65.4	44	55	123	59.1	163	60.6	32	59.3	33	66	22	48.9	558	57.7
Driving technological advances	21	22.6	12	6.7	11	13.8	20	9.62	30	11.2	5	9.26	4	8	3	6.67	105	10.9
Extending human economic activity beyond Earth	5	5.38	1	0.56	5	6.25	1	0.48	7	2.6	1	1.85	1	2	3	6.67	24	2.48
Paving the way for future settlements in space	5	5.38	5	2.79	6	7.5	3	1.44	5	1.86	0	0	1	2	5	11.1	30	3.1
Paving the way for commercial space travel	0	0	0	0	0	0	1	0.48	0	0	0	0	0	0	0	0	1	0.1
Creating opportunities for international cooperation	0	0	1	0.56	0	0	2	0.96	1	0.37	0	0	2	4	0	0	6	0.62
Maintaining our national security	8	8.6	10	5.59	1	1.25	15	7.21	6	2.23	6	11.1	2	4	1	2.22	49	5.07
Enhancing U.S. prestige	0	0	0	0	0	0	0	0	0	0	1	1.85	0	0	1	2.22	2	0.21
Inspiring young people to pursue careers in science, technology, math and engineering	7	7.53	14	7.82	7	8.75	13	6.25	23	8.55	4	7.41	5	10	4	8.89	77	7.96
Satisfying a basic human drive to explore new frontiers	15	16.1	19	10.6	6	7.5	25	12	33	12.3	4	7.41	2	4	4	8.89	106	11
DON'T KNOW	0	0	0	0	0	0	0	0	0	0	0	0	0	0	0	0	0	0
REFUSED	0	0	0	0	0	0	5	2.4	1	0.37	1	1.85	0	0	2	4.44	9	0.93

Next most important reason for space exploration in general

	Stakeholder group																Total	%
	1	%	2	%	3	%	4	%	5	%	6	%	7	%	8	%	Total	%
Expanding knowledge and scientific understanding	21	24.4	16	10.3	10	13.7	24	13.7	51	21.5	8	18.2	5	10.2	8	18.6	142	16.7
Driving technological advances	23	26.7	39	25.2	28	38.4	51	29.1	65	27.4	11	25	12	24.5	9	20.9	236	27.7
Extending human economic activity beyond Earth	1	1.16	3	1.94	3	4.11	4	2.29	7	2.95	1	2.27	4	8.16	5	11.6	28	3.29
Paving the way for future settlements in space	5	5.81	1	0.65	5	6.85	4	2.29	4	1.69	2	4.55	2	4.08	2	4.65	25	2.94
Paving the way for commercial space travel	2	2.33	0	0	0	0	0	0	2	0.84	0	0	1	2.04	2	4.65	7	0.82
Creating opportunities for international cooperation	2	2.33	5	3.23	0	0	7	4	4	1.69	2	4.55	5	10.2	2	4.65	26	3.06
Maintaining our national security	2	2.33	11	7.1	4	5.48	16	9.14	14	5.91	8	18.2	2	4.08	2	4.65	59	6.93
Enhancing U.S. prestige	6	6.98	8	5.16	0	0	1	0.57	1	0.42	0	0	1	2.04	1	2.33	18	2.12
Inspiring young people to pursue careers in science, technology, math and engineering	15	17.4	35	22.6	13	17.8	36	20.6	45	19	7	15.9	11	22.5	6	14	163	19.2
Satisfying a basic human drive to explore new frontiers	8	9.3	36	23.2	10	13.7	24	13.7	43	18.1	5	11.4	6	12.2	3	6.98	133	15.6
DON'T KNOW	0	0	0	0	0	0	0	0	0	0	0	0	0	0	0	0	0	0
REFUSED	1	1.16	1	0.65	0	0	8	4.57	1	0.42	0	0	0	0	3	6.98	14	1.65

Most important reason for human spaceflight

	Stakeholder group																		
	1	%	2	%	3	%	4	%	5	%	6	%	7	%	8	%	Total	%
Expanding knowledge and scientific understanding	5	6.33	18	17	14	18.7	22	20.2	28	15.6	7	15.9	10	23.3	5	12.2	107	16.1
Driving technological advances	15	19	8	7.55	10	13.3	13	11.9	15	8.38	5	11.4	3	6.98	5	12.2	72	10.8
Extending human economic activity beyond Earth	8	10.1	2	1.89	1	1.33	2	1.83	2	1.12	2	4.55	0	0	3	7.32	20	3
Paving the way for future settlements in space	10	12.7	10	9.43	19	25.3	12	11	24	13.4	4	9.09	9	20.9	7	17.1	95	14.3
Paving the way for commercial space travel	3	3.8	5	4.72	3	4	3	2.75	5	2.79	2	4.55	0	0	2	4.88	23	3.45
Creating opportunities for international cooperation	2	2.53	8	7.55	1	1.33	8	7.34	4	2.23	4	9.09	2	4.65	0	0	28	4.2
Maintaining our national security	2	2.53	3	2.83	1	1.33	7	6.42	6	3.35	0	0	0	0	1	2.44	20	3
Enhancing U.S. prestige	5	6.33	8	7.55	1	1.33	1	0.92	8	4.47	3	6.82	1	2.33	2	4.88	28	4.2
Inspiring young people to pursue careers in science, technology, math and engineering	12	15.2	15	14.2	16	21.3	20	18.4	35	19.6	6	13.6	7	16.3	8	19.5	118	17.7
Satisfying a basic human drive to explore new frontiers	16	20.3	28	26.4	9	12	18	16.5	50	27.9	11	25	11	25.6	6	14.6	146	21.9
DON'T KNOW	0	0	0	0	0	0	0	0	0	0	0	0	0	0	0	0	0	0
REFUSED	1	1.27	1	0.94	0	0	3	2.75	2	1.12	0	0	0	0	2	4.88	9	1.35

Next most important reason for human spaceflight

	Stakeholder group																		
	1	%	2	%	3	%	4	%	5	%	6	%	7	%	8	%	Total	%
Expanding knowledge and scientific understanding	8	11.4	1	1.16	8	11.6	14	18	15	10.1	2	5.88	4	9.52	3	7.69	55	9.87
Driving technological advances	9	12.9	16	18.6	9	13	12	15.4	27	18.2	11	32.4	11	26.2	8	20.5	102	18.3
Extending human economic activity beyond Earth	3	4.29	3	3.49	5	7.25	3	3.85	13	8.78	2	5.88	4	9.52	3	7.69	36	6.46
Paving the way for future settlements in space	10	14.3	13	15.1	5	7.25	5	6.41	11	7.43	4	11.8	4	9.52	5	12.8	54	9.69
Paving the way for commercial space travel	6	8.57	2	2.33	1	1.45	1	1.28	9	6.08	1	2.94	2	4.76	3	7.69	25	4.49
Creating opportunities for international cooperation	3	4.29	6	6.98	2	2.9	5	6.41	10	6.76	1	2.94	3	7.14	4	10.3	33	5.92
Maintaining our national security	3	4.29	3	3.49	3	4.35	1	1.28	3	2.03	2	5.88	0	0	0	0	15	2.69
Enhancing U.S. prestige	3	4.29	8	9.3	5	7.25	4	5.13	12	8.11	2	5.88	0	0	1	2.56	34	6.1
Inspiring young people to pursue careers in science, technology, math and engineering	12	17.1	18	20.9	16	23.2	19	24.4	34	23	8	23.5	11	26.2	6	15.4	121	21.7
Satisfying a basic human drive to explore new frontiers	11	15.7	15	17.4	15	21.7	11	14.1	12	8.11	1	2.94	3	7.14	4	10.3	72	12.9
DON'T KNOW	0	0	0	0	0	0	0	0	1	0.68	0	0	0	0	0	0	1	0.18
REFUSED	2	2.86	1	1.16	0	0	3	3.85	1	0.68	0	0	0	0	2	5.13	9	1.62

In your opinion, if NASA's human spaceflight program was terminated, what, if anything, would be lost?

	Stakeholder group																	
	1	%	2	%	3	%	4	%	5	%	6	%	7	%	8	%	Total	%
Knowledge and scientific understanding	11	11.7	30	15.8	36	41.9	49	21	51	18.8	7	10	13	25	5	11.6	200	19.5
Technological advances	16	17	22	11.6	25	29.1	28	12	43	15.8	9	12.9	10	19.2	11	25.6	163	15.9
Human economic activity beyond Earth	5	5.32	5	2.63	5	5.81	6	2.58	3	1.1	2	2.86	3	5.77	3	6.98	32	3.12
Future settlements in space	6	6.38	10	5.26	5	5.81	11	4.72	9	3.31	6	8.57	8	15.4	10	23.3	65	6.34
Commercial space travel	3	3.19	2	1.05	1	1.16	2	0.86	0	0	0	0	2	3.85	1	2.33	11	0.29
International cooperation	2	2.13	3	1.58	7	8.14	4	1.72	7	2.57	2	2.86	0	0	3	6.98	27	2.63
National security	6	6.38	6	3.16	4	4.65	9	3.86	10	3.68	2	2.86	3	5.77	1	2.33	41	4
U.S. prestige	34	36.2	55	29	24	27.9	43	18.5	68	25	20	28.6	14	26.9	18	41.9	270	26.3
Careers in science, technology, math and engineering	9	9.57	10	5.26	14	16.3	17	7.3	32	11.8	7	10	5	9.62	7	16.3	99	9.65
Basic human drive to explore new frontiers	14	14.9	30	15.8	26	30.2	26	11.2	48	17.7	12	17.1	9	17.3	10	23.3	175	17.1
Search for signs of life	0	0	0	0	2	2.33	0	0	2	0.74	0	0	0	0	0	0	4	0.39
Prevent threats from space	0	0	0	0	0	0	1	0.43	1	0.37	0	0	0	0	0	0	2	0.19
Humans can accomplish more than robots	2	2.13	8	4.21	4	4.65	2	0.86	11	4.04	4	5.71	0	0	3	6.98	34	3.31
Public support	8	8.51	23	12.1	6	6.98	18	7.73	42	15.4	6	8.57	5	9.62	5	11.6	112	10.9
Investment we made so far	13	13.8	21	11.1	9	10.5	15	6.44	26	9.56	4	5.71	9	17.3	7	16.3	100	9.75
Other	19	20.2	30	15.8	23	26.7	30	12.9	26	9.56	7	10	7	13.5	5	11.6	142	13.8
Nothing/Nothing would be lost	6	6.38	30	15.8	1	1.16	67	28.8	32	11.8	14	20	3	5.77	3	6.98	154	15

Do you think that NASA should focus mainly or exclusively on human spaceflight, mainly or exclusively on robotic space exploration, or a combination of both?

Stakeholder group

	1	%	2	%	3	%	4	%	5	%	6	%	7	%	8	%	Total	%
Mainly or exclusively human spaceflight	2	1.92	2	1.01	0	0	0	0	1	0.34	0	0	0	0	1	2.17	6	0.54
Mainly or exclusively robotic space exploration	20	19.2	78	39.4	14	15.6	123	46.6	108	36.7	25	35.2	9	17	1	2.17	376	34.1
A combination of both human spaceflight and robotic space exploration	80	76.9	115	58.1	75	83.3	134	50.8	182	61.9	46	64.8	44	83	41	89.1	703	63.7
DON'T KNOW	0	0	0	0	0	0	1	0.38	0	0	0	0	0	0	0	0	1	0.09
REFUSED	2	1.92	3	1.52	1	1.11	6	2.27	3	1.02	0	0	0	0	3	6.52	18	1.63

Do you think that NASA or the private sector should take the lead on each of the following activities over the next 20 years:

Space exploration for scientific research

Stakeholder group

	1	%	2	%	3	%	4	%	5	%	6	%	7	%	8	%	Total	%
NASA	98	94.2	194	98	85	94.4	246	93.2	282	95.9	68	95.8	50	94.3	40	87	1048	94.9
Private sector	2	1.92	3	1.52	3	3.33	5	1.89	6	2.04	1	1.41	2	3.77	1	2.17	22	1.99
Neither	2	1.92	0	0	1	1.11	3	1.14	3	1.02	2	2.82	1	1.89	1	2.17	13	1.18
DON'T KNOW	0	0	0	0	0	0	1	0.38	0	0	0	0	0	0	0	0	1	0.09
REFUSED	2	1.92	1	0.51	1	1.11	9	3.41	3	1.02	0	0	0	0	4	8.7	20	1.81

Human economic activity beyond Earth

Stakeholder group

	1	%	2	%	3	%	4	%	5	%	6	%	7	%	8	%	Total	%
NASA	24	23.1	30	15.2	13	14.4	37	14	48	16.3	10	14.1	9	17	11	23.9	180	16.3
Private sector	66	63.5	145	73.2	68	75.6	155	58.7	204	69.4	51	71.8	37	69.8	31	67.4	746	67.6
Neither	12	11.5	22	11.1	8	8.89	59	22.4	38	12.9	10	14.1	6	11.3	1	2.17	153	13.9
DON'T KNOW	0	0	0	0	0	0	3	1.14	0	0	0	0	0	0	0	0	3	0.27
REFUSED	2	1.92	1	0.51	1	1.11	10	3.79	4	1.36	0	0	1	1.89	3	6.52	22	1.99

Space travel by private citizens

Stakeholder group

	1	%	2	%	3	%	4	%	5	%	6	%	7	%	8	%	Total	%
NASA	3	2.88	1	0.51	1	1.11	3	1.14	6	2.04	0	0	1	1.89	1	2.17	16	1.45
Private sector	93	89.4	168	84.9	86	95.6	196	74.2	257	87.4	61	85.9	49	92.5	41	89.1	937	84.9
Neither	5	4.81	26	13.1	2	2.22	58	22	26	8.84	10	14.1	3	5.66	1	2.17	129	11.7
DON'T KNOW	0	0	0	0	0	0	0	0	1	0.34	0	0	0	0	0	0	1	0.09
REFUSED	3	2.88	3	1.52	1	1.11	7	2.65	4	1.36	0	0	0	0	3	6.52	21	1.9

Establishing an off-planet human presence

Stakeholder group

	1	%	2	%	3	%	4	%	5	%	6	%	7	%	8	%	Total	%
NASA	64	61.5	89	45	58	64.4	82	31.1	139	47.3	34	47.9	34	64.2	32	69.6	527	47.7
Private sector	15	14.4	40	20.2	15	16.7	45	17.1	72	24.5	13	18.3	12	22.6	9	19.6	216	19.6
Neither	23	22.1	66	33.3	14	15.6	126	47.7	75	25.5	24	33.8	6	11.3	1	2.17	329	29.8
DON'T KNOW	0	0	0	0	1	1.11	1	0.38	3	1.02	0	0	0	0	0	0	5	0.45
REFUSED	2	1.92	3	1.52	2	2.22	10	3.79	5	1.7	0	0	1	1.89	4	8.7	27	2.45

Which of the following best describes your views on NASA human space exploration missions beyond Low Earth Orbit?

	Stakeholder group																	
	1	%	2	%	3	%	4	%	5	%	6	%	7	%	8	%	Total	%
NASA should not conduct human space exploration missions beyond Low Earth Orbit	9	8.65	42	21.2	5	5.56	66	25	50	17	13	18.3	4	7.55	0	0	187	16.9
NASA should conduct human space exploration missions beyond Low Earth Orbit mainly or exclusively as U.S.-only missions	18	17.3	11	5.56	5	5.56	20	7.58	18	6.12	5	7.04	3	5.66	5	10.9	85	7.7
NASA should conduct human space exploration missions beyond Low Earth Orbit mainly or exclusively in collaboration with current international partners (such as ISS partners)	28	26.9	13	6.57	16	17.8	31	11.7	34	11.6	7	9.86	5	9.43	7	15.2	141	12.8
NASA should conduct human space exploration missions beyond Low Earth Orbit mainly or exclusively as part of an international collaboration that includes current partners as well as new and emerging space powers	46	44.2	127	64.1	62	68.9	132	50	186	63.3	45	63.4	40	75.5	31	67.4	656	59.4
DON'T KNOW	0	0	1	0.51	1	1.11	1	0.38	1	0.34	1	1.41	0	0	0	0	4	0.36
REFUSED	3	2.88	4	2.02	1	1.11	14	5.3	5	1.7	0	0	1	1.89	3	6.52	31	2.81

Looking beyond the very near term, consider what goals a worthwhile and feasible U.S. human space exploration program might work toward over the next 20 years. How strongly do you favor or oppose the following options for NASA, bearing in mind that these are multi-year projects and the costs given are approximate overall costs.

Continue with Low Earth Orbit flights to the International Space Station until 2020

	Stakeholder group																		
	1	%	2	%	3	%	4	%	5	%	6	%	7	%	8	%	Total	%
Strongly favor	58	55.8	70	35.4	53	58.9	98	37.1	124	42.2	33	46.5	38	71.7	26	56.5	496	44.9
Somewhat favor	28	26.9	70	35.4	28	31.1	102	38.6	104	35.4	25	35.2	10	18.9	11	23.9	373	33.8
Somewhat oppose	8	7.69	29	14.7	6	6.67	41	15.5	38	12.9	10	14.1	4	7.55	3	6.52	135	12.2
Strongly oppose	7	6.73	19	9.6	1	1.11	13	4.92	16	5.44	3	4.23	1	1.89	2	4.35	59	5.34
DON'T KNOW	0	0	0	0	1	1.11	0	0	2	0.68	0	0	0	0	0	0	3	0.27
REFUSED	3	2.88	10	5.05	1	1.11	10	3.79	10	3.4	0	0	0	0	4	8.7	38	3.44

Extend the International Space Station to 2028

	Stakeholder group																		
	1	%	2	%	3	%	4	%	5	%	6	%	7	%	8	%	Total	%
Strongly favor	53	51	53	26.8	47	52.2	83	31.4	101	34.4	27	38	34	64.2	22	47.8	412	37.3
Somewhat favor	30	28.9	53	26.8	25	27.8	76	28.8	102	34.7	25	35.2	11	20.8	13	28.3	332	30.1
Somewhat oppose	11	10.6	39	19.7	11	12.2	56	21.2	47	16	10	14.1	6	11.3	5	10.9	184	16.7
Strongly oppose	7	6.73	45	22.7	5	5.56	35	13.3	36	12.2	8	11.3	1	1.89	2	4.35	135	12.2
DON'T KNOW	0	0	1	0.51	1	1.11	0	0	2	0.68	0	0	0	0	0	0	4	0.36
REFUSED	3	2.88	7	3.54	1	1.11	14	5.3	6	2.04	1	1.41	1	1.89	4	8.7	37	3.35

Send humans to a Near-Earth asteroid in its native orbit

	Stakeholder group																		
	1	%	2	%	3	%	4	%	5	%	6	%	7	%	8	%	Total	%
Strongly favor	22	21.2	28	14.1	35	38.9	28	10.6	71	24.2	8	11.3	12	22.6	7	15.2	209	18.9
Somewhat favor	30	28.9	56	28.3	30	33.3	64	24.2	105	35.7	22	31	20	37.7	18	39.1	341	30.9
Somewhat oppose	32	30.8	47	23.7	13	14.4	77	29.2	51	17.4	23	32.4	13	24.5	9	19.6	257	23.3
Strongly oppose	17	16.4	58	29.3	10	11.1	76	28.8	54	18.4	17	23.9	6	11.3	7	15.2	243	22
DON'T KNOW	0	0	0	0	0	0	3	1.14	3	1.02	0	0	0	0	0	0	6	0.54
REFUSED	3	2.88	9	4.55	2	2.22	16	6.06	10	3.4	1	1.41	2	3.77	5	10.9	48	4.35

Return to the Moon and explore more of it with short visits

	Stakeholder group																		
	1	%	2	%	3	%	4	%	5	%	6	%	7	%	8	%	Total	%
Strongly favor	26	25	30	15.2	26	28.9	41	15.5	61	20.8	15	21.1	28	52.8	13	28.3	239	21.7
Somewhat favor	41	39.4	70	35.4	36	40	79	29.9	109	37.1	23	32.4	16	30.2	14	30.4	380	34.4
Somewhat oppose	22	21.2	45	22.7	19	21.1	72	27.3	73	24.8	18	25.4	7	13.2	10	21.7	263	23.8
Strongly oppose	11	10.6	45	22.7	7	7.78	58	22	40	13.6	15	21.1	2	3.77	4	8.7	179	16.2
DON'T KNOW	0	0	0	0	1	1.11	1	0.38	3	1.02	0	0	0	0	0	0	5	0.45
REFUSED	4	3.85	8	4.04	1	1.11	13	4.92	8	2.72	0	0	0	0	5	10.9	38	3.44

Establish outposts on the Moon

	Stakeholder group																		
	1	%	2	%	3	%	4	%	5	%	6	%	7	%	8	%	Total	%
Strongly favor	36	34.6	33	16.7	31	34.4	28	10.6	64	21.8	13	18.3	24	45.3	20	43.5	247	22.4
Somewhat favor	26	25	48	24.2	33	36.7	50	18.9	89	30.3	19	26.8	16	30.2	12	26.1	290	26.3
Somewhat oppose	22	21.2	48	24.2	15	16.7	73	27.7	74	25.2	18	25.4	9	17	4	8.7	255	23.1
Strongly oppose	16	15.4	63	31.8	8	8.89	101	38.3	58	19.7	21	29.6	4	7.55	3	6.52	271	24.6
DON'T KNOW	0	0	0	0	1	1.11	1	0.38	3	1.02	0	0	0	0	0	0	5	0.45
REFUSED	4	3.85	6	3.03	2	2.22	11	4.17	6	2.04	0	0	0	0	7	15.2	36	3.26

Conduct orbital missions to Mars to teleoperate robots on the surface

Stakeholder group

	1	%	2	%	3	%	4	%	5	%	6	%	7	%	8	%	Total	%
Strongly favor	35	33.7	52	26.3	29	32.2	69	26.1	100	34	21	29.6	26	49.1	15	32.6	345	31.3
Somewhat favor	48	46.2	70	35.4	35	38.9	88	33.3	96	32.7	24	33.8	17	32.1	17	37	385	34.9
Somewhat oppose	14	13.5	31	15.7	15	16.7	51	19.3	50	17	12	16.9	9	17	6	13	185	16.8
Strongly oppose	4	3.85	35	17.7	9	10	45	17.1	42	14.3	14	19.7	1	1.89	3	6.52	152	13.8
DON'T KNOW	0	0	0	0	0	0	0	0	3	1.02	0	0	0	0	0	0	3	0.27
REFUSED	3	2.88	10	5.05	2	2.22	11	4.17	3	1.02	0	0	0	0	5	10.9	34	3.08

Land humans on Mars

Stakeholder group

	1	%	2	%	3	%	4	%	5	%	6	%	7	%	8	%	Total	%
Strongly favor	36	34.6	40	20.2	30	33.3	29	11	68	23.1	19	26.8	27	50.9	29	63	272	24.6
Somewhat favor	29	27.9	46	23.2	36	40	42	15.9	74	25.2	18	25.4	11	20.8	6	13	258	23.4
Somewhat oppose	21	20.2	37	18.7	12	13.3	60	22.7	62	21.1	13	18.3	8	15.1	2	4.35	212	19.2
Strongly oppose	15	14.4	69	34.9	10	11.1	123	46.6	82	27.9	21	29.6	7	13.2	4	8.7	328	29.7
DON'T KNOW	0	0	0	0	0	0	0	0	2	0.68	0	0	0	0	0	0	2	0.18
REFUSED	3	2.88	6	3.03	2	2.22	10	3.79	6	2.04	0	0	0	0	5	10.9	32	2.9

Establish a human presence (base) on Mars

Stakeholder group

	1	%	2	%	3	%	4	%	5	%	6	%	7	%	8	%	Total	%
Strongly favor	21	20.2	19	9.6	17	18.9	14	5.3	25	8.5	5	7.04	20	37.7	18	39.1	137	12.4
Somewhat favor	24	23.1	34	17.2	23	25.6	22	8.33	58	19.7	17	23.9	10	18.9	11	23.9	195	17.7
Somewhat oppose	23	22.1	33	16.7	25	27.8	55	20.8	76	25.9	16	22.5	13	24.5	7	15.2	245	22.2
Strongly oppose	33	31.7	104	52.5	23	25.6	162	61.4	127	43.2	32	45.1	9	17	4	8.7	489	44.3
DON'T KNOW	0	0	0	0	0	0	0	0	3	1.02	0	0	0	0	0	0	3	0.27
REFUSED	3	2.88	8	4.04	2	2.22	11	4.17	5	1.7	1	1.41	1	1.89	6	13	35	3.17

How important is it for NASA to do each of the following over the next 20 years:

Maintain the International Space Station as a laboratory for scientific research

	Stakeholder group																	Total	%
	1	%	2	%	3	%	4	%	5	%	6	%	7	%	8	%		
Very important	47	45.2	51	25.8	48	53.3	87	33	101	34.4	26	36.6	35	66	17	37	406	36.8
Somewhat important	34	32.7	52	26.3	24	26.7	73	27.7	92	31.3	23	32.4	9	17	18	39.1	322	29.2
Not too important	12	11.5	50	25.3	13	14.4	48	18.2	67	22.8	18	25.4	5	9.43	2	4.35	213	19.3
Not important at all	4	3.85	28	14.1	2	2.22	25	9.47	21	7.14	2	2.82	2	3.77	3	6.52	84	7.61
DON'T KNOW	0	0	0	0	0	0	0	0	0	0	0	0	0	0	0	0	0	0
REFUSED	7	6.73	17	8.59	3	3.33	31	11.7	13	4.42	2	2.82	2	3.77	6	13	79	7.16

Make the investments necessary to sustain a vigorous program of human space exploration

	Stakeholder group																	Total	%
	1	%	2	%	3	%	4	%	5	%	6	%	7	%	8	%		
Very important	48	46.2	52	26.3	46	51.1	50	18.9	90	30.6	25	35.2	24	45.3	31	67.4	362	32.8
Somewhat important	30	28.9	50	25.3	26	28.9	61	23.1	92	31.3	18	25.4	21	39.6	7	15.2	300	27.2
Not too important	15	14.4	39	19.7	11	12.2	64	24.2	64	21.8	16	22.5	3	5.66	0	0	208	18.8
Not important at all	3	2.88	38	19.2	3	3.33	55	20.8	33	11.2	10	14.1	4	7.55	2	4.35	147	13.3
DON'T KNOW	0	0	0	0	1	1.11	1	0.38	0	0	0	0	0	0	0	0	2	0.18
REFUSED	8	7.69	19	9.6	3	3.33	33	12.5	15	5.1	2	2.82	1	1.89	6	13	85	7.7

Make the investments necessary to sustain a vigorous program of robotic space exploration

	Stakeholder group																	Total	%
	1	%	2	%	3	%	4	%	5	%	6	%	7	%	8	%		
Very important	50	48.1	143	72.2	58	64.4	151	57.2	218	74.2	36	50.7	41	77.4	23	50	713	64.6
Somewhat important	43	41.4	33	16.7	27	30	74	28	54	18.4	30	42.3	9	17	12	26.1	276	25
Not too important	3	2.88	4	2.02	2	2.22	6	2.27	8	2.72	3	4.23	1	1.89	3	6.52	29	2.63
Not important at all	0	0	0	0	0	0	3	1.14	1	0.34	0	0	1	1.89	1	2.17	6	0.54
DON'T KNOW	0	0	0	0	0	0	1	0.38	0	0	0	0	0	0	0	0	1	0.09
REFUSED	8	7.69	18	9.09	3	3.33	29	11	13	4.42	2	2.82	1	1.89	7	15.2	79	7.16

Limit human space exploration to Earth-orbit missions while maintaining robotic missions for exploring in and beyond the solar system

	Stakeholder group																	Total	%
	1	%	2	%	3	%	4	%	5	%	6	%	7	%	8	%		
Very important	18	17.3	60	30.3	17	18.9	87	33	99	33.7	16	22.5	17	32.1	4	8.7	318	28.8
Somewhat important	30	28.9	59	29.8	29	32.2	82	31.1	101	34.4	28	39.4	17	32.1	7	15.2	347	31.4
Not too important	24	23.1	31	15.7	31	34.4	34	12.9	54	18.4	12	16.9	10	18.9	8	17.4	200	18.1
Not important at all	24	23.1	22	11.1	8	8.89	21	7.95	17	5.78	11	15.5	7	13.2	21	45.7	128	11.6
DON'T KNOW	0	0	0	0	1	1.11	1	0.38	1	0.34	0	0	1	1.89	0	0	4	0.36
REFUSED	8	7.69	26	13.1	4	4.44	39	14.8	22	7.48	4	5.63	1	1.89	6	13	107	9.69

Maintain world leadership in human space exploration

	Stakeholder group																	Total	%
	1	%	2	%	3	%	4	%	5	%	6	%	7	%	8	%		
Very important	61	58.7	57	28.8	45	50	63	23.9	107	36.4	28	39.4	27	50.9	19	41.3	401	36.3
Somewhat important	23	22.1	60	30.3	28	31.1	82	31.1	92	31.3	23	32.4	17	32.1	14	30.4	334	30.3
Not too important	8	7.69	41	20.7	11	12.2	58	22	58	19.7	14	19.7	4	7.55	4	8.7	195	17.7
Not important at all	3	2.88	19	9.6	2	2.22	27	10.2	20	6.8	4	5.63	4	7.55	4	8.7	83	7.52
DON'T KNOW	0	0	0	0	1	1.11	0	0	0	0	0	0	0	0	0	0	1	0.09
REFUSED	9	8.65	21	10.6	3	3.33	34	12.9	17	5.78	2	2.82	1	1.89	5	10.9	90	8.15

Improve orbital technologies such as weather and communication satellites

	Stakeholder group																Total	%
	1	%	2	%	3	%	4	%	5	%	6	%	7	%	8	%		
Very important	48	46.2	123	62.1	55	61.1	179	67.8	220	74.8	41	57.8	41	77.4	12	26.1	711	64.4
Somewhat important	38	36.5	38	19.2	22	24.4	41	15.5	45	15.3	20	28.2	8	15.1	15	32.6	223	20.2
Not too important	8	7.69	14	7.07	7	7.78	10	3.79	12	4.08	5	7.04	2	3.77	10	21.7	66	5.98
Not important at all	3	2.88	5	2.53	2	2.22	1	0.38	3	1.02	2	2.82	0	0	3	6.52	19	1.72
DON'T KNOW	0	0	0	0	0	0	0	0	0	0	0	0	0	0	0	0	0	0
REFUSED	7	6.73	18	9.09	4	4.44	33	12.5	14	4.76	3	4.23	2	3.77	6	13	85	7.7

Plan for a manned mission to Mars

	Stakeholder group																Total	%
	1	%	2	%	3	%	4	%	5	%	6	%	7	%	8	%		
Very important	36	34.6	31	15.7	29	32.2	25	9.47	52	17.7	20	28.2	23	43.4	20	43.5	231	20.9
Somewhat important	22	21.2	48	24.2	36	40	46	17.4	84	28.6	18	25.4	14	26.4	13	28.3	277	25.1
Not too important	23	22.1	41	20.7	12	13.3	70	26.5	57	19.4	12	16.9	7	13.2	3	6.52	222	20.1
Not important at all	14	13.5	60	30.3	10	11.1	92	34.9	83	28.2	17	23.9	8	15.1	2	4.35	285	25.8
DON'T KNOW	0	0	0	0	0	0	0	0	1	0.34	0	0	0	0	0	0	1	0.09
REFUSED	9	8.65	18	9.09	3	3.33	31	11.7	17	5.78	4	5.63	1	1.89	8	17.4	88	7.97

Expand space exploration collaborations with other countries

	Stakeholder group																Total	%
	1	%	2	%	3	%	4	%	5	%	6	%	7	%	8	%		
Very important	32	30.8	78	39.4	43	47.8	88	33.3	150	51	29	40.9	35	66	21	45.7	470	42.6
Somewhat important	48	46.2	76	38.4	33	36.7	112	42.4	102	34.7	29	40.9	16	30.2	12	26.1	422	38.2
Not too important	12	11.5	20	10.1	10	11.1	24	9.09	24	8.16	7	9.86	1	1.89	5	10.9	103	9.33
Not important at all	4	3.85	6	3.03	0	0	8	3.03	2	0.68	2	2.82	0	0	1	2.17	21	1.9
DON'T KNOW	0	0	0	0	0	0	0	0	0	0	0	0	0	0	0	0	0	0
REFUSED	8	7.69	18	9.09	4	4.44	32	12.1	16	5.44	4	5.63	1	1.89	7	15.2	88	7.97

Currently, how involved are you in space-related work?

	Stakeholder group																	
	1	%	2	%	3	%	4	%	5	%	6	%	7	%	8	%	Total	%
Very involved	46	44.2	89	45	24	26.7	0	0	23	7.82	15	21.1	15	28.3	28	60.9	233	21.1
Somewhat involved	29	27.9	106	53.5	32	35.6	0	0	99	33.7	23	32.4	29	54.7	11	23.9	325	29.4
Not involved	27	26	0	0	33	36.7	260	98.5	171	58.2	33	46.5	9	17	4	8.7	532	48.2
DON'T KNOW	0	0	0	0	0	0	0	0	0	0	0	0	0	0	0	0	0	0
REFUSED	2	1.92	3	1.52	1	1.11	4	1.52	1	0.34	0	0	0	0	3	6.52	14	1.27

How involved are you in work related to human spaceflight?

	Stakeholder group																	
	1	%	2	%	3	%	4	%	5	%	6	%	7	%	8	%	Total	%
Very involved	30	40	25	12.8	0	0	0	0	2	1.64	3	7.89	4	9.09	15	38.5	75	13.4
Somewhat involved	29	38.7	51	26.2	12	21.4	3	1.14	22	18	14	36.8	11	25	16	41	150	26.9
Not involved	16	21.3	119	61	44	78.6	0	0	98	80.3	21	55.3	29	65.9	8	20.5	333	59.7
DON'T KNOW	0	0	0	0	0	0	0	0	0	0	0	0	0	0	0	0	0	0
REFUSED	0	0	0	0	0	0	0	0	0	0	0	0	0	0	0	0	0	0

What is the highest degree you completed?

	Stakeholder group																	
	1	%	2	%	3	%	4	%	5	%	6	%	7	%	8	%	Total	%
High school or some college	2	1.92	0	0	3	3.33	0	0	0	0	0	0	2	3.77	5	10.9	12	1.09
Bachelor's degree	28	26.9	12	6.06	25	27.8	3	1.14	0	0	1	1.41	10	18.9	18	39.1	95	8.61
Master's degree	44	42.3	16	8.08	12	13.3	25	9.47	1	0.34	15	21.1	26	49.1	9	19.6	147	13.3
Professional degree	10	9.62	8	4.04	2	2.22	12	4.55	3	1.02	8	11.3	1	1.89	3	6.52	46	4.17
Doctorate	18	17.3	160	80.8	47	52.2	219	83	289	98.3	47	66.2	14	26.4	9	19.6	791	71.7
DON'T KNOW	0	0	0	0	0	0	0	0	0	0	0	0	0	0	0	0	0	0
REFUSED	2	1.92	2	1.01	1	1.11	5	1.89	1	0.34	0	0	0	0	2	4.35	13	1.18

In what year where you born?

Year	Stakeholder group																Total	%
	1	%	2	%	3	%	4	%	5	%	6	%	7	%	8	%		
1915	0	0	1	0.51	0	0	0	0	0	0	0	0	0	0	0	0	1	0.09
1917	0	0	0	0	0	0	3	1.14	0	0	0	0	0	0	0	0	3	0.27
1918	0	0	0	0	0	0	2	0.76	0	0	0	0	0	0	0	0	2	0.18
1919	0	0	0	0	0	0	1	0.38	0	0	0	0	0	0	0	0	1	0.09
1920	0	0	0	0	0	0	3	1.14	0	0	0	0	0	0	0	0	3	0.27
1921	0	0	0	0	0	0	2	0.76	0	0	0	0	0	0	0	0	2	0.18
1922	0	0	0	0	0	0	4	1.52	0	0	0	0	0	0	0	0	4	0.36
1923	0	0	1	0.51	0	0	4	1.52	0	0	0	0	0	0	0	0	5	0.45
1924	0	0	1	0.51	0	0	5	1.89	0	0	0	0	0	0	0	0	6	0.54
1925	0	0	1	0.51	0	0	2	0.76	0	0	0	0	0	0	0	0	3	0.27
1926	0	0	2	1.01	0	0	5	1.89	0	0	0	0	0	0	1	2.17	8	0.72
1927	0	0	5	2.53	0	0	7	2.65	0	0	0	0	0	0	1	2.17	13	1.18
1928	0	0	6	3.03	0	0	8	3.03	0	0	1	1.41	0	0	1	2.17	16	1.45
1929	0	0	0	0	0	0	8	3.03	0	0	0	0	0	0	0	0	8	0.72
1930	0	0	3	1.52	0	0	8	3.03	0	0	0	0	0	0	0	0	11	1
1931	0	0	5	2.53	0	0	5	1.89	0	0	0	0	0	0	1	2.17	11	1
1932	0	0	4	2.02	0	0	6	2.27	0	0	0	0	0	0	0	0	10	0.91
1933	0	0	4	2.02	0	0	11	4.17	0	0	1	1.41	0	0	0	0	16	1.45
1934	0	0	2	1.01	0	0	6	2.27	0	0	1	1.41	1	1.89	0	0	10	0.91
1935	0	0	7	3.54	0	0	4	1.52	0	0	0	0	0	0	0	0	11	1
1936	0	0	3	1.52	0	0	7	2.65	1	0.34	1	1.41	0	0	0	0	12	1.09
1937	0	0	6	3.03	0	0	5	1.89	1	0.34	0	0	0	0	0	0	12	1.09
1938	1	0.96	9	4.55	0	0	10	3.79	0	0	3	4.23	0	0	0	0	21	1.9
1939	1	0.96	7	3.54	0	0	7	2.65	2	0.68	2	2.82	0	0	0	0	19	1.72
1940	1	0.96	6	3.03	0	0	9	3.41	1	0.34	0	0	0	0	1	2.17	18	1.63
1941	0	0	1	0.51	0	0	9	3.41	1	0.34	5	7.04	2	3.77	0	0	18	1.63
1942	2	1.92	4	2.02	0	0	6	2.27	1	0.34	0	0	0	0	0	0	13	1.18
1943	1	0.96	7	3.54	0	0	7	2.65	2	0.68	3	4.23	0	0	1	2.17	20	1.81
1944	3	2.88	4	2.02	0	0	8	3.03	4	1.36	6	8.45	1	1.89	0	0	26	2.36
1945	2	1.92	5	2.53	0	0	5	1.89	5	1.7	1	1.41	0	0	0	0	17	1.54
1946	2	1.92	4	2.02	0	0	9	3.41	6	2.04	3	4.23	0	0	0	0	23	2.08
1947	1	0.96	6	3.03	0	0	7	2.65	7	2.38	6	8.45	1	1.89	1	2.17	26	2.36
1948	0	0	9	4.55	0	0	6	2.27	4	1.36	3	4.23	0	0	0	0	22	1.99
1949	6	5.77	9	4.55	0	0	12	4.55	5	1.7	5	7.04	3	5.66	0	0	40	3.62
1950	2	1.92	10	5.05	0	0	6	2.27	9	3.06	2	2.82	2	3.77	1	2.17	31	2.81
1951	3	2.88	5	2.53	0	0	2	0.76	9	3.06	1	1.41	2	3.77	0	0	21	1.9
1952	3	2.88	6	3.03	1	1.11	11	4.17	12	4.08	4	5.63	2	3.77	2	4.35	41	3.71
1953	3	2.88	2	1.01	0	0	9	3.41	11	3.74	0	0	1	1.89	0	0	26	2.36

Year	n	%	n	%	n	%	n	%	n	%	n	%	n	%	n	%	n	%
1954	6	5.77	3	1.52	0	0	4	1.52	10	3.4	2	2.82	3	5.66	2	4.35	29	2.63
1955	4	3.85	6	3.03	0	0	5	1.89	21	7.14	1	1.41	3	5.66	1	2.17	41	3.71
1956	3	2.88	2	1.01	0	0	3	1.14	18	6.12	3	4.23	2	3.77	1	2.17	32	2.9
1957	2	1.92	5	2.53	0	0	2	0.76	22	7.48	2	2.82	1	1.89	2	4.35	35	3.17
1958	7	6.73	4	2.02	0	0	3	1.14	16	5.44	1	1.41	1	1.89	1	2.17	32	2.9
1959	3	2.88	1	0.51	0	0	3	1.14	24	8.16	2	2.82	2	3.77	1	2.17	35	3.17
1960	2	1.92	5	2.53	0	0	2	0.76	12	4.08	0	0	0	0	1	2.17	22	1.99
1961	5	4.81	2	1.01	1	1.11	2	0.76	13	4.42	3	4.23	0	0	1	2.17	26	2.36
1962	7	6.73	3	1.52	0	0	0	0	9	3.06	3	4.23	1	1.89	1	2.17	24	2.17
1963	3	2.88	0	0	0	0	0	0	12	4.08	0	0	1	1.89	0	0	16	1.45
1964	2	1.92	2	1.01	0	0	1	0.38	9	3.06	2	2.82	3	5.66	1	2.17	20	1.81
1965	3	2.88	0	0	0	0	0	0	6	2.04	1	1.41	5	9.43	0	0	15	1.36
1966	2	1.92	1	0.51	0	0	0	0	5	1.7	1	1.41	1	1.89	0	0	10	0.91
1967	4	3.85	1	0.51	1	1.11	0	0	5	1.7	0	0	0	0	1	2.17	12	1.09
1968	2	1.92	2	1.01	0	0	2	0.76	4	1.36	0	0	2	3.77	2	4.35	14	1.27
1969	4	3.85	0	0	0	0	0	0	6	2.04	0	0	2	3.77	0	0	12	1.09
1970	0	0	0	0	0	0	0	0	4	1.36	0	0	1	1.89	1	2.17	6	0.54
1971	1	0.96	1	0.51	1	1.11	1	0.38	2	0.68	0	0	0	0	0	0	6	0.54
1972	2	1.92	2	1.01	1	1.11	0	0	0	0	0	0	1	1.89	1	2.17	7	0.63
1973	1	0.96	2	1.01	0	0	0	0	0	0	0	0	3	5.66	1	2.17	7	0.63
1974	2	1.92	2	1.01	0	0	0	0	4	1.36	0	0	1	1.89	2	4.35	11	1
1976	0	0	1	0.51	0	0	0	0	2	0.68	0	0	1	1.89	0	0	4	0.36
1977	0	0	0	0	2	2.22	0	0	1	0.34	0	0	1	1.89	0	0	4	0.36
1978	0	0	2	1.01	2	2.22	0	0	2	0.68	0	0	1	1.89	0	0	7	0.63
1979	1	0.96	0	0	1	1.11	0	0	0	0	0	0	0	0	1	2.17	3	0.27
1980	1	0.96	0	0	6	6.67	0	0	0	0	0	0	0	0	1	2.17	8	0.72
1981	0	0	0	0	10	11.1	0	0	0	0	0	0	0	0	0	0	10	0.91
1982	1	0.96	0	0	4	4.44	0	0	0	0	0	0	0	0	1	2.17	6	0.54
1983	0	0	0	0	7	7.78	0	0	0	0	0	0	0	0	1	2.17	8	0.72
1984	0	0	0	0	9	10	0	0	0	0	0	0	1	1.89	2	4.35	12	1.09
1985	2	1.92	0	0	5	5.56	0	0	0	0	0	0	0	0	2	4.35	9	0.82
1986	0	0	0	0	5	5.56	0	0	0	0	0	0	0	0	0	0	5	0.45
1987	0	0	0	0	4	4.44	0	0	0	0	0	0	0	0	0	0	4	0.36
1988	0	0	0	0	7	7.78	0	0	0	0	0	0	0	0	0	0	7	0.63
1989	0	0	0	0	9	10	0	0	0	0	1	1.41	1	1.89	2	4.35	12	1.09
1990	0	0	0	0	3	3.33	0	0	0	0	0	0	0	0	0	0	4	0.36
1991	0	0	0	0	3	3.33	0	0	0	0	0	0	0	0	1	2.17	4	0.36
1992	0	0	0	0	5	5.56	0	0	0	0	0	0	0	0	1	2.17	6	0.54
1993	0	0	0	0	1	1.11	0	0	0	0	0	0	0	0	2	4.35	3	0.27
DON'T KNOW	0	0	0	0	0	0	0	0	0	0	0	0	0	0	0	0	0	0
REFUSED	3	2.88	6	3.03	2	2.22	7	2.65	6	2.04	1	1.41	0	0	2	4.35	26	2.36

Are you male or female?

	Stakeholder group																	Total	%
	1	%	2	%	3	%	4	%	5	%	6	%	7	%	8	%		
Male	93	89.4	165	83.3	66	73.3	226	85.6	267	90.8	62	87.3	36	67.9	33	71.7	936	84.8
Female	8	7.69	29	14.7	23	25.6	34	12.9	25	8.5	8	11.3	17	32.1	11	23.9	152	13.8
DON'T KNOW	0	0	0	0	0	0	0	0	0	0	0	0	0	0	0	0	0	0
REFUSED	3	2.88	4	2.02	1	1.11	4	1.52	2	0.68	1	1.41	0	0	2	4.35	16	1.45

Are you currently employed?

	Stakeholder group																	Total	%
	1	%	2	%	3	%	4	%	5	%	6	%	7	%	8	%		
Yes	99	95.2	167	84.3	77	85.6	163	61.7	292	99.3	58	81.7	51	96.2	36	78.3	928	84.1
No	2	1.92	27	13.6	12	13.3	95	36	1	0.34	12	16.9	2	3.77	8	17.4	158	14.3
DON'T KNOW	0	0	0	0	0	0	0	0	0	0	0	0	0	0	0	0	0	0
REFUSED	3	2.88	4	2.02	1	1.11	6	2.27	1	0.34	1	1.41	0	0	2	4.35	18	1.63

Which of the following best describes your current job?

	Stakeholder group																	Total	%
	1	%	2	%	3	%	4	%	5	%	6	%	7	%	8	%		
Postsecondary educator	4	4.04	73	43.7	16	20.8	72	44.2	272	93.2	10	17.2	16	31.4	3	8.33	460	49.6
Scientist in a non-teaching position	0	0	32	19.2	26	33.8	36	22.1	1	0.34	5	8.62	4	7.84	2	5.56	105	11.3
Engineer	15	15.2	11	6.59	23	29.9	10	6.13	4	1.37	13	22.4	1	1.96	6	16.7	82	8.84
Managerial or professional	79	79.8	42	25.2	0	0	36	22.1	8	2.74	30	51.7	16	31.4	19	52.8	223	24
Other, please specify:	1	1.01	8	4.79	11	14.3	8	4.91	5	1.71	0	0	13	25.5	6	16.7	52	5.6
DON'T KNOW	0	0	0	0	0	0	0	0	0	0	0	0	0	0	0	0	0	0
REFUSED	0	0	1	0.6	1	1.3	1	0.61	2	0.68	0	0	1	1.96	0	0	6	0.65

F

Acronyms and Abbreviations

ARM	Asteroid Redirect Mission
ASPCO	Asian Pacific Space Cooperation Organization
ASTP	Apollo-Soyuz Test Project
ATV	Automated Transfer Vehicle
CAIB	Columbia Accident Investigation Board
CASIS	Center for the Advancement of Science in Space
DARPA	Defense Advanced Research Projects Agency
DOD	Department of Defense
DOE	Department of Energy
DRA	Design Reference Architecture
DRM	Design Reference Mission
ECLSS	environmental, control, and life support system (or systems)
EDL	entry, descent, and landing
ELIPS	European Programme for Life and Physical Science in Space
EM	Exploration Mission
EPO	Education and Public Outreach
ESA	European Space Agency
ETD	Exploration Technology Development
EVA	Extravehicular Activity
FY	fiscal year
GAO	General Accountability Office
GCR	galactic cosmic radiation
GER	Global Exploration Roadmap
GNP	gross national product

| GPS | Global Positioning System |
| GSS | General Social Survey |

HEOMD	Human Exploration and Operations Mission Directorate
HRP	Human Research Program
HTV	H-II Transfer Vehicle

ISECG	International Space Exploration Coordination Group
I_{sp}	specific impulse
ISRU	in situ resource utilization
ISS	International Space Station

| kW | kilowatt |

L1	Lagrangian point 1
L2	Lagrangian point 2
LEAG	Lunar Exploration Analysis Group
LEO	low Earth orbit
LSAY	Longitudinal Study of American Youth

MEPAG	Mars Exploration Program and Analysis Group
MIT	Massachusetts Institute of Technology
MOL	Manned Orbiting Laboratory
MPCV	Multi-Purpose Crew Vehicle
MSL	Mars Science Laboratory
MT	metric ton
MW	megawatt

NASA	National Aeronautics and Space Administration
NCRP	National Council on Radiation Protection and Measurements
NEA	near-Earth asteroid
NEO	near-Earth object
NEP	nuclear electric propulsion
NERVA	Nuclear Engine for Rocket Vehicle Application
NRC	National Research Council
NRO	National Reconnaissance Office
NSF	National Science Foundation
NTP	nuclear thermal propulsion

| OIG | Office of the Inspector General |

SEP	solar electric propulsion
SLS	Space Launch System
SPE	solar particle event
STEM	science, technology, engineering, and mathematics
STG	Space Task Group
STMD	Space Technology Mission Directorate
Sv	Sievert

G

List of Briefings to the Committee and Panels

COMMITTEE ON HUMAN SPACEFLIGHT

December 19, 2012, Washington, D.C.

Administrator's Perspective
Charlie Bolden, NASA Administrator

Goals and Rationales for Human Space Flight
Lori Garver, NASA Deputy Administrator

NASA Science and Exploration
John Grunsfeld, Associate Administrator, NASA Science Mission Directorate

Congressional Panel: History of Request, Goals for Study, Perspective on Challenges
Ann Zulkosky and Jeff Bingham, Senate Commerce, Science, and Transportation Committee (staff)
Ed Feddeman, House Committee on Science, Space, and Technology, Subcommittee on Space and Aeronautics (staff)
Richard Obermann, House Committee on Science, Space, and Technology (staff)

Future Human Exploration and Operations
Gregory Williams, Deputy Associate Administrator for Policy and Plans, NASA Human Exploration and Operations Mission Directorate (HEOMD)

Expert Panel
• History of Human Spaceflight Policy
Roger Launius, National Air and Space Museum

- Human Spaceflight Impacts, Challenges and Opportunities (all panelists)
  Roger Launius, National Air and Space Museum
  Betty Sue Flowers, University of Texas, Austin (emeritus)
  Henry Hertzfeld, George Washington University
  Lester Lyles, The Lyles Group
  Howard McCurdy, American University

April 21-23, 2013, Washington, D.C.

Prospects and Plans Beyond LEO
William Gerstenmaier, NASA

Perspectives on Human and Robotic Spaceflight
Steve Squyres, Cornell University

Human Space Exploration in Earth Orbit
Julie Robinson and Michael Suffredini, NASA

Human Spaceflight Role in Security and International Relations
Scott Pace, George Washington University (via telecon)

Commercial Human Spaceflight
George Nield, FAA

Presidential Priorities for Human Spaceflight
John Olson, Office of Science and Technology Policy

Perspectives on the Future of Human Spaceflight
Alexey Krasnov, Roscosmos (via video)

Perspectives on the Future of Human Spaceflight
Thomas Reiter, European Space Agency (via video)

Future Robotics and Humans Partnerships in Earth Orbit
Jeff Hoffman, Massachusetts Institute of Technology

August 27, 2013 (Teleconference)

Presentation on Global Exploration Roadmap
Kathy Laurini, NASA

October 21-23, 2013, Washington, D.C.

HSF Challenges and Promise
Charlie Bolden, NASA Administrator

Lunar Pathways, Benefits, and Trade-offs
Michael Duke, Colorado School of Mines (retired)

Space Exploration as Insurance Against Existential Risk
   David Grinspoon, Denver Museum of Nature and Science

Status on HSF Plans and Challenges
   William Gerstenmaier, NASA

Delusions of Space Enthusiasts
   Neil deGrasse Tyson, Hayden Planetarium

TECHNICAL PANEL

February 4-5, 2013, Washington, D.C.

Human Exploration Program Overview
   Gregory Williams, Deputy Associate Administrator for Policy and Plans, NASA HEOMD

Exploration Systems Development (Orion, Space Launch System, and Ground Systems)
   Daniel Dumbacher, Deputy Associate Administrator for Exploration Systems Development, NASA HEOMD

Focus Session on Commercial Crew
   Philip McAlister, Director, NASA Commercial Spaceflight Development Division, NASA HEOMD
   Garrett Reisman, SpaceX
   George Sowers, ULA
   Christopher Ferguson, Boeing (via telecon)
   Michael Lopez-Alegria, Commercial Spaceflight Federation
   XCOR, Jeff Greason (via telecon)

Stepping Stones: Exploring Increasingly Challenging Destinations on the Way to Mars
   Josh Hopkins, Lockheed Martin Corporation

March 27-28, 2013, Washington, D.C.

NASA Review of Long-Term Technical Challenges for Human Spaceflight Overview
- NASA's Human Space Exploration Capability Driven Framework
  Jason Crusan, Director, Advanced Exploration Systems, NASA Headquarters
- Crew Health, Medical, and Safety: Human Resaerch Program
  Steve Davison, Human Research Program, NASA Headquarters
- Habitation Systems and Destination Systems
  John Connolly, NASA Johnson Space Center (JSC)
  Robyn Carrasquillo, NASA Headquarters
- In-Space Propulsion and Power for Human Space Flight
  Les Johnson, NASA Marshall Space Flight Center
- Robotics and Autonomous Systems
  Rob Ambrose, NASA JSC (via telecon)
- Entry, Descent, and Landing for Future Human Space Flight
  Michelle Munk, NASA Langley Research Center
- Deep Space Extravehicular Activity (EVA)
  Mike Hembree, NASA JSC

The Golden Spike Company: Extend Your Reach
  Alan Stern, Golden Spike Company (via telecon)

The Google Lunar X Prize
  Alex Hall, X Prize Foundation (via telecon)

Expanding the Final Frontier: The Bigelow Aerospace Story
  Mike Gold, Bigelow Aerospace

Asteroid Return Mission (ARM): 2012 Workshop Report and Ongoing Study Summary
  Keck Institute for Space Studies:
    Paul Dimotakis, Caltech, and Louis Friedman, The Planetary Society

Asteroid Mining and Technology
  Chris Lewicki, Planetary Resources, Inc. (via telecon)

NASA Technology Development for Human Spaceflight
  James Reuther, Deputy Associate Administrator of Programs, NASA Space Technology Mission Directorate

Technical Challenges for Piloted Missions to the Outer Planets
  Ralph McNutt, Johns Hopkins University (via telecon)

NASA's Asteroid Redirect Mission and Human Space Flight
  Michele Gates, NASA HEOMD
  Steve Stich, NASA JSC (via telecon)

April 30, 2013 (Teleconference)

Human Health Issues in Space
  Kenneth Baldwin, University of California, Irvine

June 19-21, 2013, Irvine, California

Reducing the Cost of Exploration using Near-Term Advanced In-Space Propulsion
  Roger Myers, Aerojet (via telecon)

NASA Technology Development for Human Spaceflight
  James Reuther, Deputy Associate Administrator of Programs, NASA Space Technology Mission Directorate

Technical Challenges for Piloted Missions to the Outer Planets
  Ralph McNutt, Johns Hopkins University (via telecon)

NASA's Asteroid Redirect Mission: Technical Challenges
  Michele Gates, NASA HEOMD
  Steve Stich, NASA JSC (via telecon)

August 15, 2013 (Teleconference)

NASA Technology Development for Human Spaceflight
  William Whittaker, Carnegie Mellon University

PUBLIC AND STAKEHOLDER OPINIONS PANEL

June 19, 2013, Washington, D.C.

NASA's Research on Public and Stakeholder Opinions
  Rebecca Spyke Keiser, NASA

H

List of Input Papers

The Committee on Human Spaceflight invited interested individuals and groups to submit short input papers that addressed the role of human spaceflight and its suggested future.

The request for input papers was open to any and all interested individuals and groups wishing to submit their own ideas on the role of human spaceflight and their vision for a suggested future. In developing their papers, respondents were asked to carefully consider the following broad questions.

1. What are the important benefits provided to the United States and other countries by human spaceflight endeavors?
2. What are the greatest challenges to sustaining a U.S. government program in human spaceflight?
3. What are the ramifications and what would the nation and world lose if the United States terminated NASA's human spaceflight program?

In discussing the above questions, the respondents were asked to describe the reasoning that supported their arguments and, to the extent possible, include or cite any evidence that supported their views. In considering #1 above, submitters were asked to consider private as well as government space programs.

Following is the list of lead authors and titles of the input papers submitted to in response to the committee's request. Some of the input papers represent the collective views of companies, associations, or other groups. At the time of this printing, the full text of the input papers remain available for viewing at http://www8.nationalacademies.org/aseboutreach/publicviewhumanspaceflight.aspx.

Abramson, Michael, The Roads to Space Settlement
Aguilar, Alfredo A., Jr., Edge of Creation
Akkerman, James W., Human Space Flight
Alperin, Noam, Visual Impairment and Intracranial Hypertension in Microgravity: Mismanagement of a Manageable Long-Term Spaceflight Risk
Arora, Kamal, Human Space Flight Program-Propitious or Futile
Badders, Brian D., Extrapolating Trends: Human Spaceflight Goals
Barbee, Brent W., Near-Earth Asteroids: The Next Destination for Human Explorers
Barnhard, Gary P., Human Spaceflight—Architecting the Future

Bates, William V., The Role of Human Spaceflight in the 21st Century

Baxter, David, The Need for Human Spaceflight

Beauchemin, Alyse N., Manned Spaceflight: Humanities Greatest Endeavor

Becker, Jeanne L., Input to the NRC Committee on Human Spaceflight

Bednarek, Stephanie R., Space Exploration Technologies Corp. (SpaceX) Public Comment on the National Research Council Study on Human Spaceflight

Bennett, Gary L., Human Spaceflight Observations

Bland, Joseph B., Why Continuing Human Spaceflight Is Critical for Americans

Boyle, Richard D., The Artificial Gravity Platform (AGP)—An Earth-Based Laboratory to Advance Human Exploration in Space

Bridwell, Nelson J., Myopia

Brisson, Pierre, Human Spaceflight is Needed for Mars Exploration

Brooks, Phillip, Human Space Flight, What Direction Shall We Choose?

Brown, Benjamin S., We Are the Explorers

Brown, Jeremy, Concerning a Continuation of Human Space Flight by NASA and the United States Government

Buckland, Daniel M., Early Career Researchers Views as to the Benefits and Challenges of the U.S. Government Human Spaceflight Program

Burke, James D., Long-Range Future of U.S. Human Space Flight

Burns, Neal M., Then and Now; Building Public Support for NASA Human Space Flight

Bussey, Ben, Scientific Benefits of Human Exploration

Cadorette, Normand, One Bigger Picture and Manned Space Flight

Chambliss, Joe P., Ideas on the Direction and Importance of Human Spaceflight

Chapman, Bert, More Effective Human Spaceflight and Their National Security Implications

Chapman, Philip K., The Human Future in Space

Cheuvront, David L., Addressing the Challenges to Sustaining U.S. Government Human Spaceflight with Depots

Cohen, Helen S., Occupational Therapy's Perspective on the Human Spaceflight Program

Collins, Dwight H., Questions Humanity Needs to Answer

Cooke, Douglas R., Building a Roadmap for Human Space Exploration

Cooke, Michael P., Theory for Why We Do Not Have a National Human Spaceflight Policy

Craig, Mark, What are the Greatest Challenges to Sustaining a U.S. Government Program in Human Spaceflight?

Crawford, Ian A., An Integrated Scientific and Social Case for Human Space Exploration

Crisafulli, Jim, Reaching Beyond Low-Earth Orbit: A Prescription for Cost-Effective and Sustainable Human Space Exploration

Crume, Phillip, Using a U.S. Government Human Spaceflight Program to Support Space Commerce

Dailey, Michelle K., Why America Should Sustain—Not Terminate—NASA's Human Spaceflight Program

Das, Arun C., Human Spaceflight Exploration: Benefits, Challenges and Termination Ramifications

Davies, Philip E., Response to the National Academy of Sciences Announcement of Opportunity to Submit Input to Study on Human Spaceflight

Day, Stephen, Repositioning NASA

De Vita, Mirko, Advances in Behavioural Management and Attitude Change Knowledge through Human Spaceflight Programs

DeRees, Kelly A., Pushing Our Boundaries: The Need for a Human Spaceflight Program

Dhasan, Raj A., The Wind Shuttle (or) the Wind Planet

Donahue, Benjamin B., HSF Architecture Analysis and National Goals

Donahue, Benjamin B., Exploration in Human Terms (Parts I-III)

Donahue, Benjamin B., Human Space Flight Far Term View 2025-40 Goals and Objectives

Eckert, Joy, Inspiration Is Vital to Our Future

Elifritz, Thomas L., The National Academies Committee on Human Spaceflight

Elvis, Martin, Enabling the Commercial Harnessing of Space Resources

Emken, William A., The Uncertain Future of Human Spaceflight in the United States

Espeseth, Kevin, Surface Strip Mining the Moon for Permanent Engineering Marking

Ferguson, Christopher, Human Space Exploration: Unparalleled Source of Inspiration for the Past, Present, and the Future

Ferguson, Christopher, Human Space Exploration—An Engine for Innovation

Fernhout, Paul D., The Need for Continuing Research on Human Life Support in Space via Crowsourcing Self-Replicating Space Habitats

Fornaro, John, Why Human Spaceflight Matters

Friedman, Louis D., Paving Stones for the Flexible Path into the Solar System

Friedman, Louis D., Humans to Mars, Human Telepresence Beyond

Gallegos, Zachary E., Continuing Human Space Exploration: For the Future of Our Nation and Its People

Gargani, David J., Inspiration and Cooperation—A Pathway to the Future

Geirczak, Anthony E., III, The Importance of Continued Manned Spaceflight for Improving Economic and Quality of Life Issues on Earth

Gillin, Joseph P., Human Spaceflight and the Human Future

Globus, Al, Human Space Flight, Space Industrialization, Commercialization and Settlement

Gorenstein, Paul, Human Spaceflight and a Lunar Base

Gozdecki, Jonas, Stars Entrepreneurship Closer to All

Graham, E. H., Common Thoughts

Greenhouse, Matthew A., A Values-Based Approach toward National Space Policy

Greeson, David S., Separating Human Spaceflight Policy from National Politics: The Key to Long-Term Mission Success

Grondin, Yves, Why Human Spaceflight Exploration Matters

Hamill, Doris, The Neglected Challenge: Exploiting Near Space for Human Benefit

Hana, John D., U.S. Space 2020

Harper, Lynn D., Benefits and Challenges of Human Spaceflight in the 21st Century

Harrison, Michael F., The Imperative of Human Space Flight

Hawes, Michael, NRC Inputs

Haynes, Douglas E., White Paper

Heidmann, Richard, A New American Space Exploration Initiative Is Needed! A European Point of View

Heismann, August C., The Future of Human Space Flight

Henderson, Edward M., The Importance of Human Spaceflight

Howard, Robert L., The Role of Human Spaceflight and a Vision for a Suggested Future

Huntsman, David P., Answering the Questions: Why Government and Commercial Human Spaceflight; How to Get There, How to Keep it Sustainable, and the Need for NASA Reform

Huter, Paul B., Innovators, Adventurers, Explorers, Creators, Leaders

Jensen, Dale L., Efficient Rocket Engines

Jolliff, Bradley L., Why and How the United States Must Continue to Lead the World in Human Spaceflight

Joseph, Nikolai, Human Spaceflight

Kasting, James F., Servicing of Large Space Telescopes and Geosynchronous Satellites

Kennedy, Linda F., The Role of Human Spaceflight and Vision for a Suggested Future

Keras, Kevin F., Wither NASA? Earthly Pursuits to Fund Off-Planet Exploration

Kerwin, Joseph P., Leadership in Space Exploration

Kirkpatrick, Jim, AAS Input to the National Research Council's Committee on Human Spaceflight

Kitmacher, Gary H., Reasons for Human Spaceflight; Cis-Lunar Technology Testbed

Korn, Paula, The Uncertain Future for U.S. Human Spaceflight

Krone, Bob, Philosophy for Humans in Space

Kuebler, Ulrich M., Role and Visions for Human Spaceflight

Kugler, Justin, Let Us Be Pioneers

Laine, Michael, LiftPort Lunar Space Elevator Infrastructure—Affordable Response to Human Spaceflight

Lawrence, Samuel J., The Benefits of Human Space Exploration to the United States and other Nations

Leeds, Gregory A., Countering China's Future Space Proliferation Policy

Lester, Dan F., Telepresence and the Purpose of Human Spaceflight

Lillie, Charles F., The Future of Human Spaceflight

Mandell, Humboldt C., The Future of Human Spaceflight

Maniaci, Michael T., Bold Aspirations from the Land of the Free: A Civilian's Perspective of American Spaceflight

Mapes, James M., The Philosophical Implications of Exoplanet Discovery on the Need for Human Spaceflight

Martin, Thomas N., III, Aerojet Rocketdyne Response to NRC Questions on NASA Human Spaceflight

McCain, Terence R., Switching Over: The Need to Transition from Human Wrench-Turners in Space to Advanced Robotics

McCandless, Bruce, II, Leadership through Human Spaceflight

McCarthy, Brendan, Spaceflight: A Duty of All Mankind

McMickell, M. Brett, Input to the National Research Council's Committee on Human Spaceflight

Mohanty, Ashutosh, Propulsion Technologies

Molnar, Erwin P., The Future of Human Space Flight in the United States

Monteiro, Paulo L., Work in Progress

Moulin, Nicolas, Human Space Flight

Naik, Krishna D., Human Spaceflight Is the Ultimate Human Endeavor

Neal, Clive R., Human Solar System Exploration Achieved with a "Moon First" Pathway

Ness, Peter, Human Flights Are Essential for Mineral Exploration and Mining of the Moon, Mars and Asteroids

Norton, Paul, Minimal Requirement

Nye, William S., The Goal Is Mars: The Planetary Society's Submission to the National Research Council on Human Spaceflight

Obenaus, Andre, The Importance of Human Space Flight to the Development of Medical Imaging

Okushi, Jun, A Short Review of the Necessity of Human Spaceflight Our Advocacy of a Decade

Oleson, Gary L., Toward a Thriving and Sustainable U.S. Human Spaceflight Program

Othman, Mazlan, The United Nations and Human Spaceflight

Overton, Ian M., Human Spaceflight: A Choice between Renaissance or Dark Age

Paluszek, Michael A., Human Spaceflight

Pellerin, Charles, Does Human Exploration Make Sense?

Pittman, Bruce, Human Space Flight Challenge

Pittman, Robert B., National Space Society (NSS) Response to the National Research Council (NRC) Committee on Human Spaceflight

Podnar, Gregg, We Must Explore Mars

Polk, James D., The Return on Investment of Human Spaceflight

Pomerantz, William J., The Benefits of Suborbital and Orbital Human Spaceflight

Pulham, Elliot H., Space Foundation Input to the Committee on Human Spaceflight

Raftery, Michael, Crossing This New Ocean

Rahul Arun Bagul, Capsule Project

Rahul Arun Bagul, Prototype Safety Parachute for Aerospace and Aeronautics Projects

Rahul Arun Bagul, Concept for Future Free Space Craft

Ramey, Christopher B., The Foundations of National Power and Human Spaceflight

Rice, Eric, Recommended Direction for the American Human Spaceflight Program

Richards, David L., America Needs a New Vision and Oversight Process for Its Human Spaceflight Program

Riley, Danny A., Bullet Point Responses to the 3 Questions

Riley, John T., Generating Vision with Science Fiction

Robinson, John W., The SPST Whitepaper on Space Development: The Justification for Human Space Development and Habitation Beyond Low Earth Orbit

Rodriguez, G. J., Architecting a Lunar Shipyard

Rovetto, Robert J., The Essential Role of Human Spaceflight

Sander, Michael J., Human Space Flight, Observations and Suggestions

Santos, Famar P., The Planetary Expressway

Sauvageau, Donald R., ATK Response to NRC Questions on NASA Human Spaceflight Program

Sawyer, Paul D., Why We Should Sit Out the Next Round of Manned Space Flight

Schindler, Jürgen R., The Benefits of Human Space Flight with Focus on Human Values

Schmatz, Michael J., Supremacy, Growth, and Inspiration: The Purpose of American Human Spaceflight

Schmitt, Harrison H., Deep Space Exploration: An American Imperative

Schwadron, Nathan A., Understanding and Predicting the Space Radiation Hazard—A Critical Element of
    Human Exploration Beyond Low Earth Orbit

Shaffer, Gabrielle M., The Necessity of NASA's Human Spaceflight Program for Science and Technology

Shemansky, Donald E., The NASA Human Space Exploration Program Should Be Abandoned

Sherwood, Brent, Toward a Relevant Human Spaceflight Program

Skocik, Colin R., The Moon Before Mars: Why Obama Is Wrong on Space

Skocik, Colin R., Past and Future of the Space Program

Slazer, Frank, Aerospace Industries Association Input on Human Spaceflight

Smith, O. Glenn, Challenges to Sustainability

Smith, Philippe M., Why Human Spaceflight Is Important to the United States

Smith, Stephen C., A New Economy: U.S. Human Spaceflight in the 21st Century

Smith, William L., Human Spaceflight Scenarios

Smitherman, David V., A Department of Space to Enable Human Spaceflight and the Future Settlement of Space

Spearing, Scott F., Economizing on Human Space Flight

Spudis, Paul D., Human Spaceflight: Why and How?

Srivastava, Rajesh, Other Discover Earth

Staats, Kai K., Real Heroes: A Case for Continued U.S. Involvement in Human Space Exploration

Stolc, Viktor, Bio-Electromagnetic Countermeasures Against the Risks of Spaceflight Beyond Low Earth Orbit

Strangman, Gary E., The Value of Human Spaceflight

Strickland, John K., Jr., What Should We Be Doing in Space?

Summers, Richard L., Importance of Human Spaceflight Endeavors for the Future of Biomedical Research

Sweetser, Theodore H., Mars Is the Vision, the Question Is How

Taggart, Keith A., Input to the Committee on Human Spaceflight

Thronson, Harley A., If You Set Out to Go to Mars, Go to Mars

Tran, Ricky H., An Input on Human Spaceflight for the Last Ten Years

Vail, Joel, B., The Future Is in Exploration

Van Vaerenbergh, Stefan, Need of Spaceflight Gravity from Microbiology to Geology Studies

Vartorella, William F., 'Lost Horizon'—Manned Missions Focused on Nanotechnologies, Astrobiology, Space
    Telescopes, and Advanced Materials Will Re-define NASA and NewSpace

Vedda, James A., Next Step for Human Spaceflight: Cislunar Development

Wanduragala, P., How to Overcome Limitations in Exploring Space Journeys (file unreadable)

Webber, Derek, Why We Need to Continue with Human Spaceflight

Wheelock, Terry W., "Human" Spaceflight

Whittington, Mark R., In Pursuit of Space Power

Wilkins, Richard T., The Value of Human Space Flight on Higher Education and the Minority Communities

Winn, Laurence B., The Case for a New Diaspora: Cosmic Dispersal

Woodard, Daniel, Practical Benefits for America

Woodcock, Gordon R., Human Space Flight

Woodlyn, Sjon O., Space Exploration: Foundation to Build a Global Response

Zimpfer, Douglas J., Future of Human Spaceflight: Draper Laboratory Response

I

Committee, Panel, and Staff Biographies

COMMITTEE

MITCHELL E. DANIELS, Jr., *Co-Chair,* is the president of Purdue University. Immediately prior to this appointment, he served two terms as the 49th Governor of the State of Indiana. Previously, he has been the CEO of the Hudson Institute and president of Eli Lilly and Company's North America Pharmaceutical Operations. In the political arena, he also served as chief of staff to Senator Richard Lugar, senior advisor to President Ronald Reagan, and director of the Office of Management and Budget under President George W. Bush (January 2001 to June 2003). He is the author of the book *Keeping the Republic: Saving America by Trusting Americans.* Mr. Daniels earned his B.S. from the Woodrow Wilson School of Public and International Affairs at Princeton University and his J.D. from the Georgetown University Law Center.

JONATHAN I. LUNINE, *Co-Chair,* is the director of the Center for Radiophysics and Space Research and the David C. Duncan Professor in the Physical Sciences at Cornell University. Dr. Lunine is interested in how planets form and evolve, what processes maintain and establish habitability, and what the limits of environments capable of sustaining life are. He pursues these interests through theoretical modeling and participation in spacecraft missions. He works with the radar and other instruments on the Cassini Saturn Orbiter and was part of the science team for the Huygens landing on Saturn's moon Titan. He is co-investigator on the Juno mission to Jupiter, launched in 2011, and an interdisciplinary scientist for the James Webb Space Telescope. Dr. Lunine has contributed to or led a variety of mission concept studies for solar system probes and space-based detection of planets around other stars. He has chaired or served on a number of advisory and strategic planning committees for NASA and the National Science Foundation (NSF). He is the winner of the Harold C. Urey Prize of the DPS/American Astronomical Society, the Macelwane Medal of the American Geophysical Union (AGU), the Zeldovich Prize in Commission B of Committee on Space Research (COSPAR), and the Basic Science Award of the International Academy of Astronautics. He is a member of the National Academy of Sciences (NAS) and a fellow of the AGU and American Association for the Advancement of Science (AAAS). Dr. Lunine received a B.S. in physics and astronomy from the University of Rochester and an M.S. and a Ph.D. in planetary science from the California Institute of Technology. Dr. Lunine has served on several National Research Council (NRC) committees, including as co-chair for the Committee on the Origins and Evolution of Life and the Committee for a Review of Programs to Determine the Extent of Life in the Universe, and as a member of the Committee on Decadal Survey on Astronomy and Astrophysics 2010.

BERNARD F. BURKE is the William A.M. Burden Professor of Astrophysics, emeritus, at the Massachusetts Institute of Technology (MIT). He is also a principal investigator (PI) at the MIT Kavli Institute for Astrophysics and Space Research. His research career has covered a wide range of activities, including the co-discovery of Jupiter radio bursts and the discovery of the first "Einstein Ring," a manifestation of the warping of space-time by matter that was predicted by Albert Einstein in his general theory of relativity. Dr. Burke was president of the American Astronomical Society and served as a member of the National Science Board. He is a member of the NAS and the American Academy of Arts and Sciences, a fellow of AAAS, and a recipient of the NASA Group Achievement Award for Very Long Baseline Interferometry. He earned a Ph.D. in physics from MIT. Dr. Burke has served on numerous NRC committees, including the Committee on the Assessment of Solar System Exploration, the U.S. National Committee for the International Astronomical Union, and the International Space Year Planning Committee.

MARY LYNNE DITTMAR is president and executive consultant for Dittmar Associates, Inc., an engineering and consulting firm in Houston, Texas. Previously, Dr. Dittmar managed International Space Station (ISS) Flight Operations and Training Integration for the Boeing Company and later served as chief scientist and senior manager for Boeing's Commercial Space Payloads Program before advising on business development and strategic planning for the company's Space Exploration group. More recently she has acted as senior advisor to executives in a variety of aerospace companies, at NASA, and at the Center for the Advancement of Science in Space (CASIS) where she was instrumental in developing the strategic plan for utilization of the ISS National Laboratory. Dr. Dittmar's areas of practice focus on strategic planning, public/private partnerships, strategic communications, systems engineering, and change management. She is published in a variety of fields including operations, engineering, artificial intelligence, human factors, communications, and business, and has authored several papers on the impact of regulatory frameworks and investor engagement with emerging sectors such as the commercial spaceflight industry. She also has served as a member of the Federal Aviation Administration Commercial Space Transportation Advisory Committee's Space Operations Working Group and holds a number of industry and academic awards, including Meritorious Inventions and the Chief Technology Officer's Award for Technical Excellence from Boeing and NASA's "Silver Snoopy" Award for significant contributions to human spaceflight, and is a fellow of Sigma Xi, the research honor society of scientists and engineers. Dr. Dittmar earned a Ph.D. in human factors from the University of Cincinnati.

PASCALE EHRENFREUND is research professor of space policy and international affairs at the Space Policy Institute of George Washington University. During the past 15 years Dr. Ehrenfreund has contributed as PI, co-investigator, and team leader to experiments in low Earth orbit and on the ISS, as well as to various European Space Agency (ESA) and NASA space missions, including astronomy and planetary missions. She is a lead investigator with the NASA Astrobiology Institute, a virtual institute that integrates research and training programs, and her research experience and interests range from biology to astrophysics. Dr. Ehrenfreund served as the project scientist of NASA's O/OREOs satellite, the first mission of the NASA Astrobiology Small Payload program currently in orbit. She has served on several committees dealing with space strategy issues, including the European Space Science Committee, ESA's Life and Physical Science Advisory Committee, and ESA's Life Science Working Group. Since 2010, she has chaired the Panel on Exploration of COSPAR. Dr. Ehrenfreund is president of IAU Commission 51 (Bioastronomy) and a full member of the International Academy of Astronautics. She serves as member of the NRC Committee on Astrobiology and Planetary Sciences and served on the Committee on Planetary Science Decadal Survey: 2013-2022. Dr. Ehrenfreund was a member of the FP7 Space Advisory Group of the European Commission and has been reelected in the Horizon2020 Space Advisory Group and started her term in November 2013. Since 2013 she has served as president of the Austrian Science Fund. She holds a master's degree in molecular biology from the University of Vienna (Austria), a Ph.D. in astrophysics from the University Paris VII/University Vienna (Austria), a Habilitation in astrochemistry from the University of Vienna (Austria), and a master's degree in management and leadership from Webster University (Netherlands). She authored and co-authored more than 300 publications.

FRANK G. KLOTZ[1] (USAF, retired) is the under secretary of energy for nuclear security and administrator of the National Nuclear Security Administration. He is a former senior fellow for Strategic Studies and Arms Control at the Council on Foreign Relations (CFR) and a former commander of Air Force Global Strike Command, where he established and then led a new 23,000-person organization that merged responsibility for all U.S. nuclear-capable bombers and land-based missiles under a single chain-of-command. Earlier in his military career, General Klotz served as the defense attaché at U.S. Embassy Moscow during a particularly eventful period in U.S.-Russian relations. Later, as the director for nuclear policy and arms control on the National Security Council staff, he represented the White House in talks that led to the 2002 Moscow Treaty to reduce strategic nuclear weapons. General Klotz also served as the vice commander of Air Force Space Command. He was awarded the prestigious General Thomas D. White Trophy by the Air Force Association for the most outstanding contribution to progress in aerospace in 2006. General Klotz has spoken extensively on defense and space topics to audiences throughout the United States, as well as abroad. He is the author of *Space, Commerce and National Security* (1998) and *America on the Ice: Antarctic Policy Issues* (1990). He served as a White House fellow at the State Department and as a military fellow at CFR. He is a member of CFR, the International Institute for Strategic Studies, and the Secretary of State's International Security Advisory Board. A distinguished graduate of the U.S. Air Force Academy, General Klotz attended Oxford University as a Rhodes Scholar, where he earned an M.Phil. in international relations and a D.Phil. in politics. He is also a graduate of the National War College in Washington, D.C. He currently serves as a member of the NRC Committee on International Security and Arms Control.

JAMES S. JACKSON (IOM) is the director and research professor of the Institute for Social Research at the University of Michigan. He is also the Daniel Katz Distinguished University Professor of Psychology, a professor of health behavior and health education in the School of Public Health, and a professor of Afroamerican and African studies. He is past director of the Program for Research on Black Americans and the Center for Afroamerican and African Studies. His research focuses on issues of racial and ethnic influences on life course development, attitude change, reciprocity, social support, and coping and health among blacks in the diaspora. He is past chair of the Section on Social, Economic, and Political Sciences of the AAAS. He is a former national president of the Black Students Psychological Association, the Association of Black Psychologists, and the Society for the Psychological Study of Social Issues. He served on the councils of the National Institute of Mental Health and the National Institute on Aging. He is a fellow of the Gerontological Society of America, the Society of Experimental Social Psychology, the American Psychological Association, the Association of Psychological Sciences, AAAS, and the W.E.B. Du Bois Fellow of the American Academy of Political and Social Science. He received numerous awards, including the Distinguished Career Contributions to Research Award of the Society for the Psychological Study of Ethnic Minority Issues, the James McKeen Cattell Fellow Award for Distinguished Career Contributions in Applied Psychology of the American Psychological Association, and the Medal for Distinguished Contributions in Biomedical Sciences of the New York Academy of Medicine. He is a fellow of the American Academy of Arts and Sciences. Dr. Jackson earned a Ph.D. in social psychology from Wayne State University. He is currently a member of the NRC Board on Behavioral, Cognitive, and Sensory Sciences and has served as a member of the Committee on International Collaborations in Social and Behavioral Research, the Committee to Study the National Needs for Biomedical, Behavioral, and Clinical Research Personnel, and the Committee on U.S. Competitiveness: Underrepresented Groups and the Expansion of the Science and Engineering Workforce Pipeline.

FRANKLIN D. MARTIN is president of Martin Consulting, Inc. His interests include independent review services for NASA spaceflight projects. Over the past decade, he has taught nearly 100 team development workshops for organizations and flight projects across NASA while working as a subcontractor to 4-D Systems. He has more than 40 years of experience with space science, space systems, engineering, and management. His experience covers robotic, remote sensing, and human spaceflight. His career with NASA and Lockheed Martin includes the following: Science Mission Operations for Apollo 16 and Apollo 17; director, Solar Terrestrial and Astrophysics at

---

[1] General Klotz resigned from the committee on April 10, 2014, to take up an appointment as Under Secretary of Energy for Nuclear Security and Administrator of the National Nuclear Security Administration.

NASA Headquarters; director for Space and Earth Science, Goddard Space Flight Center; NASA deputy associate administrator, Space Station; NASA assistant administrator, Human Exploration; and director of Space Systems and Engineering in Civil Space for Lockheed Martin, with responsibility for the Hubble Servicing Missions, Space Infrared Telescope Facility (Spitzer), Lunar Prospector, and the Relativity Mission (Gravity Probe-B). Dr. Martin also served as assistant editor of *Geophysical Research Letters* and worked as a physicist with the Naval Oceanographic Office. He resigned from NASA in 1990 at senior executive service (SES) Level ES-6 and retired from Lockheed Martin in 2001. He currently serves on the NASA Innovative Advanced Concepts External Council for the NASA Chief Technologist Office. Dr. Martin received NASA's Exceptional Service Medal, an Outstanding Leadership Medal, and the SES Presidential Ranks of Meritorious Executive and Distinguished Executive. He is a fellow of the American Astronautical Society (AAS). He earned a B.A. with majors in physics and in mathematics from Pfeiffer College (aka Pfeiffer University) and a Ph.D. in physics from the University of Tennessee. Dr. Martin has served as a member of the NRC Committee on Human Spaceflight Crew Operations, the Committee on NASA's Suborbital Research Capabilities, and the Committee on Science Opportunities Enabled by NASA's Constellation System.

DAVID C. MOWERY is the William A. and Betty H. Hasler Professor of New Enterprise Development (Emeritus) at the Walter A. Haas School of Business at the University of California, Berkeley, and a research associate of the National Bureau of Economic Research. Dr. Mowery's research interests include the impact of technological change on economic growth and employment, the management of technological change, and international and U.S. trade policy. His academic awards include the Raymond Vernon Prize from the Association for Public Policy Analysis and Management, the Economic History Association's Fritz Redlich Prize, the Business History Review's Newcomen Prize, the Cheit Outstanding Teaching Award, and the Distinguished Scholar award from the Academy of Management. He received his undergraduate and Ph.D. degrees in economics from Stanford University and was a postdoctoral fellow at the Harvard Business School. Dr. Mowery has served on several NRC committees, including as vice chair of the Committee on Competitiveness and Workforce Needs of United States Industry and as a member of the Committee to Review the National Nanotechnology Initiative and the Committee to Assess the Capacity of the U.S. Engineering Research Enterprise.

BRYAN D. O'CONNOR is an independent aerospace consultant and former Marine pilot and NASA senior executive. He previously served as NASA's chief of the Office of Safety and Mission Assurance where he led an extensive restructure of system safety, reliability, quality, and risk management organizations throughout the agency in response to the findings of the Columbia Mishap Investigation Board. He was previously director of engineering for the Futron Corporation, providing system safety engineering and risk management consulting to the Department of Defense, the Department of Energy, the Federal Aviation Administration (FAA), NASA, and industry. Prior to that, he was the director of the Space Shuttle Program. Mr. O'Connor served as a Marine Corps test pilot and as pilot of the STS-61B Space Shuttle mission and as commander of the STS-40 mission. He also served in a variety of RDT&E functions in support of the first test flights of the space shuttle. He is the recipient of several awards, including the Naval Test Pilot School Distinguished Graduate Award, the Distinguished Flying Cross, the NASA Distinguished Service Medal, and the International Association for the Advancement of Space Safety's Jerome Lederer Space Safety Pioneer Award. He currently serves on NASA's Aerospace Safety Advisory Panel, which directly observes NASA operations and evaluates and advises NASA on its safety performance; the Panel submits an annual report to Congress and to the NASA Administrator. Mr. O'Connor earned an M.S. in aeronautical systems from the University of West Florida. He previously served as chair of the NRC Committee on Space Shuttle Upgrades.

STANLEY PRESSER is a Distinguished University Professor at the University of Maryland where he teaches in the Sociology Department and the Joint Program in Survey Methodology. His research interests include questionnaire design and testing, the accuracy of survey responses, and the nature and consequences of survey nonresponse. Dr. Presser is a past president of the American Association for Public Opinion Research, a fellow of the American Statistical Association, and a recipient of the Paul F. Lazarsfeld Award for a career of outstanding contributions to methodology in sociology. He earned his Ph.D. in sociology from the University of Michigan. Dr. Presser has served as a member of the NRC Committee to Review the Bureau of Transportation Statistics' Survey Programs and the NRC Panel to Review USDA's Agricultural Resource Management Survey, and he currently serves as a member of the NRC Panel on Measuring Civic Engagement and Social Cohesion to Inform Policy.

HELEN R. QUINN (NAS) is a professor of particle physics and astrophysics (emeritus) at the SLAC National Accelerator Laboratory and co-chair of Stanford University's K12 Initiative. Dr. Quinn is a theoretical physicist who holds numerous honors for her research contributions, including the prestigious Dirac (Italy) and Klein (Sweden) medals. She has had a long-term engagement in education issues and has worked at the local, state, and national level on them. Her interests range from science curriculum and standards to the preparation and continuing education of science teachers. She is a member and former president of the American Physical Society. She received her Ph.D. in physics from Stanford University. She currently chairs the NRC Board on Science Education and serves as a member of the Committee on a Framework for Assessment of Science Proficiency in K-12. She has also chaired the NRC Committee on Conceptual Framework for New Science Education Standards and served as a member of the Committee on Physics of the Universe; the Astro 2010 decadal Survey and many other NRC committees.

ASIF A. SIDDIQI is an associate professor of history at Fordham University. He specializes in the history of modern science and technology and has authored numerous books and articles on the history of spaceflight. His book *Challenge to Apollo: The Soviet Union and the Space Race*, 1945-1974, published by NASA in 2000, was the first major work on the history of the Soviet space program and was based on evidence revealed after the end of the Cold War. *The Wall Street Journal* named it one of the five best books published on space exploration. His writings extend beyond the Russian/Soviet space program to such topics as Asian space initiatives, military space research, and the historiography of American space exploration. His most recent book, *The Red Rockets' Glare: Spaceflight and the Soviet Imagination, 1857-1957*, focused on the cultural roots of space enthusiasm in Russia and was published in 2010. He has served as a visiting scholar at MIT and was one of the co-authors of *The Future of Human Spaceflight*, a report presented to members of Congress and NASA. In 2013-2014, Dr. Siddiqi will serve as the Charles A. Lindbergh Fellow in aerospace history at the National Air and Space Museum in Washington, D.C. Dr. Siddiqi has a B.S. and M.S. in electrical engineering from Texas A&M University and a Ph.D. in history from Carnegie Mellon University.

JOHN C. SOMMERER is a retired senior fellow in the Director's Office of the Johns Hopkins Applied Physics Laboratory (APL), the largest of the Department of Defense University Affiliated Research Centers. He also holds a permanent appointment as Daniel Coit Gilman Scholar at Johns Hopkins University. Until January 1, 2014, he led the APL Space Sector, with responsibility for all APL contributions to military, intelligence community, and civil space programs, including the MESSENGER mission to Mercury, the New Horizons mission to Pluto, Van Allen Probes, Solar Probe Plus, the MDA Precision Tracking Space System, and ORSTech 1 and 2. Prior to 2008, he held a number of other senior executive positions at APL, including director of science & technology, chief technology officer, and director of the Milton S. Eisenhower Research Center, and he led a number of enterprise-level task forces, strategic plans, and other initiatives. Dr. Sommerer received bachelor's and master's degrees in systems science and mathematics from Washington University in St. Louis, a master's degree in applied physics from Johns Hopkins University, and a Ph.D. in physics from the University of Maryland. Dr. Sommerer has served on a number of advisory bodies for the U.S. government, including terms as chair and vice chair of the Naval Research Advisory Committee, senior technical advisory committee to the Secretary of the Navy, Chief of Naval Operations, and Commandant of the Marine Corps. He has also served on numerous NRC boards and committees. He is a full member of the International Academy of Astronautics.

ROGER TOURANGEAU is a vice president and associate director at Westat, Inc., one of the largest survey firms in the United States. Before joining Westat, he was research professor at the University of Michigan's Survey Research Center and the director of the Joint Program in Survey Methodology at the University of Maryland. He has been a survey methodologist for nearly 30 years. Dr. Tourangeau is an author on more than 60 research articles, mostly on survey methods topics. He is also the lead author of a new book on web survey design (*The Science of Web Surveys*) with Fred Conrad and Mick Couper. His earlier book, *The Psychology of Survey Response*, with Lance Rips and Kenneth Rasinski, received the 2006 Book Award from the American Association for Public Opinion Research. He was elected as a fellow of the American Statistical Association in 1999. Dr. Tourangeau has a Ph.D. in psychology from Yale University. He recently served on the NRC Committee on National Statistics and has previously served as chair of the Panel on Research Agenda for the Future of Social Science Data Collection.

ARIEL WALDMAN is the founder of Spacehack.org, a directory of ways to participate in space exploration, and is the global instigator of Science Hack Day, an event that brings together scientists, technologists, designers, and people with good ideas to see what they can create in one weekend. Ms. Waldman is also a fellow at Institute For The Future and was the recent recipient of an honor from the White House as a Champion of Change in citizen science. In 2012, Ms. Waldman authored a paper on democratized science instrumentation for the Science and Technology Policy Institute. Previously, she worked at NASA's CoLab program, whose mission was to connect communities inside and outside NASA to collaborate. Ms. Waldman has also been a sci-fi movie gadget columnist for *Engadget* and a digital anthropologist at VML. She has keynoted O'Reilly's Open Source Convention (OSCON) and DARPA's 100 Year Starship Symposium, appeared on the SyFy channel, and regularly gives talks to a variety of global audiences. In 2008, she was named one of the top 50 most influential individuals in Silicon Valley. Ms. Waldman earned a B.S. from the Art Institute of Pittsburgh in graphic design.

CLIFF ZUKIN is a professor of public policy and political science at Rutgers University's Edward J. Bloustein School of Planning and Policy and the Eagleton Institute of Politics. He is also a senior research fellow at the Rutgers University John J. Heldrich Center for Workforce Development. Dr. Zukin's research interests include public opinion, survey research, mass media, and political behavior. He was founding director of the Rutgers University Center for Public Interest Polling and the Star-Ledger/Eagleton-Rutgers Poll, a quarterly opinion survey. He also served as a consultant to the NBC News Election Unit and for 15 years was on the board of advisers for the Pew Research Center. Dr. Zukin is a member of the *Public Opinion Quarterly* editorial board and past president of the American Association for Public Opinion Research. His recent book *A New Engagement? Political Participation, Civic Life, and the Changing American Citizen* (co-authored with S. Keeter, M. Andolina, K. Jenkins, and M.X. Delli Carpini) uses survey data and the authors' own primary research to examine generational differences in political participation. Dr. Zukin has a Ph.D. in political science from Ohio State University.

PUBLIC AND STAKEHOLDER OPINIONS PANEL

ROGER TOURANGEAU, *Chair. See the committee listing above.*

MOLLY ANDOLINA is associate professor in the department of political science at DePaul University. She researches civic practices and attitudes of the millennial generation toward politics, volunteerism, and community involvement. She co-authored the book *A New Engagement? Political Participation, Civic Life and the Changing American Citizen* and the chapter, "A Conceptual Framework and Multi-Method Approach for Research on Political Socialization and Civic Engagement," in the *Handbook of Research on Civic Engagement in Youth*. In the past, she was survey director at the Pew Research Center for the People & the Press, where she conducted public opinion polling on attitudes toward public policy issues and co-directed a survey of political elites. Prior to joining the Pew Research Center, she wrote and designed qualitative interview guides and conducted stakeholder interviews with public officials and citizens. Dr. Andolina has a Ph.D. in government from Georgetown University.

JENNIFER L. HOCHSCHILD is the Henry LaBarre Jayne Professor of Government, professor of African and African American Studies, and Harvard College Professor at Harvard University. She also holds a lectureship in the Harvard Kennedy School. Dr. Hochschild studies the intersection of American politics and political philosophy and works on issues in public opinion and political culture. Her current research interests include citizens' use of factual information in political decision-making, and the political or ideological developments around genomic science. Dr. Hochschild is an expert in qualitative research methodologies, especially in-depth interviews and elite interviews. She is the author or co-author of numerous books, including *What's Fair: American Beliefs about Distributive Justice*, a classic study based on in-depth interviews. She was founding editor of *Perspectives on Politics*, published by the American Political Science Association, and until recently was a co-editor of the *American Political Science Review*. She is a fellow of the American Academy of Arts and Sciences, a former vice-president of the American Political Science Association, a former member and vice-chair of the board of trustees of the Russell Sage Foundation, and a former member of the board of overseers of the General Social Survey. Dr. Hochschild has a Ph.D. in political science from Yale University. She served two terms on the National Academies Division Committee for the Behavioral and Social Sciences and Education.

JAMES S. JACKSON (IOM). *See the committee listing above.*

ROGER D. LAUNIUS is associate director for collections and curatorial affairs at the Smithsonian Institution's National Air and Space Museum, where he previously served as a senior curator in the Space History Department. From 1982-1990 he worked as a civilian historian with the U.S. Air Force (USAF). He then became chief historian of NASA until 2002. Dr. Launius has written or edited more than 20 books on aerospace history, and pursues research in other historical areas as well. He is frequently consulted by the media for his views on space issues and has been a guest commentator on National Public Radio and major television networks. He is a member of the American Historical Association, the Organization of American Historians, the Society for History in the Federal Government, the National Council on Public History, the History of Science Society, and the Society for the History of Technology. He also is a fellow of AAAS, the American Astronautical Society, and the International Academy of Astronautics, and associate fellow of the American Institute for Aeronautics and Astronautics. He served as chair of the history and education panel of the U.S. Centennial of Flight Commission between 1999 and 2004. In 2003 he served as a consultant to the Columbia Accident Investigation Board. He is a recipient of the Exceptional Service Medal and the Exceptional Achievement Medal from NASA. He holds a Ph.D. in American history from Louisiana State University.

JON D. MILLER is research scientist and director of the International Center for the Advancement of Scientific Literacy, Institute for Social Research, at the University of Michigan. He has measured public understanding of science and technology in the United States for the past three decades, and has examined the factors associated with the development of attitudes toward science and science policy. He is currently the director of the Longitudinal Study of American Youth, which is a longitudinal study of the development of attitudes toward science, mathematics, and citizenship among two cohorts of students. In the past Dr. Miller was the PI for a national study of science policy leaders and space policy leaders. Another one of his recent studies, which involved in-depth interviews with leading scholars in the popularization of science and technology, developed recommendations for a longer-term research program to enhance our understanding of how various segments of the public views, conceptualizes, and understands scientific research. He is author of the book The American People and Science Policy: The Role of Public Attitudes in the Policy Process. He is a member of the Planetary Protection Subcommittee of NASA's Advisory Council. Dr. Miller has a Ph.D. in political science from Northwestern University. Previous NRC experience includes service on the Committee on Assessing Technological Literacy in the United States and the Committee on NASA Education Program Outcomes Study.

STANLEY PRESSER. *See committee listing above.*

CLIFF ZUKIN. *See committee listing above.*

TECHNICAL PANEL

JOHN C. SOMMERER, *Chair. See committee listing above.*

DOUGLAS S. STETSON, *Vice Chair,* is founder and president of Space Science and Exploration Consulting Group, a network of senior advisors and experienced individuals drawn from NASA, national laboratories, industry, and universities. He is a consultant specializing in innovative mission and system concepts, strategic planning, decision analysis, proposal development, and university and industry partnerships. Prior to becoming a consultant, Mr. Stetson spent 25 years at the Jet Propulsion Laboratory (JPL) in a variety of technical and management positions, including several assignments at NASA Headquarters. At JPL he was most recently the manager of the Solar System Mission Formulation Office, where he was responsible for development of all new planetary mission and technology strategies and programs. Earlier in his career, Mr. Stetson played key roles in the design and development of several major planetary missions, including Cassini and Galileo, and he was the leader of many planetary advanced studies and proposals. He received his M.S. in aeronautics and astronautics from Stanford University. He is a veteran of two NRC studies, most recently as a member of the Committee on Planetary Protection Standards for Icy Bodies in the Outer Solar System.

ARNOLD D. ALDRICH is an aerospace consultant. He joined the NASA Space Task Group at Langley Field, Virginia, in 1959, 6 months after the award of the contract to build the Mercury Spacecraft and 4 months following the selection of the seven original astronauts. He held a number of key flight operations management positions at Langley and at the NASA Johnson Space Center during the Mercury, Gemini, and Apollo programs. Subsequently, he served as Skylab deputy program manager; Apollo Spacecraft deputy program manager during the successful Apollo Soyuz Test Project with the Soviet Union; Space Shuttle Orbiter project manager, where he oversaw 15 successful flights as well as the construction of the orbiters Discovery and Atlantis; and as space shuttle program manager. Following the space shuttle *Challenger* accident, Mr. Aldrich was appointed director of the National Space Transportation System (Space Shuttle Program) at NASA Headquarters where he led space shuttle program recovery and return-to-flight efforts. Subsequently, Mr. Aldrich was appointed NASA associate administrator for Space Systems Development, overseeing the Space Station Freedom program, development of the Space Shuttle Super Lightweight External Tank, and other space system technology initiatives, including single-stage-to-orbit concepts and feasibility. He also led political and technical initiatives with Russia, leading to the incorporation of the Russian Soyuz spacecraft as the on-orbit emergency rescue vehicle for the ISS. In 1994, Mr. Aldrich left NASA and joined Lockheed Missiles and Space Company in Sunnyvale, California, where he served as vice president of commercial space business development and subsequently as vice president of strategic technology planning. With the merger of Lockheed and Martin Marietta, he joined Lockheed Martin corporate headquarters in Bethesda, Maryland, where he oversaw X-33/Venturestar single-stage-to-orbit program activity. Later, he became director of program operations and pursued a broad array of initiatives to enhance program management across the Corporation. Mr. Aldrich has received numerous honors during his career, including the Presidential Rank of Distinguished Executive and the NASA Distinguished Service Medal. He is an honorary fellow of the AIAA. Mr. Aldrich holds a B.S. in electrical engineering from Northeastern University. He is a member of the NRC's Aeronautics and Space Engineering Board.

DOUGLAS M. ALLEN is an independent consultant. Mr. Allen has more than 30 years of experience in advanced aerospace technology research, development, and testing. He is an expert in space power technology; his achievements include leading the successful first flight of multi-junction solar cells, leading the successful first flight of modular concentrator solar arrays, teaching AIAA's Space Power Systems Design short course, leading development of high specific energy batteries, managing development of nuclear space power systems, and leading development of solar power systems designed to survive hostile threats. Mr. Allen was awarded AIAA's Aerospace Power Systems Award for outstanding career achievements. Previously, he worked for the Schafer Corporation for 18 years. Mr. Allen led multiple modeling and simulation efforts for the Air Force Research Laboratory and the National Air and Space Intelligence Center. He was Schafer's chief engineer for a NASA contract that included developing a concept for Moon and Mars exploration and conceptual design of a Crew Exploration Vehicle. Prior to that, Mr. Allen managed launch vehicle and power technology programs for the Department of Defense's Strategic Defense Initiative Organization. Mr. Allen received an M.S. in mechanical engineering/energy conversion and a B.S. in mechanical engineering from the University of Dayton. His previous NRC membership service includes the Committee on Radioisotope Power Systems, the Committee on Thermionic Research and Technology, and the NASA Technology Roadmap: Propulsion and Power Panel.

RAYMOND E. ARVIDSON is the James S. McDonnell Distinguished University Professor in the Department of Earth and Planetary Sciences at Washington University in St. Louis. He is also a fellow of the McDonnell Center for the Space Sciences. He directs the Earth and Planetary Remote Sensing Laboratory (EPRSL), which is involved in many aspects of NASA's planetary exploration program, including developing science objectives and plans for missions, participating in mission operations and data analysis, and archiving and distributing data (NASA PDS Geosciences Node) relevant to characterizing and understanding planetary surfaces and interiors. Dr. Arvidson has participated in the Viking Lander (Image Team), Mars Global Surveyor, Odyssey (interdisciplinary scientist), Mars Exploration Rover (Spirit and Opportunity as deputy principal investigator), Phoenix Mars Lander (Robotic Arm co-investigator), Mars Reconnaissance Orbiter (CRISM Team), Mars Science Laboratory (Curiosity mobility scientist) and the European Space Agency's Mars Express missions (OMEGA Team). He received his Ph.D. in planetary sciences from Brown University. His NRC experience includes previously serving as chair of the Committee on Data Management and Computation and as a member of the Planetary Science Decadal Survey: Mars Panel, the Panel on Remote Sensing, and the Space Studies Board.

RICHARD C. ATKINSON (NAS/IOM) is president emeritus of the University of California (UC) and professor emeritus of cognitive science and psychology at UC, San Diego. He has also served as president of the UC system. His tenure was marked by innovative approaches to admissions and outreach, research initiatives to accelerate the UC's contributions to the state's economy, and a challenge to the country's most widely used admissions examination—the SAT—that paved the way to major changes in the way millions of America's youth now are tested for college admissions. Before becoming president of the UC system, Dr. Atkinson served for 15 years as chancellor of UC, San Diego, where he led that campus's emergence as one of the leading research universities in the nation. Dr. Atkinson has also served as director of NSF, as president of AAAS, and as a long-term member of the faculty at Stanford University. His research has been concerned with problems of memory and cognition. He is a member of the National Academy of Education and the American Philosophical Society. Dr. Atkinson is the recipient of many honorary degrees and the Vannevar Bush Medal of the National Science Board. He received his Ph.D. in mathematical psychology in Indiana University and a B.A. in mathematical psychology the University of Chicago. Dr. Atkinson's previous NRC service includes chair of the Division Committee for the Behavioral and Social Sciences and Education and the Board on Testing Assessment and member of the Committee on the Fiscal Future of the United States: Analysis and Policy Options, National Forum on Science and Technology Goals: Harnessing Technology for America's Economic Future, and the Committee on Science, Engineering, and Public Policy.

ROBERT D. BRAUN serves as the David and Andrew Lewis Professor of Space Technology in the Daniel Guggenheim School of Aerospace Engineering at the Georgia Institute of Technology. As director of Georgia Tech's Space Systems Design Laboratory, he leads an active research program focused on the design of advanced flight systems and technologies for planetary exploration. Dr. Braun has worked extensively in the areas of entry system design, planetary atmospheric flight, and space mission architecture development and has contributed to the design, development, test, and operation of several robotic space flight systems. In 2010 and 2011, he served as the first NASA chief technologist in more than a decade. In this capacity, he was the senior agency executive responsible for technology and innovation policy and programs. Earlier in his career, Dr. Braun served on the technical staff of the NASA Langley Research Center. He is an American Institute of Aeronautics and Astronautics (AIAA) fellow and the principle author or co-author of more than 200 technical publications in the fields of planetary exploration, atmospheric entry, multidisciplinary design optimization, and systems engineering. Dr. Braun has a B.S. in aerospace engineering from the Pennsylvania State University, an M.S. in astronautics from George

Washington University, and a Ph.D. in aeronautics and astronautics from Stanford University. He previously served as co-chair of the NRC's Committee on Review of the NASA Institute for Advanced Concepts and as a member of the Committee on the Planetary Science Decadal Survey: Mars Panel and the Committee on New Opportunities in Solar System Exploration.

ELIZABETH R. CANTWELL is the director for economic development at the Lawrence Livermore National Laboratory. She previously served as the deputy associate laboratory director for the National Security Directorate at Oak Ridge National Laboratory. Prior to joining Oak Ridge, Dr. Cantwell was the division leader for the International, Space, and Response Division at Los Alamos National Laboratory. Her career began in building life support systems for human spaceflight missions with NASA. She received an M.S. in mechanical engineering from the University of Pennsylvania, an M.B.A. in finance from Wharton School, and a Ph.D. in mechanical engineering from the University of California, Berkeley. Dr. Cantwell has extensive NRC experience including current memberships on the Space Studies Board and the Division on Engineering and Physical Sciences Board; co-chair of the Committee on Decadal Survey on Biological and Physical Sciences in Space; and member of the Committee on NASA's Bioastronautics Critical Path Roadmap, the Review of NASA Strategic Roadmaps: Space Station Panel, the Committee on Technology for Human/Robotic Exploration and Development of Space, and the Committee on Advanced Technology for Human Support in Space.

DAVID E. CROW (NAE) is professor emeritus of mechanical engineering at the University of Connecticut and retired senior vice president of engineering at Pratt and Whitney Aircraft Engine Company. At Pratt and Whitney he was influential in the design, development, test, and manufacturing in support of a full line of engines for aerospace and industrial applications. He was involved with products that include high-thrust turbofans for large commercial and military aircraft; turboprops and small turbofans for regional and corporate aircraft and helicopters; booster engines and upper stage propulsion systems for advanced launch vehicles; turbopumps for the Space Shuttle; and industrial engines for land-based power generation. His involvement included sophisticated computer modeling and standard work to bring constant improvements in the performance and reliability of the company's products, while at the same time reducing noise and emissions. Dr. Crow received his Ph.D. in mechanical engineering in from the University of Missouri-Rolla, his M.S. in mechanical engineering from Rensselaer Polytechnic Institute, and his B.S. in mechanical engineering from University of Missouri-Rolla. Dr. Crow's current NRC service includes chair of the Panel on Air and Ground Vehicle Technology and as a member on the Army Research Laboratory Technical assessment Board. His previous membership service with the NRC is extensive and includes the Committee on Examination of the U.S. Air Force's Aircraft Sustainment Needs in the Future and its Strategy to Meet those Needs, the Board on Manufacturing and Engineering Design, the Committee for the Evaluation of NASA's Fundamental Aeronautics Research Program, the Committee on Analysis of Air Force Engine Efficiency Improvement Options for Large Non-Fighter Aircraft, the Committee on Air Force/Department of Defense Aerospace Propulsion and the NASA Technology Roadmap: Propulsion and Power Panel.

RAVI B. DEO is president and founder of EMBR, an aerospace engineering and technology services company specializing in strategic planning, business development, program management and structural engineering. Dr. Deo formerly served as the director of the technology, space systems market segment at Northrop Grumman Corporation's Integrated Systems Sector. He has worked as a program and functional manager for government-sponsored projects on cryotanks, integrated airplane and space vehicle systems health management, and structures and materials, thermal protection systems, and software development. He has extensive experience in road mapping technologies, program planning, technical program execution, scheduling, budgeting, proposal preparation, and business management of technology development contracts. Among his significant accomplishments are the NASA-funded Space Launch Initiative, Next Generation Launch Technology, Orbital Space Plane, and High Speed Research programs, where he was responsible for the development of multidisciplinary technologies. Dr. Deo is the author of more than 50 technical publications and is the editor of one book. He has served on the Scientific Advisory Board to the Air Force Research Laboratories. Dr. Deo received a B.S. in aeronautical engineering from the Indian Institute of Technology and an M.S. and Ph.D. in aerospace engineering from Georgia Institute of Technology. His NRC service includes membership on the Aeronautics and Space Engineering Board, the Committee on the NASA Technology Roadmap, the Panel C: Structures and Materials and the Committee on Assessment of NASA Laboratory Capabilities

ROBERT S. DICKMAN is an independent consultant for RDSpace, LLC. Prior to retirement, he served 7 years as the executive director of the AIAA. He is also a USAF major general (retired), having served 34-years as an USAF officer. His military career spanned the space business from basic research in particle physics to command of the 45th Space Wing and director of the eastern range at Cape Canaveral, Florida. He served as the USAF director of space programs, the DOD Space Architect, and the senior military officer at the National Reconnaissance Office (NRO). He retired from active duty as a major general. Prior to joining the AIAA, he was deputy for military space in the office of the undersecretary of the U.S. Air Force. Major General Dickman has been a member of the U.S. Department of Transportation's Commercial Space Transportation Advisory Committee and has served on the Air Force Scientific Advisory Board and the NRO's Technical Advisory Group. He is a fellow of the AIAA and a Corresponding Member of the International Academy of Astronautics. Major General Dickman earned a B.S. in physics from Union College, an M.S. in space physics from the Air Force Institute of Technology, and an M.S. in management from Regina College. In addition, he is a distinguished graduate of the Air Command and Staff College and the Naval War College. He previously served as a member on the NRC Committee for the Reusable Booster System: Review and Assessment.

DAVA J. NEWMAN is a professor in the Department of Aeronautics and Astronautics and Engineering Systems at MIT. She also serves as affiliate faculty in the Harvard-MIT Health Sciences and Technology Program; MacVicar Faculty Fellow and director of the Technology and Policy Program and the MIT Portugal Program at MIT. She specializes in investigating human performance across the spectrum of gravity. Dr. Newman has served as PI on four spaceflight experiments, and she is an expert in the areas of extravehicular activity (EVA), human movement, physics-based modeling, human-robotic cooperation, and design. Currently she is working on advanced spacesuit design and biomedical devices, especially to enhance locomotion implementing wearable sensors. Her exoskeleton innovations are now being applied to "soft suits" to study and enhance locomotion on Earth for children with Cerebral Palsy. Dr. Newman's finite element modeling work provided NASA the first three-dimensional representation of bone loss and loading applicable for long-duration missions. Her teaching focuses on engineering design, aerospace biomedicine, and leadership, all involving active learning, hands-on design, and information technology to enhance student learning. Dr. Newman has more than 175 research publications, including an Engineering Design text and CDROM. She was named one of the Best Inventors of 2007 for her BioSuit™ system by *Time Magazine*, which has been exhibited at the MET, Boston's Museum of Science, Paris' Le Cité des Sciences et Industrie, London's Victoria and Albert Museum, and the American Museum of Natural History. She serves on the NASA Advisory Council (NAC) Committee on Technology and Innovation. Dr. Newman received a B.S. in aerospace engineering from the University of Notre Dame, an M.S. in both aeronautics and astronautics and technology and policy, and a Ph.D. in aerospace biomedical engineering from MIT. Her prior NRC service includes membership for two terms on the Aeronautics and Space Engineering Board, the Steering Committee on the NASA Technology Roadmaps, the Decadal Survey on Biological and Physical Sciences in Space, the Committee on Engineering Challenges to the Long-Term Operation of the International Space Station, the Committee on Advanced Technology for Human Support in Space, and the Committee on Full System Testing and Evaluation of Personal Protection Equipment Ensembles in Simulated Chemical and Biological Warfare Environments.

JOHN R. ROGACKI is associate director of the Florida Institute for Human and Machine Cognition (IHMC) (Ocala). Prior to joining IHMC, Dr. Rogacki served as director of the University of Florida's Research and Engineering Education Facility (REEF), a unique educational facility in Northwest Florida supporting U.S. Air Force (USAF) research and education needs through graduate degree programs in mechanical, aerospace, electrical, computer, industrial, and systems engineering. Under his leadership, REEF grew into a highly capable and internationally respected research and education facility. Among Dr. Rogacki's past experiences, he served as the NASA's deputy associate administrator for space transportation technology (in charge of the Space Launch Initiative);

program director for the Orbital Space Plane and Next Generation Launch Technology Programs; co-chair of the NASA/DOD Integrated High Payoff Rocket Propulsion Technology Program; director of NASA's Marshall Space Flight Center's Space Transportation Directorate; director of the propulsion directorate for the USAF Research Laboratory; director of the USAF Phillips Laboratory Propulsion Directorate; and deputy director of the Flight Dynamics Directorate of the USAF Wright Laboratory. An accomplished pilot, Dr. Rogacki has logged more than 3,300 flying hours as pilot, instructor pilot, and flight examiner in aircraft ranging from motorized gliders to heavy bombers. He has served as primary NASA liaison for the National Aerospace Initiative; co-chair of the DOD Future Propulsion Technology Advisory Group; co-chair of the DOD Ground and Sea Vehicles Technology Area Readiness Assessment Panel; member of the National High Cycle Fatigue Coordinating Committee; and senior NASA representative to the Joint Aeronautical Commanders Group. Later, Dr. Rogacki became associate professor of engineering mechanics (and chief of the materials division) at the USAF Academy. In 2005, he graduated from the Senior Executives Program in National and International Security at Harvard's John F. Kennedy School of Government. Dr. Rogacki earned his Ph.D. and M.S. in mechanical engineering from the University of Washington and his B.S. in engineering mechanics from the USAF Academy. He previously chaired the NRC NASA Technology Roadmap: Propulsion and Power Panel.

GUILLERMO TROTTI is president of Trotti and Associates, Inc. (TAI), a firm specializing in sustainable architecture and design for extreme environments such as remote islands, the Antarctic, space, and underwater environments in Cambridge, Massachusetts. He is an internationally recognized architect and industrial designer with more than 35 years of experience designing space habitats and structures, architectural projects for the eco-tourism, entertainment, medical and education sectors. Previously, Mr. Trotti was the president of Bell and Trotti, Inc. (BTI), a design and fabrication studio that specialized in space architecture and exhibit design. His experience includes designing diverse elements of the ISS for NASA and leading aerospace companies. He co-founded Space Industries Inc. for the purpose of building a privately owned space station. His lunar base design is included in the Smithsonian Air and Space Museum's permanent collection. He and his students won the NSF design competition for the U.S. South Pole Station. He has worked with NASA's Institute of Advanced Concepts on revolutionary mission architecture concepts for exploring the Moon with habitable and inflatable rovers. His Extreme Expeditionary Architecture: Mobile, Adaptable Systems for Space and Earth Exploration research proposed a revolutionary way for humans and machines to explore the lunar surface. Currently, TAI collaborates with MIT leading the design of the BioSuit™ System, an advanced mechanical counterpressure space suit for lunar and Mars planetary exploration; and the design of novel EVA injury protection and countermeasure devices for astronaut safety. Mr. Trotti's teaching experience includes architecture and industrial design at the University of Houston (UH) and the Rhode Island School of Design, respectively. At UH, he co-founded the Sasakawa International Center for Space Architecture. He received a B.A. in architecture from the University of Houston and an M.A. in architecture from Rice University. Mr. Trotti has served on the NRC Committee to Review NASA's Exploration Technology Development Programs and the Decadal Survey on Biological and Physical Sciences in Space: Translation to Space Exploration Systems Panel.

LINDA A. WILLIAMS is a program manager for the Wyle Aerospace Group. She leads a team of analysts for a major government customer's Cost Assessment and Analysis Group. She is also a Wyle subject matter expert in cost estimating and analysis, with more than 30 years of experience with space system cost estimating at Wyle, RCA Astro Electronics (now Lockheed Martin), Futron, Harris and L-3 Communications. She has developed numerous space system cost models, collected and normalized data, conducted price-to-win analyses, participated in satellite industry demand-based forecasts, and developed economic and strategic planning analyses. She has provided support to national, civil, and commercial space programs. Some of the projects she has lead or supported include development of cost and technical trade studies for the Space Station Work Package 3 study, development of a demand-based forecast for the NASA Reusable Launch Vehicle 2 project, participated in the development of a demand-based forecast for a major commercial communications satellite operator, and conducted cost trades and estimates for satellite systems including Defense Meteorological Satellite Program (DMSP), Television and Infrared Operational Satellite (TIROS), Mars Observer, Earth Observing System (EOS), Mobile User Objective System (MUOS), Global Positioning System (GPS), and many commercial satellite programs. She has co-authored several papers focused on the commercial satellite market and price analyses. She has an M.B.A. from Rider University and a B.A. in economics from Rutgers University. Ms. Williams is a certified cost estimating analyst (CCEA) and a project management professional (PMP). For the past 4 years she has provided an annual training course and problem-solving workshop at the International Cost Estimating Analysts Association (ICEAA) annual conference in support of the industry certification exam. This course covers all aspects of cost estimating and economic/program analysis.

STAFF

Committee and Technical Panel Staff

SANDRA J. GRAHAM, *Study Director*, has been a senior program officer at the National Research Council's Space Studies Board (SSB) since 1994. During that time Dr. Graham has directed a large number of major studies, many of them focused on space research in biological and physical sciences and technology. Studies in other areas include an assessment of servicing options for the Hubble Space Telescope, a study of the societal impacts of severe space weather, and a review of NASA's Space Communications Program while on loan to the NRC's Aeronautics and Space Engineering Board (ASEB). More recently, she directed the work of the committee and seven panels to develop the comprehensive decadal report *Recapturing a Future for Space Exploration—Life and Microgravity Sciences Research for a New Era.* Prior to joining the SSB, Dr. Graham held the position of senior scientist at the Bionetics Corporation, where she provided technical and science management support for NASA's Microgravity Science and Applications Division. She received her Ph.D. in inorganic chemistry from Duke University, where her research focused primarily on topics in bioinorganic chemistry, such as rate modeling and reaction chemistry of biological metal complexes and their analogs.

MICHAEL MOLONEY is the director for Space and Aeronautics at the Space Studies Board and the Aeronautics and Space Engineering Board of the National Research Council of the U.S. National Academies. Since joining the ASEB/SSB. Dr. Moloney has overseen the production of more than 40 reports, including four decadal surveys—in astronomy and astrophysics, planetary science, life and microgravity science, and solar and space physics—a review of the goals and direction of the U.S. human exploration program, a prioritization of NASA space technology roadmaps, as well as reports on issues such as NASA's Strategic Direction, orbital debris, the future of NASA's astronaut corps, and NASA's flight research program. Before joining the SSB and ASEB in 2010, Dr. Moloney was associate director of the Board on Physics and Astronomy (BPA) and study director for the decadal survey for astronomy and astrophysics (Astro2010). Since joining the NRC in 2001, Dr. Moloney has served as a study director at the National Materials Advisory Board, the BPA, the Board on Manufacturing and Engineering Design, and the Center for Economic, Governance, and International Studies. Dr. Moloney has served as study director or senior staff for a series of reports on subject matters as varied as quantum physics, nanotechnology, cosmology, the operation of the nation's helium reserve, new anti-counterfeiting technologies for currency, corrosion science, and nuclear fusion. In addition to his professional experience at the National Academies, Dr. Moloney has more than 7 years' experience as a foreign-service officer for the Irish government—including serving at the Irish Embassy in Washington and the Irish Mission to the United Nations in New York. A physicist, Dr. Moloney did his Ph.D. work at Trinity College Dublin in Ireland. He received his undergraduate degree in experimental physics at University College Dublin, where he was awarded the Nevin Medal for Physics.

ALAN C. ANGLEMAN has been a senior program officer for the Aeronautics and Space Engineering Board (ASEB) since 1993, directing studies on the modernization of the U.S. air transportation system, system engineering and design systems, aviation weather systems, aircraft certification standards and procedures, commercial supersonic aircraft, the safety of space launch systems, radioisotope power systems, cost growth of NASA Earth and space science missions, and other aspects of aeronautics and space research and technology. Previously, Mr. Angleman worked for consulting firms in the Washington area providing engineering support services to the DOD and NASA Headquarters. His professional career began with the U.S. Navy, where he served for 9 years as a nuclear-trained submarine officer. He has a B.S. in engineering physics from the U.S. Naval Academy and an M.S. in applied physics from the Johns Hopkins University.

ABIGAIL A. SHEFFER joined the Space Studies Board (SSB) in fall 2009 as a Christine Mirzayan Science and Technology Policy Graduate Fellow to work on the report *Visions and Voyages for Planetary Science in the Decade 2013-2022*. She continued with the SSB to become an associate program officer. Dr. Sheffer earned her Ph.D. in planetary science from the University of Arizona and her A.B. in geosciences from Princeton University. Since joining the SSB, she has worked on several studies, including *Defending Planet Earth: Near-Earth Object Surveys and Hazard Mitigation Strategies*, *Assessment of Impediments to Interagency Collaboration on Space and Earth Science Missions*, and *The Effects of Solar Variability on Earth's Climate: A Workshop Report*.

AMANDA R. THIBAULT, research associate, joined the Aeronautics and Space Engineering Board in 2011 and left in January 2013. Ms. Thibault is a graduate of Creighton University where she earned her B.S. in atmospheric science in 2008. From there she went on to Texas Tech University where she studied lightning trends in tornadic and non-tornadic supercell thunderstorms and worked as a teaching and research assistant. She participated in the VORTEX 2 field project from 2009-2010 and graduated with a M.S. in atmospheric science from Texas Tech in August 2010. She is a member of the American Meteorological Society.

DIONNA J. WILLIAMS is a program associate with the SSB, having previously worked for the National Academies' Division of Behavioral and Social Sciences and Education for 5 years. Ms. Williams has a long career in office administration, having worked as a supervisor in a number of capacities and fields. Ms. Williams attended the University of Colorado, Colorado Springs, and majored in psychology.

F. HARRISON DREVES was a Lloyd V. Berkner Space Policy Intern for the SSB during the study and is now a communications/media specialist for the Division on Engineering and Physical Sciences. Mr. Dreves recently received a B.A. degree from Vanderbilt University with concentrations in the communication of science and technology and Earth and environmental sciences. His academic interests include science policy, climate science, and science communication through video. At Vanderbilt, he served as a senior video producer for student media. Mr. Dreves hopes to pursue a career in science journalism or science policy, working to translate between the scientific community and the public. He is interested in combining his lifelong passion for space exploration (attending Space Camp in Huntsville at age 11) with his interest in science policy at the Space Studies Board, especially to gain insight into the political and economic structures behind space science programs. As a future space science communicator, Mr. Dreves would like to explain how research is funded, how a research target is selected, and, most importantly, why space science research funding matters.

JINNI MEEHAN, Lloyd V. Berkner Space Policy Intern, Fall 2013 (SSB) is a Ph.D. student at Utah State University in the department of physics. Her research is directed toward alleviating space weather effects on the Global Navigation Satellite System (GNSS) by better characterizing the ionosphere, which can improve forecast models. Ms. Meehan developed an interest for science policy when she spent a summer with the American Meteorological Society as a policy fellow working on space weather policy issues. She has authored several publications and presentations and has been a contributing author to numerous workshop reports for the space sciences community. She is passionate about the societal impacts due to space weather effects on GNSS and understands the importance of effective communication between scientists and government and she plans to pursue a career in the field when she completes her Ph.D.

CHERYL MOY, Christine Mirzayan Science and Technology Policy Graduate Fellow, Fall 2012, received her Ph.D. in chemistry from the University of Michigan. In her graduate work, Dr. Moy focused on elucidating the interactions that drive the formation of unique materials categorized as molecular gels. During her graduate career, she also helped design and implement a class-project centered on students editing Wikipedia pages as means of improving science education and the public's access to science. She holds a B.A. from Willamette University, and her experiences there led to an interest in bridging the gap between scientists and the general public. In 2011, she interned at the Office of Science and Technology Policy. Dr. Moy is excited for the opportunity to be a Mirzayan Fellow to learn how to connect scientific discoveries with everyone who can benefit outside of the research atmosphere—from consumers to government to industry.

SIERRA SMITH, Lloyd V. Berkner Space Policy Intern, Fall 2013 (SSB), recently graduated from James Madison University with an M.A. in history. The research for her master's thesis focused on the sociopolitical context of the search for extraterrestrial intelligence and its broader relationship to space sciences. While working at the National Radio Astronomy Observatory, she conducted research on the evolution of radio astronomy in the United States. She plans to continue her studies by pursuing a Ph.D. in the history of science. An internship with the Space Studies Board presents her with an exciting opportunity to experience the real-time development of space policy.

PADAMASHRI SURESH, Christine Mirzayan Science and Technology Policy Graduate Fellow, Winter 2014 (SSB/ DEPS), is currently finishing her Ph.D. in electrical engineering at Utah State University. Ms. Suresh is a NASA Earth and Space Science Fellow working on understanding the effects of space weather. Her dissertation focuses on studying the effects of solar storms on Earth's upper atmosphere. Her other research interests include instrumentation for CubeSats and sounding rocket missions. She is also a member of the student government working as the graduate research director and serving as the graduate student liaison across various research and student welfare committees. Ms. Suresh's interest in the Mirzayan fellowship stems from her interest to pursue a career in space program management. As a Mirzayan fellow, she hopes to learn how the different stakeholders of the space industry interface when deciding system-level problems and making enterprise-level decisions. Originally from Bangalore, India, Ms. Suresh obtained her B.S. in electrical engineering from Visveswaraiah Technological University. Following which, she worked with IBM as a systems engineer and architect for 2 years and then moved to Utah to pursue a master's with a focus on space systems.

Public and Stakeholder Opinions Panel Staff

KRISZTINA MARTON is a senior program officer with the Committee on National Statistics (CNSTAT). She is currently serving as study director for the Panel on Addressing Priority Technical Issues for the American Community Survey and has lead CNSTAT's contribution to the Committee on Human Spaceflight. Previously, she was the study director for the Panel on the Statistical Methods for Measuring the Group Quarters Population in the American Community Survey, the Panel on Redesigning the Commercial Buildings and Residential Energy Consumption Surveys of the Energy Information Administration, the Workshop on the Future of Federal Household Surveys, and an expert meeting on more efficient screening methods for the Health and Retirement Study of the National Institute on Aging. Prior to joining CNSTAT, she was a survey researcher at Mathematica Policy Research (MPR) where she conducted methodological research and oversaw data collections for the National Science Foundation, the Department of Health and Human Services, the Agency for Healthcare Research and Quality, the Robert Wood Johnson Foundation, and other clients. Previously, she was a survey director in the Ohio State University Center for Survey Research. She has a Ph.D. in communication with an interdisciplinary specialization in survey research from the Ohio State University.

CONSTANCE F. CITRO is director of the Committee on National Statistics (CNSTAT), a position she has held since May 2004. She previously served as acting chief of staff (December 2003-April 2004) and as senior study director (1986-2003). She began her career with CNSTAT in 1984 as study director for the panel that produced *The Bicentennial Census: New Directions for Methodology in 1990.* Dr. Citro received her B.A. in political science from the University of Rochester and an M.A. and Ph.D. in political science from Yale University. Prior to joining CNSTAT, she held positions as vice president of Mathematica Policy Research, Inc. and Data Use and Access Laboratories, Inc. She was an American Statistical Association/National Science Foundation/Census research fellow in 1985-1986, and she is a fellow of the American Statistical Association and an elected member of the International Statistical Institute. For CNSTAT, she directed evaluations of the 2000 census, the Survey of

Income and Program Participation, microsimulation models for social welfare programs, and the NSF science and engineering personnel data system, in addition to studies on institutional review boards and social science research, estimates of poverty for small geographic areas, data and methods for retirement income modeling, and a new approach for measuring poverty. She coedited the 2nd–5th editions of *Principles and Practices for a Federal Statistical Agency* and contributed to studies on measuring racial discrimination, expanding access to research data, the usability of estimates from the American Community Survey, the National Children's Study research plan, and the Census Bureau's 2010 census program of experiments and evaluations.

JACQUELINE R. SOVDE has been a program associate with the Committee on National Statistics since December 2011. Before joining CNSTAT, she was with the Committee on Population. Prior to joining the Academies, she worked for the National Museum of Women in the Arts and for Alice Smith & Associates. Ms. Sovde received her B.A. from the University of California, Santa Cruz, in 2003, where she had an independent major in writing and communication.